Philosophy, God and Mot

In the post-Newtonian world motion is assumed to be a simple category which relates to the locomotion of bodies in space, and is usually associated only with physics. *Philosophy, God and Motion* shows that this is a relatively recent understanding of motion and that prior to the scientific revolution motion was a much broader and more mysterious category, applying to moral as well as physical movements.

Simon Oliver presents fresh interpretations of key figures in the history of western thought including Plato, Aristotle, Aquinas and Newton, examining the thinkers' handling of the concept of motion. Through close readings of seminal texts in ancient and medieval cosmology and early modern natural philosophy, the book moves from antique to modern times investigating how motion has been of great significance within theology, philosophy and science. Particularly important is the relation between motion and God. Following Aristotle, traditional doctrines of God have understood the divine as the 'unmoved mover' while more recent theology and philosophy has suggested that, in order for God to be involved in the cosmos, the divine must in some way be subject to motion. Simon Oliver argues that, while God is beyond all qualifications of change, motion is nevertheless a means of creation's perfection and participation in the dynamic eternal life of God. *Philosophy, God and Motion* therefore suggests that there may be an authentically theological, as well as a natural scientific, understanding of motion.

This volume will prove a major contribution to theology, the history of Christian thought and to the growing field of science and religion.

Simon Oliver is Lecturer in Systematic Theology at the University of Wales, Lampeter and an Anglican priest.

Routledge Radical Orthodoxy Series
Edited by John Milbank, Catherine Pickstock
and Graham Ward

Radical orthodoxy combines a sophisticated understanding of contemporary thought, modern and postmodern, with a theological perspective that looks back to the origins of the Church. It is the most talked-about development in contemporary theology.

1 Philosophy, God and Motion
Simon Oliver

Previous titles to appear in the Routledge Radical Orthodoxy Series include:

Radical Orthodoxy
edited by John Milbank,
Catherine Pickstock
and Graham Ward

Divine Economy
D. Stephen Long

Truth in Aquinas
John Milbank
and Catherine Pickstock

Cities of God
Graham Ward

Liberation Theology after
the End of History
Daniel M. Bell, Jr

Genealogy of Nihilism
Conor Cunningham

Speech and Theology
James K. A. Smith

Culture and the Thomist
Tradition
Tracey Rowland

Being Reconciled
John Milbank

Augustine and Modernity
Michael Hanby

Truth in the Making
Robert Miner

Philosophy, God and Motion

Simon Oliver

 Routledge
Taylor & Francis Group

LONDON AND NEW YORK

First published 2005 by
Routledge
2 Park Square, Milton Park, Abingdon, Oxon OX14 4RN

Simultaneously published in the USA and Canada
by Routledge
711 Third Avenue, New York, NY 10017

Routledge is an imprint of the Taylor & Francis Group

© 2005 Simon Oliver

Typeset in Garamond by
Taylor and Francis Books

First issued in paperback 2013

British Library Cataloguing in Publication Data
A catalogue record for this book is available from the British Library

Library of Congress Cataloging-in-Publication Data
A catalog record for this book has been requested

ISBN 0-415-36045-5
ISBN13: 978-0-415-84918-0 (pbk)

T&F informa

Taylor & Francis Group is the Academic Division of T&F Informa plc.

For Jayne and Benedict

But oh, how poor and feeble, speech and word!
    How weak to my conception! For what was shown
    Me there, to call this slight, would be absurd!
O Light eternal, self-indwelt alone,
    Only self-understood, self-understanding,
    And understanding, love and smile on your own!
Those rings that had appeared to me as banding
    Of light reflected, as my eyes conceived,
    Those moments that my vision held their stranding,
Within itself, in self-colour, seemed relieved
    With human image; seeing which, my eye
    Was wholly dedicated, and there it cleaved.
As the geometer who means to try
    Squaring the circle but cannot find,
    Think as he may, the principle to apply,
Such, at this new sight, was I in mind;
    I would perceive how image could inhere
    Within the circle yet remain defined.
But my own wings were not sufficient gear
    For that – had not a flash then come to me
    By which what I had wished was made quite clear.
Here, power failed for this high phantasy;
    But, now, as a smooth wheel – without any jars –
My will and my desire turned evenly
    Through Love that moves the sun and other stars.

Dante Alighieri, *La Divina Commedia*, *Paradiso*, Canto 33
(trans. Peter Dale)

Contents

Acknowledgements

I should like to thank two bodies who funded the research contained within this essay: the Wordsworth Fund of the Faculty of Divinity, Cambridge, and the Arts and Humanities Research Board of the British Academy.

Portions of chapters 3 and 4 have previously appeared in an earlier form in Vivarium 42.2 (2004) and The International Journal of Systematic Theology 7.1 (2005). I am grateful for permission to reproduce the material here.

I am also indebted to the Diocese of Ely, and most particularly to the Rt Revd John Flack, the Revd Christopher Boulton, the Revd George Westhaver and the people of St Andrew's, Cherry Hinton, and All Saints', Teversham. In their various ways, they provided practical support and a prayerful Eucharistic community. More recently, the Fellows, students and staff of Jesus College, Cambridge and Hertford College, Oxford have generously provided congenial and friendly environments in which to work. I am particularly grateful to six former students for their questions and conversation: Matthew Bullimore, Joel Cabrita, John Hughes, Samuel Klein, Catriona Laing and Vincent Vitale.

Many people have given me the benefit of their learning and wisdom during the execution of this project. Amongst the many who have kindly offered comments and criticisms of portions of this essay, I should like to thank Luc Brisson, Timothy Jenkins and Andrew Louth. I am also grateful for suggestions and criticism from the members of seminars at whose meetings portions of this work have been read: the 'D' Society, the Systematics Seminar and the Science and Religion Research Group in the Faculty of Divinity, Cambridge; and the Doctrine Seminar in the Faculty of Theology, Oxford. I have learned much from many others. In particular, I should like to mention Michael Hanby, Mary-Jane Rubenstein, Stephen Snobelen, Janet Soskice, Tony Street, Denys Turner and Graham Ward. I am particularly indebted to John Milbank for his continual guidance and friendship: his influence pervades every part of this work. My greatest intellectual debt is to Catherine Pickstock. I have benefited immeasurably from her learning, intense theological insight, ceaseless encouragement, perspicuous criticism and friendship. The vagaries and shortcomings of this essay remain entirely my own.

Family and friends have offered great encouragement and wonderful company, and I should like particularly to thank my parents, Stephen and Hilary Oliver, my brother, Adam Oliver, Gwen and Geoff Ellis, Peter and Beatrice Groves and Simon Jones. Finally, greatest thanks are due to my wife, Jayne, and our son, Benedict, whose love and companionship I treasure above all things. They keep life very firmly in perspective.

The Feast of the Transfiguration
2004

Introduction

This essay is about motion. We commonly think of motion as the subject of the science of physics. It is a category which apparently refers to bodies and their spatial locomotions. We have come to accept Newton's three laws of motion and his stipulation that motion is a simple category well known to all. However, for those philosophers and theologians prior to the scientific revolution of the seventeenth century, motion tends to be presented as a more mysterious category which is not confined to spatial or local motion. Rather, it may apply to moral as well as physical movements: learning, growing, ripening and thinking count as motion, just as much as the movement of bodies through space. The cosmos is seen as saturated with motions of many kinds, and this apparently renders nature difficult to grasp: natural beings continuously move in their various ways. What causes motion? Why do things move? When I throw a ball into the air, what preserves it in motion after it has left my hand? Is motion orientated towards a particular goal? Are there different qualities of motion? Meanwhile, within the traditions of apophatic theology, God is frequently understood negatively as beyond all qualifications of motion. The divine reality is the first unmoved mover, the impassible and ineffable source of all created, moving being. So in addition to questions concerning the nature and purpose of motion which have concerned philosophers and scientists for centuries, a central theological question arises: what is the relation between a universe whose very nature is, to paraphrase Aristotle, a principle of motion and rest, and God who is wholly beyond motion?

Given this latter question, it might be thought that motion is something profoundly negative to be assuaged or overcome, marking the boundary between the ontological stability and certainty of the unchanging eternal, in which motion is absent, and the capricious cosmos which is saturated with motion. Perhaps motion does render things ultimately ungraspable, thereby constituting the cosmos as a mere diversion from truth, for any created thing is never absolutely identical with itself from one moment to the next. Such a negative view of the cosmos is traditionally ascribed to Plato. Similarly, one might suggest that modern science is an attempt to stop or slow down nature within the laboratory in order to make fixed and more certain observations. This is perhaps one of the reasons why contrived experiment did not fit well with Aristotelian natural philosophy: once one removed or manipulated motion – the essential characteristic of any natural entity – one was observing something wholly *unnatural*.

Just as motion might constitute the boundary between the unchanging eternal and the cosmos, so motion might also constitute the boundary between the discourses which refer to these realms. So we might think that the study of motion is properly and primarily confined to the science of physics, which deals with nature, while it is no concern of metaphysics or theology. Motion might therefore be understood as the boundary limit in a dualistic ontology: moving nature stands over and against the eternal stability of the Forms, the first unmoved mover or the unchanging, simple God.

This essay offers a different narration of the concept of motion. I seek to demonstrate that traditionally motion is a broad and complex concept applied analogically in various discourses, and that there can be an authentically theological, as well as natural philosophical, account of motion. For those theologians and philosophers prior to the rise of modernity, motion does not constitute the separation of discourses, but is a means of their unity. For Plato, in wholly undualistic fashion, motion is the very means by which we come to know all that is contained in the eternal stability of the Forms; it is, for Aristotle, the means of our passage from potency to actuality; it is, for Grosseteste, the means of the propagation of the universe from the simple, eternal light; it is, for Aquinas, the means of our participation in the eternal dynamism of the Trinitarian life of God. It is only later, when the concept of motion is narrowed dramatically to encompass only spatially quantified change, that the study of motion is confined solely to one discourse and comes to constitute a boundary between natural science, on the one hand, and metaphysics and theology, on the other.

I begin with Plato's *Timaeus*, a seminal text in western thought which exerted a great influence on the natural philosophers of the sixteenth and seventeenth centuries. Here one finds a cosmology which relates the motion of the realm of becoming to the eternity and reality of the Forms through the metaphysics of participation. For Plato, because cosmology is discourse about a universe of motion, it must to some degree be subject to the same transience as its subject matter; there is no objective or supra-cosmic stance from which one can discern the nature of the cosmos. However, this is not to say that cosmology is pure caprice for Plato, nor that motion is to be eschewed in order to secure certainty. On the contrary, I argue that motion makes possible human knowing and participation in the realm of the Forms because the eternity of being is known successively in the motions of the cosmos. It is precisely the motions of the cosmos, particularly the most perfect circular orbits which are more unitary and complete, which mediate to us the unity and stability of the Forms. So rather than constituting the division between being and becoming in dualistic fashion, I argue that for Plato it is precisely motion which enables the participation of becoming in being. This text also instructs us in pre-Newtonian qualities of motion: we learn that motion is subject to subtle hierarchical differentiation.

In Chapter 2, I examine the natural philosophy of Aristotle, for his principal works were central to the medieval understanding of motion which was apparently supplanted by early modern science. In particular, I focus on Aristotle's *Physics* and examine his understanding of motion as that which hovers between potency and act while remaining related to a particular *telos*. Motion, for Aristotle, is not indifferent and self-explanatory, for motion can be natural (tending towards a being's *telos*) or violent (tending away from a being's *telos*), and each and every motion involves a mover and something that is moved in such a way that motion is a relational category. But motion for Aristotle not only applies to the physics of moving bodies. Motion is also applied analogically to the human ethical life to describe the way in which humanity moves and is moved towards its *telos*. In particular, in the sphere of human virtues one finds principles of motion towards the Good, which is the defining limit of all motion. Therefore, motion is not just an indifferent transfer of a density of being from one place to another; it is the means of the perfection of things within nature, and as motion is the continual actualisation of beings, I argue that motion is ecstatic for Aristotle, for any being in motion is constantly exceeding its own limits as it strives towards its goal and fulfilment.

Having described two principal accounts of motion within ancient philosophy, I turn to examine the influence of this ancient philosophy upon certain medieval thinkers. Initially focusing on the work of Robert Grosseteste, one of the first Latin commentators on Aristotle, I examine his cosmogony of light and motion, and his understanding of method in natural philosophy, particularly in his philosophical treatises *De Luce* and *De Veritate*. For Grosseteste, God is light and all things are a more or less rarefied form of light. Therefore, all things may be analogically related through the mediation of light. I argue that for Grosseteste cosmology is made possible because the universe is a motion of emanation from the one, true, eternal light of God. Therefore, to study light is to study God and all things in relation to God. Here we find a theological rendering of the qualitative and hierarchical motion of Plato and Aristotle.

The thought of Grosseteste has been regarded as one of the origins of experimental practice, for he appears to advocate the verification and falsification of hypothesis by contrived observation. In addition to examining Grosseteste's view of motion and his cosmogony of light, I also investigate the role of the *experimentum* in his thought and suggest that he advocates such practice not in order to mitigate a proto-Humean inductive scepticism, but rather to assuage the effects of the Fall. For Grosseteste, there is no dark chasm of un-illuminated nature which inductive reasoning is unable to traverse and which must therefore be surmounted by the constant repetition of experiment. Rather, all knowledge is a participation in God's illumination of all things, and therefore the motions of experimental practice perform a similar role to cosmological analysis in Plato's *Timaeus*: the patterns of non-identical repetition and the modulations of temporal succession of the motion of the cosmos make clear to our clouded, fallen minds all that is known in the eternal simplicity of the divine light.

Grosseteste's immediate successor, Roger Bacon, extends the method of the *experimentum* into a *scientia experimentalis* while also advocating the study of mathematics. Does one find in Bacon's work an early intimation of the scientific practice which will come to characterise investigations of motion from the sixteenth century onwards? I argue that, in a number of respects, Bacon does indeed anticipate modern scientific concerns. I also investigate the way in which Bacon anticipates a characteristic feature of later medieval and early modern thought, namely an understanding of knowledge as representation. Knowledge for Bacon is not a visceral motion whereby a known object comes to reside in a different mode via the mysterious protocols of non-identical repetition in the mind of the knower. Rather, experience is the mere occasion for knowledge, that knowledge instead emerging from a separate agent intellect which is 'stamped' onto the mind. A new distrust of our experience of the world emerges from this view, because sensations are no longer reliable; God could, by a simple act of will, determine which representations assert themselves in the mind without there being a created source of that representation. Thus begins a scepticism, a distrust, concerning moving nature which places the identical repetition of experiment as the condition of possibility for certain knowledge at the heart of natural philosophy.

Having examined one of the earliest attempts to assimilate Aristotelian natural philosophy into theology, I move on to study the Platonic and Aristotelian synthesis in the work of Thomas Aquinas. I argue that questions of motion appear at all levels in the hierarchy of sciences addressed by Aquinas and that, ultimately, all motion is analogically related to the eternal emanations of the persons of the Trinity. This forms a reassessment of the doctrine of God as the 'first unmoved mover'. Many philosophers and theologians reject this supposedly Aristotelian view which is associated particularly with Aquinas's so-called 'Five Ways' of proving God's existence. Critics argue that such a view renders God static and divine action restricted to a 'first push' or merely efficient causality. This paves the way for deism. In response to the notion that God is the first motionless mover, an emphasis has been placed more recently on the passible nature of divinity. One thinks particularly of 'process theology' and the focus on suffering as the form of kinship between God and humanity. By contrast, I demonstrate that God does not, and could not, 'become' for Aquinas (there is no potentiality in the divine), but by means of a Neoplatonic hierarchical understanding of causation in which God 'touches' every motion, one can attain a more nuanced understanding of God as 'unmoved mover' in which the dynamism of Trinitarian love is mediated via motion to the cosmos.

From the importance of motion for Aquinas's doctrine of God and understanding of creation, I examine the importance of motion in his ethics, most particularly the analogy between the fulfilment of bodies in nature through motion and the attainment of beatitude in God's moving of humanity through the law and virtue. The incarnation is, I suggest, a crucial moment for Aquinas, for here one finds the origin and *telos* of human motion in the midst of the way. This is a revelation which reorientates us towards a goal which of our own power we could not discern. The justification and passion wrought by Christ's death and resurrection removes the obstacle of sin which prevents humanity from achieving its end. The salvation of Christ is continually mediated through the sacraments, and most particularly the Eucharist where, I suggest, one finds the unification of all cosmic motion in a single point of rest in the midst of the way, though 'rest' itself is here also to be seen in qualitative and hierarchical terms.

Having outlined one of the most dominant visions of the medieval period, I examine a different tradition which finds a focus in the work of the Persian Islamic philosopher Avicenna. I argue that, in contrast to Aquinas, for whom the various sciences are analogically related to each other through the application of concepts such as motion, Avicenna anticipates the separation of physics from the realm of metaphysics and thus suggests what will later become the restriction of questions of motion to an autonomous physics. For example, for Avicenna the question of the proof of God's existence, which for Aquinas belongs both to physics and metaphysics, is confined purely to metaphysics. Motion, which is the subject of physics, no longer points beyond itself to a divine origin and end. Moreover, the Plotinian emanationist cosmology espoused by Avicenna suggests a substantial cosmic hierarchy between the absolute One and moving creation in such a way that motion, rather than providing the means of relating creation to God or becoming to being, defines and determines the boundary between the two. By contrast, for Aquinas, God's 'touch' is intimate to every motion.

In addition to Avicenna's separation of physics and metaphysics, I examine another medieval tradition relating to motion, namely the theory of impetus. In various presentations of this theory, a number of recent commentators have seen a precursor of Newton's principle of inertia. I argue that, through impetus theory, motion becomes a more exclusively quantitative category, and that certain motions, such as those of the heavenly spheres, are related to God not because of their exalted cosmological status, but because a being is now required which possesses a sufficient quantity of power to sustain such motion. Although impetus theory remains Aristotelian in many important respects, in others it anticipates the physics and voluntarist theology of Newton.

Having examined some principal understandings of motion in medieval theology and natural philosophy, I address the revolution in physics brought about by the publication of Sir Isaac Newton's *Principia*. I argue that, although Newton produces laws of motion and places these at the heart of his treatise, nevertheless motion becomes a simple category and is supplanted by force as the central subject of physics. Moreover, because motion is no longer teleologically ordered and is understood as a state, this category is no longer ontological and tells us nothing about the being of a moving body.

In addition to an examination of Newton's physics, I explore the relationship of his science of motion to his theology. Whereas for Aquinas all motion is relational and analogically related to the emanations of the persons of the Trinity, because Newton maintains an Arian, non-Trinitarian position, he is unable to find a source and ground for a relational cosmos at the highest ontological level. How can a motionless God, construed by Newton in voluntaristic terms, now be related to a cosmos in motion? For Aquinas, God's relation to the cosmos is understood through motion's participation in the emanation of the persons of the Trinity, through which God created the universe. By contrast, Newton conceives of an absolute space begotten of God (the infamous *sensorium dei*) in and through which God creates and acts in the world. I argue that, curiously, this absolute space takes on the characteristics of a more orthodox Christ in Newton's theology.

This is the view of motion bequeathed to the eighteenth century. In the concluding section of this essay I comment on more recent understandings of the nature of motion and suggest that mechanistic cosmologies were already thought to be dubious, even at times by Newton himself. I examine briefly the rise of electromagnetism and thermodynamics, suggesting that these innovations rely on a form of analogical reasoning which bears an interesting resemblance to the natural philosophy which was the subject of the first four chapters of this essay. Moreover, I suggest that motion is understood once again as relational rather than idealised as the state of a body in a vacuum. The second law of thermodynamics, for example, is interpreted as the donation or sharing of energy by all created beings in such a way that relations are constantly being re-established between creatures. Finally, I suggest, following Wolfgang Smith, that a Thomistic ontology which includes substantial form and an ontological hierarchy with a varied understanding of motion is more fitting to some of the most recent developments in physics, particularly the peculiarities of quantum theory.

Although motion is thought to be the exclusive purview of physics, this essay, through the examination of key texts, seeks to argue that this is a recent view and that theology, rather than averring to scientific accounts of motion in a straightforward fashion, may be able to offer its own authentic approach. While this is not an attempt to supplant scientific notions of motion, it seeks to suggest that these apparently opposed discourses can be brought into closer proximity once again through the sharing of a category which is applied analogically, and ultimately to its divine origin and end.

This essay may therefore be placed very broadly within the ambit of what has become known as 'science and theology'. Science is often assumed to express literally what theology expresses through myth, narrative or analogy. Alternatively, science might apparently provide criteria of rationality to which theology should aspire, or science is thought to offer a kind of natural proof of theology's claims in such a way that theologians are free to pursue their particular enquiries in isolation, resting secure in their vindication by science's rigour. Or theology simply fills in the ever shrinking gaps left by science. However, this essay seeks a shift in emphasis by examining the genealogy of a concept which might once again be shared analogically by the different *scientia*. This is not arbitrary or alien to science or theology, for this essay argues that both rely on analogical predication, namely discourse which examines the multifarious, complex, non-univocal relations between things which conjoin, by shared participation in a transcendent origin and goal, to form a *uni*-verse. Analogy implies more than a series of resemblances of the kind which might lead to the claim that 'motion' in physics happens to resemble 'motion' understood theologically in an innocuously pleasing fashion. Rather, it is the contention of this essay that science and theology are related, like all things, by their shared participation in the *scientia divina*, that participation being made possible by the various analogically related motions of created being and, more particularly, of human knowing.

1 Plato's *Timaeus* and the soul's motion of knowing

The architects of the modern scientific revolution of the seventeenth century onwards, from whom we inherit our common understanding of such concepts as motion, have frequently been seen to build upon a tradition whose foundations lie some two millennia earlier in ancient Greece, and particularly in the Platonic treatise *Timaeus*.[1] It is in this work, written in the fourth century BC during the later part of Plato's career,[2] that he sketches the very purpose and limits of scientific enquiry. The cosmos is, perhaps for the first time, investigated and described through the establishment of axioms from which emerge mathematical proportion and the 'harmonic music of the heavens'.[3] In terms later adopted by the founders of modern science, Plato seeks explanation of phenomena as diverse as the motion of the heavenly planets and stars, the behaviour of the elements and the workings of the human body by a description of their origins, structure, place and behaviour within the ordered universe. No longer are the happenings of nature the result of the activities of recalcitrant deities. Instead, through one of the earliest instances of a method that might be broadly recognised as 'scientific' in the modern sense, Plato describes an orderly cosmos which is constituted according to certain identifiable proportions and which behaves in observable and predictable patterns.[4]

The *Timaeus* is in part a polemic against the *physiologoi* of ancient Greece, those philosophers who believed that the universe originated by chance rather than design or 'art', and that its activity is the result of the natural activity of soulless bodies.[5] This immanentism was anathema to Plato.[6] Rather, cosmology in the *Timaeus* is fundamentally a 'theological' enterprise which seeks to identify the *telos* of nature. It is theological in the sense that the cosmos is described in terms of its participation in a transcendent origin and purpose: it is not the self-explanatory and closed system of the *physiologoi*. The cosmos is a realm not of static being but of change and becoming which finds its explanation only with reference to its origin and continued participation in the unchanging, eternal and transcendent realm of the Forms and, ultimately and crucially, in its relation to the Good.[7] This change and becoming manifests itself in terms of motion, and therefore motion is a central concern in Plato's cosmological treatise.

Because this change and becoming which are manifest in movement are defining aspects of the cosmos investigated in the *Timaeus*, this chapter will examine Platonic cosmology through a study of the nature and purpose of motion as expounded in the treatise. It will be seen that Plato does not have a 'theory' of motion in the modern sense of the provision of a predictive model, but that motion is understood teleologically as, on the one hand, a purposeful ordering of the cosmos and, on the other hand, the means of nature's fulfilment and the a priori condition for the understanding of the universe in all its diversity. Crucially, motion will be understood as emerging from the persuasion of the pre-existent realm of 'necessity' from chaos to teleological order by the craftsmanship of *nous*. In addition, it will be seen that the study of cosmology, and particularly the study of motion in astronomy, has its own pedagogical *telos*. Finally, I will suggest ways in which the understanding of motion in Plato's *Timaeus* hints at the very purpose and nature, or movement, of natural and political philosophy.

The nature of the cosmos

The status of cosmological accounts

The *Timaeus* tells the story of the formation of the universe. Plato would not have intended this to be a definitive cosmology or a final and exhaustive account of the universe. This is because 'truth' does not lie within the universe. For Plato, truth lies in the realm of the unchanging Forms, that truth being illuminated by the Form of the Good. The visible realm of becoming, whose origin Plato is seeking to investigate, only discloses truth in so far as it participates in the unchanging and intelligible world of Forms. To the extent that the visible world fails to participate in the Forms, it becomes unintelligible and corresponds to the realm of mere 'opinion'. Because Plato's treatise is itself a discourse *within* the realm of becoming and change and also a discourse *about* that realm, it will to some extent be subject to the same demur as this realm of mere opinion. The cosmology presented by Plato cannot fully escape or transcend the limits described by its subject matter. So the *Timaeus*, unlike many modern speculative scientific cosmologies, is not an account aimed at definitive statements, because there cannot be any such neutral exhaustive, or in any alleged fashion, supra-cosmic statements about the universe. Why? Because the universe is the visible realm of becoming. Rather, the *Timaeus* presents a mythical story about a universe which is itself 'mythical'.

However, this is not to say that the *Timaeus* presents mere baseless fiction or a pure invention of Plato's mind. For Plato, philosophic discourse consists in affirmative and negative statements about Forms. The philosopher discerns the true structure of the realm of the Forms, what each Form is in itself and how it differs from other Forms. A false judgement is understood as the mistaking of one Form for another, or mistaking the visible realm of becoming and the intelligible and unchanging realm of the Forms.[8] Therefore, Plato is not concerned to give a purely immanent account of the universe; such an account, if it is to be true to its subject matter, is not possible. His treatise will only be true to the extent that it comes close to an understanding of the universe in its relation to the realm of the Forms: this is the means by which the *Timaeus* can receive an element of the truth which resides in that transcendent realm. Plato's cosmology is an attempt to see the universe in its most fundamental being; that is, in its relation to the Forms. Ultimately, the Form of the Good is the source of that being and therefore Plato's cosmology will be truthful to the extent that it can correctly identify the universe in its relation to the Good.[9]

Because the realm of becoming is an 'image' or 'likeness' of the Living Creature which resides in the realm of the Forms, Plato is comparing a model with its copy. The universe was created by the Demiurge[10] according to a transcendent ideal model, and just as a copy cannot be understood apart from its relation to, and origin in, the model from which it emanates or to which it is aligned, so neither can cosmology be undertaken without recourse to the model from which the universe emanates and in which it participates.[11] So Timaeus says, 'If then, Socrates, in many respects concerning many things – the gods and the generation of the universe – we prove unable to render an account at all points entirely consistent with itself and exact, you must not be surprised'.[12]

If Platonic cosmology is concerned to identify the universe in its relation to the Forms, and fundamentally the relation of everything to the Form of the Good, how is the natural philosopher to identify and know anything of the transcendent realm where truth resides? To put this question in the familiar terms of the allegory of the cave in *The Republic*, Book VII, how does one escape from the cave to see everything in the light of the Good? In *Timaeus*, Plato outlines a view which regards motion as a central means for the realm of becoming to participate in being, for the identification of the Forms and therefore the key to a truthful account of the universe.

The World Soul and the body of the universe

How does Plato conceive the origin of the universe? He begins with a discordant and disordered realm that is brought into order by the Demiurge.[13] The visible world is a living creature made after the likeness of an eternal original by the Demiurge, having soul in body and reason in soul. In keeping with the Living Creature, the model after which the universe is formed embraces all the intelligible Forms of living creatures. So too its visible image, the universe, embraces all the visible living creatures within itself.[14] For Plato, the universe is unique with nothing lying outside.[15] It is therefore whole and cannot suffer external assault. Neither can the universe suffer internal dissolution because this would entail the supremacy of a part over the whole.[16] It contains within its body the four primary bodies of

constant proportion to each other.[34] The seven parts become the distances from the earth and therefore the proportion of the motions of the seven heavenly spheres known to Plato (the sun, the moon, Mercury, Venus, Mars, Jupiter and Saturn).[35] These motions were observed by astronomers as constant and therefore rational. This is a cosmology which is taken directly from the mathematical theory of music known to the Pythagoreans. The identifiable proportions which enabled the description of the rational and harmonious movement of music are transposed to describe the rational and harmonious movement of the heavenly bodies and thus identify their participation in the symmetry, proportion, order and beauty bestowed by the Good. The motion of the stars, which is directly bestowed by the circle of the Same, and the motion of the planets, which is bestowed by the circle of the Different, is thus rendered rational. This is the movement of the music of the spheres transmitted through the soul of the universe.[36]

Cosmological knowing

Having established this mediatory character of the composition of the World Soul, the perfection of the motion of the Same, the unifying character of that motion and the proportions and symmetry in the motions of the Same and the Different, Plato can further explain why this motion is so important for our knowledge of the intelligible realm and therefore for cosmology as a whole. First, the rational movement of the World Soul, which is transmitted throughout the body of the cosmos and which is in mathematical proportion and therefore symmetry, makes cosmology possible. True knowledge is possible only of the Forms, and these have beauty, proportion, symmetry and order bestowed upon them by the Good, thus constituting them as rational and intelligible. It is through the mediation of the World Soul, which partakes of this symmetry through the form and proportions of its motion, that cosmology, the *eikos logos* or verisimilar account, is made possible. Secondly, and in a similar vein, Plato comments that

> whenever she [the World Soul] is in contact with anything that has dispersed existence or with anything whose existence is indivisible, she is set in motion all through herself and tells in what respect precisely, and how, and in what sense, and when, it comes about that something is qualified as either the same or different with respect to any given thing, whatever it may be, with which it is the same or from which it differs, either in the sphere of things that become or with regard to things that are changeless.
>
> (Plato, *Timaeus*, 37a–b)

As was stated above, cosmology for Plato involves correctly identifying the universe in its relation to the Forms and ultimately in its relation to the Good. The World Soul can be seen as the condition of possibility for this understanding precisely because of its characteristic motions. However, as regards the understanding of the cosmos, these motions are not primarily physical. They are instead the motions of 'wish, reflection, diligence, counsel, opinion true and false, joy and grief, cheerfulness and fear, love and hate'.[37] Whereas for modern science it is a commonplace that motion is attributed primarily to bodies and only metaphorically to the emotions and intelligence, Plato is here describing the latter as primary, for they reside in the soul which is the very source of motion. It now becomes clear that it is by participating in the ordered motion of the World Soul that we come to identify correctly the realms of being and becoming and their relation to each other. It is by our ordered movements of wishing, reflecting, being diligent, forming true opinions and expressing appropriate joy, grief and love that we correctly distinguish the real from its image. Disordered wanderings of the soul produce the confusion which misunderstands the Forms and fails correctly to identify the realms of becoming and being. This is discussed by Plato in relation to the motions of the soul in children. F. M. Cornford comments that

> in infancy the motions of the soul-circles in human beings are perturbed and distorted by the inflow of nourishment and of sense-impressions, and 'when they meet with something outside that falls under the Same or the Different they speak of it as "the same as this" or "different from that" contrary to the true facts, and show themselves mistaken and foolish'.
>
> (Cornford, *Plato's Cosmology*, p. 96; Plato, *Timaeus*, 44a)

Once the individual becomes older, the motion of the soul begins to govern the body and the right names are given to what is the same and what is different, in such a way that the individual becomes rational in accordance with the rational movement of the World Soul. It will be noticed that Plato is not giving an intellectualist account of what it is to be rational and to possess rational and ordered motion. The World Soul itself is not pure intelligence. Owing to its existence with a perceptible body, it may be said to possess emotions and feelings in the same way as the human soul experiences such things as they arise from an encounter with the body.[38]

One may suspect that Plato is again erring towards a dualistic account of the role of soul and body within cosmology. This might be the case were it not for the body's own exaltation through its production of time. Time, for Plato, is the moving image of eternity.[39] The Living Being is eternal, and by necessity this characteristic cannot be fully conferred on anything generated, so Plato believes that the Demiurge created an everlasting likeness which moves according to number, 'that to which we give the name Time'.[40] It might be thought that the nearest one could come to the unchanging realm of being would be a situation of rest or *stasis*, for this would remove the corruption of change. However, it is through time that the realm of becoming attains a greater approximation to the realm of being because the realm of becoming is then *most fully itself*, just as the realm of being is always most fully realised as itself. Within the realm of becoming, *stasis* actually denies the very nature of that realm as a realm of movement, change and therefore of becoming. In a sense, *stasis* would be a mere parody of the realm of being. Therefore, Plato suggests that the visible realm of becoming, the universe, more fully participates in the realm of being by the circular motion of time. Time is circular because it is coterminous with the motion of the celestial clock, namely the circular movement of the heavenly bodies. This circular time, the wheel of becoming – birth, growth, maturity, decay, death, rebirth – is a non-identically repeated cycle marked by the movement of the years which are in turn the movement of the heavenly bodies. Curiously, this cyclical motion incorporates the eternal in its repetitive nature. Yet time can never be a tedious or repetitive motion because its non-identically repetitive nature indicates that the eternal, in which time seeks to participate, can never be exhaustively disclosed. This point can be made with reference to the seasons. The realm of being comprehends the seasons together in a complete fashion. However, it is better for the realm of becoming to comprehend the seasons one by one through the circular motion of time rather than contemplate, for example, only a perpetual and static winter. Therefore, far from being denigrated, it is the physical motion of time which is exalted to the status of disclosing what is only fully comprehended in the eternity of the realm of the Forms. It is also in the physical motion that the realm of being is most fully itself as the realm of change and does not lapse into a parodic form of being. It is important to remember that Plato does not view the *telos* of the cosmos as finally entering into being. The *telos* of the cosmos is to be the cosmos most properly in relation to that from which it emanates. The physical motion of time is a crucial means of deepening that participation.

Therefore, motion for Plato originates in the World Soul, and reaches its most perfect and rational form in the movement of the circle of the Same. This motion is dispersed throughout the cosmos through the circular motion of the Different. The myriad souls of the universe partake of this motion, as also does the physical realm by means of the motion of the heavenly bodies. Because the World Soul has a composition of Existence, Sameness and Difference which is intermediate between the realm of being and becoming, between the realm of the Forms and the cosmos, it acts as a mediating element in the visible realm's comprehension both of the invisible and of itself. The motions of the soul are primary for Plato, while the physical motions are dependent on those of the soul. However, this does not drive a division between the soul and the body, for it is by the circular motion of the heavenly bodies, a motion that constitutes time, that the realm of the visible seeks to comprehend all that is held together in the realm of the intelligible. Through the motion of time the cosmos becomes more fundamentally itself, that is a realm of change and becoming which emanates and seeks to participate in the realm of the Forms. In this sense, the motion of the world's body is a kind of self-transcendence as its operational capacity (the means of comprehending the eternal realm of being) has a higher potency than its material being (the definition of the circle of change and time).

Reason, necessity and the power of rhetorical persuasion

It is thus that Plato describes the relation between the World Soul and the body of the universe, so establishing the nature and possibility of cosmological enquiry. The primary concern in the *Timaeus* thus far has been to elucidate the purpose of creation. Timaeus has spoken of the benevolence of the divine craftsman, 'the best of causes', and the desire to fashion a realm 'that is the best of things that have become'.[41] It is the craftsmanship of reason, seen most particularly in the Demiurge and manifest in the potent motion of the World Soul, that has been the focus of consideration in the first part of this treatise. Perhaps unexpectedly, Plato now makes a new beginning and returns to consider that upon which reason works, namely the realm of 'necessity'.[42] To what is Plato referring in his consideration of the realm of 'necessity'? To answer this question, one must realise the full importance of teleology in Plato's cosmology. This is well illustrated in Plato's middle to late dialogue, the *Phaedo*, where we find a dramatisation of Socrates's last day and execution.

Phaedo of Elis narrates a conversation between Socrates and two visitors to Athens, Simmias and Cebes. For a brief period, during a consideration of the intellectual biography of Socrates, the conversation settles on the subject of enquiry into nature and its causes.[43] Socrates recalls an encounter with someone reading from a book by the fifth-century Athenian philosopher Anaxagoras in which it is claimed that intelligence orders all things. At first, Socrates delights in the thesis that intelligence is the reason for everything, that *nous* is the ordering principle behind the cosmos. He believes that intelligence must situate and order the visible realm in the best possible way so that any enquiry into an element of nature will ask how it is best for that thing to exist, or act, or be acted upon. Socrates claims that explanations should always refer to 'the best, the highest good'. So, for example, Socrates expects Anaxagoras to explain the shapes and positions of the earth and planets in terms of what is best and the finest exemplar of the benevolent cause of the cosmos.

> For I never imagined that, when he said they [the sun, the moon and the stars] were ordered by intelligence, he would introduce any other cause for these things than that it is best for them to be as they are. So I thought when he assigned the cause of each thing and of all things in common he would go on to explain what is best for each and what is good for all in common.
>
> (Plato, *Phaedo*, 98a–b)[44]

However, to his great disappointment, Socrates soon finds that Anaxagoras explains the phenomena of nature in immanent and mechanistic terms. He mocks the way in which Anaxagoras might explain his sitting with the visitors to Athens in terms of the bones, sinews, joints, sockets and muscles of the body, or their conversation in terms of the operation of auditory sensations and vocal sounds. These explanations are wholly insufficient for Socrates. His explanation for his sitting and conversing with the visitors refers to his condemnation by the Athenian authorities and his decision that

> it was best for me to sit here and that it is right for me to stay and undergo whatever penalty they order. For, by the Dog, I fancy these bones and sinews of mine would have been in Megara or Boeotia long ago, carried thither by an opinion of what was best, if I did not think it was better and nobler to endure any penalty the city might inflict rather than to escape and run away.
>
> (Plato, *Phaedo*, 98e)

A proper explanation requires reference to a *telos* reaching far beyond merely mechanistic causal explanation. Socrates explains his sitting and conversing with reference to justice, and ultimately with reference to what is best, to the way in which his actions relate to the Good. He makes a crucial distinction between 'a cause ... and the thing without which the cause could never be a cause'.[45] The mechanistic or immanent explanation of Socrates's position is necessary, but wholly insufficient, to a full understanding of his circumstances.

This distinction between what constitutes a necessary and a sufficient causal explanation informs Plato's consideration of 'reason' and 'necessity' in the *Timaeus*. For Plato, 'necessity' refers to the realm of chaos which preceded the creation of the cosmos by the Demiurge. It is referred to as 'necessity' for two reasons. First, because it is indeed a necessary cause of the existence of an ordered and visible realm of becoming (just as the bones and sinews were a necessary cause of Socrates sitting and talking with the visitors to Athens). The Demiurge required raw materials with which to work in fashioning the universe after the Form of the Living Creature. Secondly, this realm of 'necessity' is one of mechanistic causes in which elements interact 'of necessity' but with no reference to a *telos* which transcends this mechanism. In contrast to modern accounts of causation, Plato, like Aristotle who followed him,[46] associates 'necessity' with 'chance'. On this view, phenomena are the result of causes which cannot act other than the way they do, which are necessary but not ordered to any transcendent end and which can therefore give rise to chance. 'Necessity' is contrasted with 'purpose'. To use Aristotle's example, rain may be said to fall because of the necessary cooling of evaporated moisture, its necessary condensation and eventual fall to the ground. However, in a cosmos which is explained with no reference to final causes, this rain may be said to fall for a multitude of purposes – to flood houses, to drown, or to grow crops. On this view of the cosmos, rain has no 'purpose', and its coincidence with any other phenomena of nature would be regarded as mere 'chance'.

Plato regards the realm of 'necessity' with which the Demiurge begins as one of mechanical causes with no distinct purpose; it manifests chance. These mechanical causes are not arranged in any order, and it is for this reason that this realm is properly described as 'chaotic'. The realm of 'necessity' is called the 'errant cause' by Plato. It is a necessary, but wholly insufficient, cause of the universe. They are things which are 'incapable of any plan or intelligence or purpose'.[47] So, given this atavistic disorder, how does 'reason' in the form of the Demiurge act upon this realm of 'necessity' and 'chance' in order to give rise to a purposeful and ordered cosmos?

Plato employs a highly suggestive phrase in describing the interaction between 'reason' and 'necessity': 'this universe was fashioned in the beginning by the victory of reasonable persuasion over necessity'.[48] In understanding this phrase, one must keep in mind the image of the Demiurge as craftsman. A good craftsman knows the characteristics and possibilities inherent in the materials which are available. No good craftsman will attempt to make a saw from wood. In other words, a good craftsman selects material best adapted to his purpose and cannot 'force' those materials to take on forms contrary to their natures. All the causal elements behind the production of an artefact combine to effect their product: the purpose of the craftsman, the work of his hands and tools, the nature of the materials to hand. All this is highly suggestive of 'persuasion' as opposed to 'force' or 'compulsion'. Plato is emphasising that reason and necessity work in co-operation in such a way that the universe is not the result of an arbitrary or violent imposition of the will of the Demiurge. Rather, the Demiurge 'persuades' the realm of necessity, the realm of chance and chaos, to yield an ordered and purposeful cosmos.

In considering the action of the Demiurge as 'the victory of reasonable persuasion over necessity', Plato is suggesting that the power which formed the universe is analogous to the power of rhetorical persuasion. One might therefore learn something of Plato's understanding of the persuasive power behind the cosmos by a consideration of his understanding of the practice and purpose of rhetoric.[49] Plato's mature understanding of rhetoric is most clearly expressed in the *Phaedrus* where we find Socrates outside the walls of Athens on the banks of the River Ilissus.

Together with his interlocutor, Phaedrus, Socrates commences a consideration of the art of rhetoric. Phaedrus reads a speech delivered by Lysias concerning the advantages of a relationship with a non-lover over a relationship with a lover.[50] Socrates expresses a dissatisfaction with the speech because of its repetitious and ill-formed character.[51] However, it is Lysias's inclusion of copious points which most impresses Phaedrus, for this apparently renders the former's account of the subject exhaustive.[52] Socrates, in his displeasure at Lysias's speech, claims that he can deliver a speech quite different from that of Lysias and yet just as good. Phaedrus delights in the possibility of another speech offering yet more points on this same subject, and so persuades Socrates to give his own oration concerning the favour to be expressed towards the non-lover rather than the lover.[53] Yet this attempt by Phaedrus to elicit another speech from Socrates is more indicative of compulsion or extortion than proper persuasion.[54]

Having delivered his speech favouring the non-lover over the lover, Socrates realises that his recitation on this subject was 'foolish and somewhat impious'.[55] He remembers the divinity of Love and therefore offers a recantation of this first speech and begins a consideration of the advantages of the lover over the non-lover. Socrates's first speech is now seen as an example of the deception that is possible in rhetoric when the subject matter is one about which the speaker and hearers are not clear, in this case the character and purpose of love.[56] Therefore, Socrates delivers a second speech in praise of the lover, one which ends with a prayer to Love for the acceptance of his words and the turning of Phaedrus and Lysias towards love and philosophical discourses.[57] This second speech of Socrates features the characteristics of proper rhetoric and the art of persuasion upon which he will focus in his conversation with Phaedrus. To what view of rhetoric does this initial exchange lead?

Socrates's speech, and the art of rhetoric, begins with a consideration of the soul, for the goal of a speech is 'to lead souls by persuasion'.[58] Therefore, if one is effectively to persuade one's hearers, it is necessary to know the nature of the soul, whether it is one or multiform and varied, to what end it acts and the manner of its being acted upon.[59] Given this knowledge of the nature of the soul, the rhetorician 'will classify the speeches and the souls and will adapt each to the other, showing the causes and effects produced and why one kind of soul is necessarily persuaded by certain classes of speeches, and another is not'.[60] Socrates identifies the essence of soul as self-movement. The soul never ceases to move, it is the source and beginning of all motion and is therefore immortal. That which moves itself, says Socrates, must be the ungenerated and indestructible source of motion, for otherwise all the heavens and all generation must fall into ruin, stop and never again have a source of motion or an origin.[61]

Socrates claims that the characterisation of the soul as immortal is as much as can be said before resorting to the allegory of the charioteer as an explanation of the form of the soul. He likens the soul to a winged chariot and its pilot, pulled by two horses. It is claimed that when the soul is perfect and in possession of full wings, it rises up to the heavens, partakes of the nature of the divine and is nourished by 'beauty, wisdom and goodness', while any soul which falls below is destroyed by 'the opposite qualities, such as vileness and evil'.[62] It is the souls of the gods, likened to chariots pulled by two good and obedient horses, which can mount up when at 'a feast or a banquet' to the outer surface of the heavens to be carried around by the revolution of the heavens to behold the eternal realm which 'is visible only to the mind, the pilot of the soul'.[63] However, the souls of embodied humans, unlike those of the gods, are hindered in their flight by unruly passions, represented by an errant horse pulling the chariot. The charioteer is constantly distracted by the need to compensate for the unruly behaviour of this errant horse. Thus the chariot finds great difficulty in rising to contemplate the eternal and unchanging realm, but is instead pulled back towards the vicissitudes of the realm of opinion, there to be deprived of those good things which nourish the wings of the soul.[64]

Through the mouth of Socrates, Plato claims that the soul may rise to partake of the revolution of the heavens to make that perfect motion its own. Once again, as in the *Timaeus*, this motion of the soul is contrasted to the wandering which ensues when the unruly passions drag the soul down to the confusion and conflict of the realm of opinion. The crux of Plato's argument appears to be that the soul must rise to the point where its own motion is indistinguishable from what might be described, in the terms of the *Timaeus*, as the motion of the World Soul. This attainment of the motion of the World Soul is termed a kind of 'madness' because the life of the soul exceeds its own solitary capabilities in ways that might be regarded as strange and unexpected.[65] This participation in the motion of the World Soul is thus a kind of self-abandonment to partake in a greater beauty and perfection, and yet in that self-abandonment the soul realises its own inherent capabilities all the more. In Socrates's speech, this participation in the more perfect motion of another is contrasted with Lysias's preference for the wilful self-control of the soul by a refusal to exercise the soul's proper desire, its nourishment in Love. The affection of the non-lover, says Socrates, will only cause the soul to be 'a wanderer upon the earth',[66] forever confined to its own solitary innate capabilities.

Having put forward his elaborate allegory of the charioteer, Socrates can identify the goal of rhetoric as the persuasion of the soul towards the Good by means of the soul's ever intensifying participation in the perfect revolution of the World Soul. In other words, rhetoric can bring to a halt the soul's wanderings. Therefore, any speech must itself be ordered and in proper proportion. So Socrates clearly states that

> every discourse must be organised, like a living being, with a body of its own, as it were, so as not to be headless or footless, but to have a middle and members, composed in fitting relation to each other and to the whole.
>
> (Plato, *Phaedrus*, 264c)

The protocol of ordering and arranging is a process of dialectic, of dividing and collecting things by classes, trying not to break up any part 'after the manner of a bad carver',[67] and these processes of division and bringing together are regarded by Socrates as 'aids to speech and thought' by which one can be led to the truth.[68] He also claims that a speech must be in right proportion to its subject matter and that the rhetorician must know the proportions of his art, just as the musician is aware of the harmony which is contained in the proportions of the notes which lie between the highest and the lowest.[69] Finally, it is conviction in the truth, which is, of course, coterminous with the path of the Good, which is the ultimate *telos* of the art of rhetoric.[70]

It is now possible to delineate the similarities between the art of rhetoric as described by Socrates in the *Phaedrus* and the triumph of the Demiurge in persuading the realm of necessity to yield an ordered and rational universe. Like the rhetorician in the *Phaedrus*, the Demiurge works with, and not against, the materials to hand: there is no arbitrary imposition of a will alien to the cosmos or the hearer of a speech. Socrates claims that the deliverer of any speech must know both the subject matter at hand and the souls to which the speech is addressed so that all might be in harmonic proportion. Likewise, the intelligence which persuades necessity into order is aware of the possibilities within this realm as a craftsman is aware of the possible forms which his materials might possess. It is motion, however, which forms a crucial link between the creative and persuasive arts in the *Phaedrus* and *Timaeus*. As was stated above, Plato claims, through the second speech of Socrates, that the soul may rise on its wings to sit on the outer regions of the heavens, making the rotation of the World Soul its own and thus being in such ordered and perfect motion (one might almost describe this as a 'motionless motion')[71] that a contemplation of the eternal realm of the Forms is then possible without the distractions of wandering change. Similarly, once Plato has described the formation of the World Soul and the nature of its priority over body in the *Timaeus*,[72] he then begins to describe how the Demiurge persuades the realm of necessity away from chance wanderings to assume, in differing intensities, the perfect motion of the World Soul as its own.[73] Yet the intelligence which persuades the realm of necessity and the speech which persuades the soul towards truth begin to be eclipsed by the brilliant light of the Good which is their common end. The goal of the persuasive activity of the Demiurge is not the Demiurge himself; rather, it is the deepening participation in the Good through ordered motion. The same is true of the goal of the rhetorician in the *Phaedrus*. Whereas for Phaedrus and Lysias the goal of speech writing was the praise of the writer,[74] the goal for Socrates is the praise of the divine truth, expressed, for example, in the hymn to Love at the end of the second speech, and the drawing of the soul toward the Good. Just as a sculptor 'persuades' a sculpture to emerge from a block of marble, so too the Socratic dialectic and the essentially dialectical process of bringing order from the realm of necessity draw out what is latent within the materials to hand. Meanwhile, both the rhetorician and the craftsman fade before the new beauty which emerges from the dialectical motion of persuasion.

This dialectic of persuasion is also latent within the second part of what was intended – along with the *Timaeus* and a putative third treatise – to be a trilogy. That second treatise, the unfinished *Critias*, begins to describe human history, and particularly the formation and government of the mythical city of Atlantis. Through the character of Critias, Plato describes how, 9,000 years earlier, the gods had divided up the earth into kingdoms, 'not according to the results of strife' but by the acceptance of the just allocation of land. The gods looked after the inhabitants of each land not by using any physical means of control, but rather 'they directed from the stern where the living creature is easiest to turn about, laying hold on the soul by persuasion, as by a rudder, according to their own disposition'.[75]

The land of Atlantis came to Poseidon. Critias tells of how Poseidon fathered five sets of male twins by Cleito, the daughter of two inhabitants of Atlantis. These ten were to become the kings of the regions of the island. Critias goes on to describe the city and buildings of Atlantis. These take a form and proportion that is reminiscent of the formation of the World Soul and cosmos by the Demiurge in the *Timaeus*. For example, the capital city of Atlantis is surrounded by a series of concentric rings of canals and land which are formed in strict proportion to each other. These rings of water mediate between the sea and an irrigated plain at the foot of a protective range of mountains upon which food was grown and on whose rivers timber and other goods were transported.[76] Atlantis is thereby depicted as a realm of co-operation between its constituent parts, both natural and human, which is achieved not by force but by persuasion. This was not, however, a realm of wholly perfect or entirely civil motions. Because of this laws were required, which were exercised with gentleness and wisdom 'in dealing with the changes and chances of life and in their [the regions of Atlantis] dealings with one another'.[77] These laws thus acted as a means of the perpetual luring of the constituent parts of Atlantis into the movement of friendly co-operation. In addition, the ritual of the pillar made from orichalcum (a mineral peculiar to Atlantis) at the temple of Poseidon in the centre of the capital formed an important part of the balanced and ordered government of the island. This ritual was celebrated every fifth and sixth year around the pillar on which was inscribed the law of Poseidon. The law maintained the harmonious balance between the regions. At this ritual, the ten kings of the regions gathered and 'took counsel about public affairs and gave judgement'.[78]

Through Critias, Plato therefore describes an order of harmonious balance and right motion which results not from force, but from the right use of persuasion through the rhetoric of law and government. It was only when that harmonious balance was upset, 'when the portion of divinity that was within them [the inhabitants] was ... becoming weak through being ofttimes blended with a large measure of mortality' that they 'lost their comeliness'.[79] In charting the course of human history, Plato describes how the now lost city of Atlantis began to desire its wealth for its own sake and adopt an attitude of lawless ambition. Along with the *Timaeus*, the first part of this trilogy is therefore pointing to what was to be an organic whole in which cosmology, human history and civil government are understood within the ambit of harmony, symmetry and proportion given through persuasion into a right motion, all aimed at drawing the elements of becoming into a mutual participation of friendship in the beauty of the Good. Thus every aspect of the realm of becoming is understood in terms of its orientation towards its teleological fulfilment.

The pedagogy and ethics of cosmology[80]

Having established the *telos* of the realm of becoming to be a deepening participation in the realm of the Forms, and particularly the Form of the Good, by means of its rhetorical persuasion into ordered motion, one may ask if the cosmological task itself participates in this movement to the *telos*. Or is the *Timaeus*, like so many of its Ionian predecessors and modern successors, an example of mere *curiositas*?

Within the process of the creation of the universe, Plato focuses on the relation of the mind of the Demiurge with the Forms.[81] However, it is not the Demiurge alone who comes to know the Forms. Plato clearly states that 'true belief, we must allow, is shared by all mankind, intelligence only by the gods and a small number of men'.[82] How may mankind come to share in belief and even intelligence? As was seen above, the World Soul is the vital ontological condition of possibility for man's knowledge of the Forms. More particularly, it is the motion of the World Soul, which is mathematical, symmetrical and rational and which orders the motions throughout the body of the universe, which prevents the 'wanderings' of the mind from taking men into the realm of mere basest opinion. The form and symmetry of the rational motion of the World Soul is particularly manifest to us in the motions of the stars and planets. Plato comments that, 'Sight, then, in my judgement is the cause of the highest benefits to us in that no word of our present discourse about the universe could have been spoken, had we never seen stars, Sun, and sky'.[83] Plato goes further to say that the opportunity to investigate the nature of the world is the greatest gift that has ever come to mortal man from heaven. Yet one must do more than merely gaze upon the heavens. It is necessary too that we use sight in order to 'observe the circuit of intelligence in the heaven and profit by them for the revolutions of our own thought, which are akin to them'.[84] Plato is here recommending the study of astronomy, that is the exercise of intellectual powers in the learning of the mathematical proportions which constitute the music of the heavens, as a means of reproducing within ourselves the unerring motion of the heavenly gods so as to 'reduce to settled order the wandering motions in ourselves'.[85] However, this study of astronomy is not even an end in itself, for the purpose is to instil in ourselves the motion of reason which is ordered towards a deepening knowledge, and therefore participation, in the Forms and most particularly the Form of the Good.[86]

In seeking this deepening participation in the realm of being, Plato's utmost concern is with the health, and therefore happiness, of the soul. Towards the end of the *Timaeus*, he claims that if man is concerned only with the satisfaction of appetites (one might here include the appetites of curiosity) and ambitions, his thoughts will be merely mortal and will pass away. However,

> if his heart has been set on the love of learning and true wisdom and he has exercised that part of himself above all, he is surely bound to have thoughts immortal and divine, if he shall lay hold upon truth, nor can he fail to possess immortality in the fullest measure that human nature admits ... he must needs be happy above all.
>
> (Plato, Timaeus, 90b–c)

One must care for the soul by 'giving it nourishment and motions proper to it'.[87] These motions, which are akin to the divine aspect in mankind, are those of the thoughts and revolutions of the universe. For Plato, it is by studying these motions that one might correct the motions in the head which were distorted at birth in such a way that man might bring the intelligence into the likeness of that which intelligence discerns and thereby win the fulfilment of the best life set by the gods before mankind.[88]

It might be thought that Plato is claiming that the study of cosmology, and astronomy in particular, is an enterprise exclusively for those who are, or who could become, philosophers. It is by studying the order and rationality of the motions of the World Soul manifest in the heavens with their harmonious and mathematical proportions that such order and rationality is instilled in the human soul. As Plato claims when discussing the motions of the soul in infancy,[89] it is only by preserving a rational and therefore ordered circular motion of thought that one comes to know anything of the eternal Forms. It is through 'bringing the intelligence into the likeness of that which intelligence discerns'[90] that one finds fulfilment in the life proposed as the human *telos*. Whereas elsewhere Plato stresses the affinity between the Forms and the human soul, in the *Timaeus* it is seen that mankind finds particular affinity and fulfilment in participating in the motions of the World Soul and the cognitive life of the Demiurge. In this fashion, Plato is maintaining that the realm of the Forms and the light of the Good are mediated by intermediate gods in the guise of the Demiurge, World Soul and heavenly ensouled beings. It is by the imitation of the eternal (which is mediated by the gods) through its regular and rational moving image that one comes to know anything of being and thus find fulfilment.

However, Plato does exhort all people to the study of cosmology and astronomy because of its immediate effect in producing an ordered and therefore more ethical intellect.[91] Although it may not be possible for all to know the Forms, it would be possible to learn of regular and rational motion by some observation and study of the motions of the World Soul seen in the heavens. However, another benefit, namely health, may be obtained by the study of the workings of the cosmos. Plato describes how ill health occurs owing to an imbalance between soul and body and that health can be restored by appropriate care and exercise of each part. He states that as our body is heated and cooled within by things that enter it, and again is dried and moistened by what is outside, and suffers affections consequent upon disturbances of both these kinds, if a man surrenders his body to these motions in a state of rest, it is overpowered and ruined.

(Plato, *Timaeus*, 88d)

So Plato maintains that one must, in the first instance, imitate the motion of the Receptacle of Becoming so as to shake the parts of the body into some kind of order.[92] As an aid to more general health, Plato observes that 'of motions the best is that which is produced in oneself by oneself, since it is most akin to the movement of thought in the universe'. So he recommends gymnastic exercise as a means of maintaining and promoting health. This recommendation, rather than emerging from a detailed biology, is made on the basis of what constitutes a proper and orderly motion, that inference being made from a study of the workings of the cosmos.

It has been seen that Plato's cosmology as expounded in the *Timaeus* is fundamentally 'theological' in character. The cosmos is identified as that which participates in and obtains its being from a transcendent source. That participation is achieved through persuasion from necessity and chance to an ordered motion, and more particularly through the acceptance of the motion of the World Soul. The creative dialectic of the Demiurge is alike in character to the art of the rhetorician as expressed in the *Phaedrus*. Both matter and the soul are persuaded to yield a beauty which is already inherent within themselves, and yet this can be achieved only through a participation in the motion of that which is greater, namely that of the World Soul. That motion constitutes time, the 'moving image of eternity', by which the temporal realm of becoming comes to know, in the succession of time, the Forms which are comprehended in total in their eternity. Thus for Plato cosmology cannot be a totalising discourse. The non-identically repetitive nature of time reveals the truth that the universe neither is 'presence' itself, nor is it fully absent. The universe cannot be fully grasped because it is never wholly still. Instead, motion is continuous and there is always more to be bestowed and seen within time's own movement. In addition, the very composition of the World Soul and its rational motion indicates the stability, beauty, symmetry and proportion of the Good. By the study of this motion in astronomy allied to the metaphysical category of music, the *Timaeus* outlines the pedagogical and ethical nature of cosmology. In a sense, therefore, the *Timaeus* is a discourse offering a theological ethics as well as a mere investigation into the workings of the cosmos. Perhaps even more unexpected, however, is that Plato's cosmology also provides a health regime based on the study of motion.

Throughout this treatise, motion is seen to be the subject matter of cosmology, the means of the mediation of the being which resides in the Forms, and the very condition of possibility for human knowledge of the cosmos, a knowledge which will see the realm of becoming in its relation to the eternal realm of being. Yet the unifying and crucial aspect of this Platonic motion is that it is teleologically orientated towards human fulfilment through a deepening participation in the Forms and particularly the Good. Motion is the means of mediating the Forms to the cosmos; it is the means by which humans may know and therefore imitate and participate in the rational and intelligible; it is therefore the means of the fulfilment in happiness through the right motions of both intellect and body. The *Timaeus* is far from mere curiosity. Through the subject of motion Plato writes a theological, ethical and pedagogical cosmology. Because motion is so central to this treatise it can be seen that the most excellent motion is recommended as the very motion of the human psyche and body, not only the immediate human body but also the body politic. This is seen particularly in the consonance between the principles enunciated in the *Timaeus* and those expressed in the successor work, the *Critias*. In the latter, the same principles of beauty and harmony which characterised the motion and unity of the cosmos also characterise the motion and unity of the political realm. These principles help to explain the course of human history in the descent of Atlantis into a disordered wandering.

The motion which characterises the realm of becoming, therefore, is both beautiful and symmetrical; it is a rational revolution which expresses the harmonic balance of the Good. Thus Platonic cosmology is not an acquisitive exercise which seeks the attainment of facts, after the fashion of an inexorable forward motion towards an indeterminate end. Rather, motion as presented in the *Timaeus* is the perfect synthesis of the ideal 'end' with the embodied means of fulfilment in that end.

2 Aristotle
Ecstasy and intensifying motion

In the previous chapter I outlined the way in which the category of motion is central to Plato's cosmology in the *Timaeus*; motion was seen to be the embodied means of gaining a genuine understanding of the universe. Such an understanding views the realm of becoming in relation to the eternal and transcendent Forms from which it receives its measure of being and truth by means of participation. In addition, within the realm of cosmology Plato alludes to the importance of motion within arenas as apparently diverse as ethics and music. Most generally, motion is central to Plato's philosophical task of referring all things to their transcendent origin and end.

In devoting a dialogue to cosmological enquiry, Plato set an example for philosophical speculation and emphasised not an escape from the uncertainty of becoming, but rather the importance of an investigation into the nature of the cosmos and its central characteristic, motion. As well as establishing an understanding of the realm of becoming, this enterprise was also to be the means of coming to know something of the unchanging reality of the Forms as this is embodied in a universe which is the moving image of eternity. Plato was soon to be followed by Aristotle in this investigation of nature. However, the student was not merely to repeat the thoughts of his teacher as these had been expressed in the *Timaeus*: the differences and disagreements between Plato and Aristotle in matters of cosmology and metaphysics have been well documented.[1] Yet despite Aristotle's revision and development of many of Plato's cosmological doctrines, we will see that they shared the crucial conviction that motion is a central concept within philosophical investigation as well and the embodied means of enquiry into the nature of the universe and its relation to a transcendent eternal.

In the present chapter a number of the themes outlined in the discussion of Plato's cosmology will be developed with reference to Aristotle's understanding of motion. I will begin with an examination of the centrality of motion in the definition of nature given in the *Physics*. Aristotle will be seen to understand motion not as a feature of individual, subsistent and self-explanatory bodies, but rather as a network of relations between natural beings as they partake in the motion of the cosmos. Yet the order of nature and motion is not mechanistic for Aristotle, but fundamentally teleological.

This theme of motion's orientation to exterior ends will lead to a discussion of motion as an intensifying and ecstatic self-transcendence in an approach towards a *telos* or goal. The distinction between motion (*kinesis*) and actuality (*energeia*) will be seen as important: just as for Plato there is no dualistic separation of being and becoming, so it seems that for Aristotle there is no formal division between the potency of *kinesis* and the actuality of *energeia*. However, there will be cause for reservation, for it will be argued that, unlike Plato, Aristotle, by postulating a single unmoved mover, forecloses any sense of relationality at the highest ontological level, and hence maintains at least a latent division between relational motion in the universe and the ultimate divine goal of that universe.

Returning to the central theme of teleology, an important issue will be discussed concerning the 'direction' or 'orientation' of motion. If Aristotle claims that motion is passage from potency to act, could motion in some way be the means not of the realisation or intensification of being, but also the means of the destruction of being in passage from act to potency? Is motion determined towards the perfection of being? This will lead to a consideration of Aristotle's wider concept of motion within the realm of ethics. I will suggest that, for Aristotle as for Plato, all motion is orientated to a fundamental goal in the Good. This orientation of motion towards the Good will be discussed in relation to the human ethical life which Aristotle understands as an ecstatic motion in the attainment of the virtues towards a life which is worthy of being its own end: the potential becomes actual and the eternal is glimpsed in time's movement as motion is intensified as actuality.

The *Physics* and nature's motion

How does Aristotle define nature and motion? At the beginning of Book 2 of the *Physics*, nature is identified by distinguishing this realm from art (the works of human contrivance) and chance (an unintended, accidental or irrational meeting of two bodies or phenomena).[2] Nature is 'the principle and cause of motion and rest to those things, and those things only, in which she inheres primarily, as distinct from incidentally'.[3] By inhering 'primarily' rather than 'incidentally', Aristotle is drawing a distinction between things which have within themselves the principles of their own making, and things which are essentially manufactured by another. A plant or a human being has within itself primarily the principle of movement (or change) and rest. A bed or a garment (to use Aristotle's examples) only have these things incidentally, by virtue of another. As well as being a primarily inhering principle of motion and rest, nature is identified in two further ways: first, as an active principle in things, namely a source of motion; and secondly as a passive principle, the ability to be the recipient of motion. Whereas the early Greeks had identified nature as residing in matter which they believed constituted the substance and the active principle of behaviour, Aristotle claims that true matter is antecedent to substance and is a pure potentiality. One would not attribute the term 'art' to what is only potentially a sculpture and has not yet received the form of a sculpture. Likewise, matter which is only potentially flesh and bone 'has not yet the "nature" of flesh until it actually assumes the form indicated by the definition that constitutes the thing in question, nor is this potential flesh or bone as yet a product of nature'.[4] Therefore, the term 'nature' does not primarily apply to an inchoate underlying matter, but rather to form which realises the matter as, for example, flesh and bone. Form and matter do not thereby stand over and against one another; rather, form is a realisation of matter as a particular kind of being. Thus Aristotle clarifies his definition of nature to be 'the distinctive form or quality of such things as have within themselves a principle of motion, such form or characteristic property not being separable from the things themselves, save conceptually'.[5]

Within this definition of nature as an inherent principle of motion, Aristotle is considering not only local motion, but also qualitative and quantitative movement.[6] Most generally, motion is

> the actualization of what potentially is, as such; for example the actual progress of qualitative modification in any modifiable thing *qua* modifiable; the actual growing of a thing or shrinking ... of anything capable of expanding or contracting; the process of coming into existence or passing out of it of that which is capable of so coming and passing; the actual moving of a physical body capable of changing its place.
>
> (Aristotle, *Physics*, III.1.201a)

In short, motion is passage from potency to act. Once again, using Aristotle's own examples, building materials pass from being potentially building materials to actuality as building materials in the motion of building. Similarly, one can consider the motions of growing, learning, healing, jumping, maturing, ageing and so on.[7] Furthermore, Aristotle is insistent that all motion always pertains between two poles or opposites, from something to something. As for Plato, motion for Aristotle is never a random 'wandering' without an identifiable origin and terminus. The following are examples of the contraries between which motion may take place: 'in substantive existence ... form and shortage of form; in quality, white and black; in quantity, the perfectly normal and an achievement short of perfection'.[8]

It is important to note that Aristotle's definition identifies nature not only as a principle of motion, but also as a principle of rest. By rest is meant not merely the absence of motion nor motion reduced quantitatively to zero, but the positive possession of fulfilment by that which was previously in motion. For Aristotle, all motion implies an attainment or fulfilment of something. As is suggested by Aristotle's insistence that motion always takes place between contraries, the identification of a *telos* is integral to his understanding of motion. If there were no *telos*, there would be no motion, for the *telos* is the reason for the motion. Motion is never for its own sake, but for the attainment of some end or goal. Therefore, if nature is the principle of motion and rest and these terms 'motion' and 'rest' refer to a fulfilment and the means of that fulfilment, nature can never be understood merely as a series of undirected discrete processes. In fact, such aimless activity would be regarded by Aristotle as unintelligible chaos. Moreover, Aristotle is not here referring only to the motion of those items in nature capable of self-movement, that is those in possession of the active principle of motion. He is also thinking of those remaining non-living phenomena which possess only the passive principle of motion. On this view, heavy or light objects, for example, have a teleologically defined motion: light bodies tend to rise up to their proper place, heavy bodies tend to fall to their proper place. This teleo-logical view, in being one aspect of Aristotelian cosmology which contrasts with modern frameworks, requires further explanation. In the following section, nature will be seen to be saturated with a purposeful motion which, in a fashion reminiscent of Plato, makes cosmological enquiry possible.

Teleological motion and the possibility of cosmology

In being thoroughly teleological, Aristotle's account of nature and motion can be distinguished from mechanistic accounts. In particular, Aristotle is writing against Empedocles's theory of 'love and strife' and the Atomists' natural philosophy, particularly those of Democritus and Anaxagoras.[9] In the latter, orderly patterns of behaviour are described in terms of necessary material interactions. For example, with reference to modern biology, a dog's systematic use of its sense of smell in hunting prey may be understood through the genetic inheritance or neurology of the dog. On the other hand, with regard to teleology, the dog's orderly use of its sense of smell may be understood in terms of the goal or end at which such behaviour aims, namely the satisfaction of the desire and need for food. However, as Jonathan Lear points out, Aristotle would reject the suggestion that mechanism within nature is a basis for teleology.[10] Such a view regards mechanism as a ground for goal-orientated behaviour in nature in such a way that teleolog-ical explanation is always thereby reducible to mechanistic explanation. However, following Lear, it is possible to clarify this distinction between mechanistic and teleological accounts of nature and motion through a discussion of Aristotle's understanding of chance and spontaneity. Aristotle regards mechanistic behaviour not as a ground for goal-orientated behaviour, but as a form of necessity and spontaneity.[11]

'Spontaneity' (*to automaton*) and 'chance' (*tuchē*) feature prominently in Aristotle's cosmology as instances of *apparent* teleology.[12] A spontaneous event is one in which 'any causal agency produces a significant result outside its aim'.[13] Such an event would have an external cause and it could have taken place for the sake of some end, but in fact did not. For example, if a man were struck on the head by a falling rock which happened to roll down a hill, Aristotle would describe this as a spontaneous event. The rock falls because it seeks its natural place downwards and is unhindered in seeking that place; the rock does not fall *in order* to hit the man on the head.[14] A chance event is a variety of spontaneity: it is an apparently purposive action by a person capable of exercising deliberate choice.[15] Aristotle describes the case of a man who goes to a market to buy goods and happens to meet someone who is due to repay him a loan on that day.[16] It *appears* that the man went to market to recover his debt, yet this was an unintended or chance outcome of a journey to the market which in fact had the purpose of buying food. In other words, a chance event is an unintended result of a deliberate action which in fact had another purpose or goal. However, it is important that neither spontaneity nor chance constitute a break in the natural order; they are merely a meeting of two phenomena which *appear* to belong together in the order of formal and final causes. If the man had not walked past the hill the rock would nevertheless have fallen to its natural place; if the debtor had not been present at the market, his creditor would nevertheless have made the journey in order to buy his food.

It is now possible to see that for Aristotle a mechanistic account of nature and motion regards phenomena as instances of necessity and spontaneity rather than genuine teleology. If the motions of nature were merely the result of necessary material interactions whose purposes and ends were decided after their occurrence, those ends would be incidental to the phenomena concerned. Aristotle claims that if, for example, teeth grew in their characteristic pattern due to mechanistic necessity, and those teeth persisted which happened to be useful for a given purpose (namely masticating certain foods), such 'natural selection' would not be an instance of genuine purposive teleology, but of necessity coupled with spontaneity; this would be, then, an instance of what merely *appears* to be goal-orientated behaviour.[17] In the case of the growth of teeth understood mechanistically, teeth arise as the fulfilment of a necessary mechanical process and only *appear* to have the mastication of certain foods as their causal end. On this understanding, it would be mere chance that teeth were useful in the consumption of foods which sustain the body.

However, for Aristotle teleology cannot be merely apparent in this way. The chance and spontaneity of a mechanistic nature, far from grounding a teleological order, actually presuppose a teleological order: 'Clearly then luck itself, regarded as a cause, is the name we give to causation which incidentally inheres in deliberately purposeful action taken with respect to some other end but leading to the event we call fortunate.'[18] To understand the extent to which Aristotle understands nature as infused at every level with substantive rather than merely apparent teleological order, it is necessary to return to the distinction between matter and form.

As was stated above, matter is merely potential and requires form in order to be realised as a particular being. Just as matter is allied to potency, so form is allied to actuality. Therefore, the necessary interactions of material alone could not account for the orderly motions of nature *from* potency *to* act, for material is merely potential; some account of form is required. Moreover,

Aristotle is also committed to the idea that every composite natural being is in fact a hierarchy of form and matter, and this provides another element of natural order for which material mechanism could not account: the transition from one level of matter and form to the next. The human body, for example, contains both matter and form, but it is also composed of the body's different parts (hands, feet and so on) which, if considered individually, might have their own individual matter and form; just as in the case of the human body, so order is understood more generally as an expression of form throughout nature, from top to bottom. Yet Aristotle believes that some account must be given of how each level of natural organisation gives rise to the next level; how, for example, do the form and matter of limbs and a torso, which are only potentially a human being, come together as that which transcends their individuality, namely an actual human body which itself possesses both form and matter *qua* human body? For Aristotle, it is no more possible for the material interactions of the flesh and bone of the parts of a human body to give rise to human life than it is possible for the material order of a paint box to give rise to a painting. Within mechanistic and necessary order there is no basis for a developmental motion from potency to act. Aristotle therefore believes in the basic ontological reality of form as that which provides the 'why' of nature's motion from potency to act. Such formal causation is also an instance of final causation, for the form is the goal and purpose of any motion; any natural being by its very nature desires its own realisation by the attainment of the form proper to that being.[19]

An important corollary of Aristotle's view is that the order of being is the opposite of the order of generation. In *Parts of Animals*, it is suggested that, for example, the matter and form of the individual parts of an animal come together first in time, but only for the sake of that which is ontologically greater or more complete, namely the animal in question. Aristotle states that, 'The order of generation is the opposite of that of being: things prior in generation are posterior in nature, and the primary thing is last in generation'.[20] There is thus a genuine hierarchy of cause allied to a hierarchy of being in Aristotle's natural teleology.

However, this conception of teleological explanation has sometimes been construed as attributing a 'conscious' purposefulness to nature and its motions in such a way that phenomena are regarded anthropomorphically. Purposive order is subsequently understood to be 'imposed' on nature by the human mind.[21] Such a view searches for some kind of psychological basis for natural teleology. However, for Aristotle the case is quite the reverse. The ability of the human mind to impose purposeful order onto natural phenomena in, for example, the manufacture of artefacts, is itself the result of nature and motions which are inherently teleological and of which human intentionality is a part. Humanity does not read teleology into nature; rather, nature's order itself gives rise to humanity's purposiveness.[22] Thus the purposeful ordering of mind is not, as it were, an instance of spontaneity or chance within nature, but rather an integral part of the teleological being of nature.

It is now possible to outline an important congruence between Plato's and Aristotle's understandings of motion in relation to cosmology. As was seen in the previous chapter, for Plato the motions of the human mind are a reflection of, and participation in, the motions of the cosmos in such a way that nature becomes to some degree intelligible by intimating through time's motion the eternal plenitude and inherent intelligibility of the Forms. Although Aristotle rejects the view that forms exist in an independent transcendent realm, nevertheless form is an ontological reality that does not exist merely in the mind. Forms are instances of that to which nature tends as its *telos*, in such a way that motion is the result of an intricate weaving of formal, final, efficient and material causation.[23] Why does form in particular render nature intelligible? Because it is form which identifies something as an instance of one particular natural being rather than another, for example a tree rather than a flower. The form is 'the *logos* of the essence'[24] of something, for it instantiates a definitive order and being into matter so that it can be defined and understood. Nature is therefore a principle of intelligible order,[25] which is to say that nature is an inherent principle of motion from potency to actuality in form. For Aristotle, cosmology is a part or reflection of this motion, for 'all men naturally desire knowledge'.[26] In other words, because humanity is not understood as outside the natural order, the motion of humanity from potency to actuality with regard to knowledge is part of nature's motion from potency to intelligibility in act. Crucially, this motion from ignorance to understanding would not be possible were it not for nature's intelligibility which arises from its infusion with a teleologically ordered motion. Thus cosmology is made possible for both Plato and Aristotle because of the ontological kinship between the motion of human knowing and the motion of the universe which are both instances of a teleological reduction of potency to actuality.

The integrated motions of nature

Having established the extent of the teleological order of nature, it is now possible to outline the way in which Aristotle differentiates between qualities of motion and, furthermore, how he describes not a wholly flattened universe, but one which is hierarchically unified. As has been seen, nature is the principle of motion towards an end and a state of fulfilment in rest. As natural phenomena have determinate characteristics expressed in their form and matter (in what they are 'by nature')[27], so they have determinate motions and formal ends. Thus magnesium does not behave like copper and goats are not born from cows. Likewise, as well as the active principle in nature's motion having a determinate end, so too those phenomena which have a merely passive principle are openly receptive to certain kinds of motion which will provide the means of their fulfilment. So a rock has a natural passive receptivity of downward motion to its proper place in the universe. Therefore, if nature as both an active and passive principle has a particular *telos*, and motion is the means of fulfilment in that *telos*, this implies that natural phenomena will be the subjects of motions which enable or hinder fulfilment in a *telos*, what one may call 'natural' or 'violent' motions.[28] Natural motions are those characteristic patterns of behaviour which are produced by a being in a given environment. One should also add that the being in question may have a certain intrinsic receptivity for 'natural' motion. For example, the fall of a heavy body to earth is a natural motion. By contrast, a violent motion is one in which there is no intrinsic intentionality of that motion within or by the being itself. Such violent or non-natural motions may be due to human control, chance or extrinsic force. Motion is not here an indifferent category: there can be good or bad, better or worse, proper or improper motions.

Allied to this distinction between natural and violent motion, which was seen to be based in turn upon a fundamentally teleological understanding of nature, is another characteristic principle of the Aristotelian understanding of motion which underlies the discussion of the first unmoved mover in *Physics* Book 8 and *Metaphysics* Book 12: 'everything which is moved is moved by another'.[29] James Weisheipl has argued that this statement has been the subject of considerable misinterpretation both by scholastic and contemporary readers of Aristotle.[30] For example, W. D. Ross interprets the principle to mean that everything that is moving must be moved by something here and now conjoined to the moving body, or at least something that is moving the body continuously: 'One body can be in movement as a result of the influence of another only so long as the other body is continuing to act on it, and is in fact still in contact with it'.[31] However, Weisheipl argues that Aristotle did not wish to claim that what is here and now in motion is moved here and now by something; he was not claiming that the mover and the thing moved are in constant conjunction. Neither does the principle claim that whatever moves, or is in motion, is moved *per se* by another with respect to the motion at hand, for it is clear that for Aristotle there are things which move without being moved in such a fashion, namely living beings within nature.[32] Thus, for example, it is not claimed that a painter is painted or a builder built. Rather, this is a logical principle which merely claims that if there is something which is moved, then it is moved by a mover which is other than itself. In other words, no one thing can be both the mover and the thing moved.

It should be noted that the so-called motor-causality principle that 'whatever is moved is moved by another' is most obviously true of the violent motion of inanimate bodies. Such motion cannot originate in the object moved because it has no intrinsic tendency to that motion. Thus the movement of a heavy body upwards is a non-natural motion which must have an external origin. The moved, namely the heavy body, must be moved by another.[33] However, in the case of the natural motion of inanimate beings, such as the fall of a heavy body, it is less clear where to make the distinction between mover and moved. When a rock falls to the ground, what is moving the rock? Where does one find a differentiation between mover and moved? Is the rock a self-mover? Some interpreters suggest that, despite Aristotle's occasional statements to the contrary, he is bound to the idea that simple bodies are in a sense self-moving.[34] At first sight, this may appear to concur with Aristotle's statement that nature is an internal source of motion and rest, and therefore an efficient cause of motion. Since in some sense the form of the rock is its realised nature, one might be led to the conclusion that inanimate objects move themselves, by their natures, to their proper places. In the case of a rock, one may be tempted to suggest that it is the form of heaviness which moves the rock.

However, Aristotle does not consider nature to be an efficient cause of the motion of inanimate bodies. He accounts for the natural motion of inanimate bodies, for example the fall of a rock downwards, by distinguishing two different senses of potentiality from which motion takes place to actuality.[35] In order to understand this distinction, consider the case of someone who is not yet capable of speaking a language. That person possesses the language potentially in a different and more radical sense than someone who already possesses the language but is not currently exercising their ability to speak the language. Aristotle claims that it is one thing to be unable to speak a language, while it is another to be able to speak a language without exercising that ability. We might call the former an instance of first potentiality, and the latter an instance of second potentiality. In relation to the motion of the inanimate elements, it is possible to see that air is in first potentiality to moving downwards because it must first be given the form of heaviness in being changed into earth. On the other hand, earth which is held up by some impediment is in second potentiality to motion downwards. Such a heavy body, in being in second potentiality to move downwards, has the nature necessary to move downwards. Such a body gains that nature from its generator and the only reason for it not exercising its nature in moving downwards is that it is impeded from so doing by another. Therefore, Aristotle claims that in the case of the motion of inanimate bodies there are two movers which can be identified in any such motion. The first is only an accidental mover, namely that which removes any impediment which prevents the natural motion of an inanimate body (for example, someone who removes a supporting column from beneath a weight). The second and more proper cause of the natural motion of an inanimate body is said to be the generator of the natural body in question. The generator gives all that is required for a natural body to move in its characteristic way. For Aristotle, this generator must be in initial contact with that which it generates; but once this new and independent substance is generated, it has everything required to move in its characteristic way – in other words, it has a nature which is a principle (*archē*) but not a cause (*aitia*) of motion.[36] Therefore, it seems that the generator is the efficient *per se* cause of the natural motion of inanimate bodies, yet the generator or mover does not have to be in constant contact with the body moved. When reading

Aristotle's writings on motion, according to Weisheipl it is therefore very important that the motor-causality principle is not understood in Ross's sense described above to mean that whatever is in motion is *here and now* moved by something by being in constant contact with a mover.

We will return to a discussion of the 'motor-causality principle' in Chapter 4 when considering Aquinas's view of the matter. There, we will find a subtle alteration to Aristotle's understanding of the natural motion of inanimate bodies. Contrary to Weisheipl's stipulations, it will be seen that Aquinas does maintain the need for a constant mover in any motion, yet such a mover emerges from a more Neoplatonic metaphysics of the divine ideas as the foundation of every motion in creation. This will draw a distinction between Aquinas's understanding of such motion and that of early modern science and the principle of inertia. The latter, of course, rejected the need for a constant mover in any motion and, in ascribing a similar view to Aristotle and Aquinas, Weisheipl is attempting to indicate a consonance between medieval and modern physics. For the moment, however, one can note that Aristotle's account of the motion of natural bodies has been regarded as a particular weakness in his physics.[37] There is no concept of 'gravity' in Aristotle's thought, and it was not until Galileo that a principle of inertia was articulated as an attempt to account for the motion of bodies. One might think that Aristotle comes near to a concept of gravity in the notion that place (*topos*) is a cause of the natural motion of inanimate bodies. However, it must be remembered that gravity refers to the mutual attraction of bodies,[38] whereas Aristotle is concerned with the motion of bodies to their proper and determined *topos* within the cosmos. In what way does *topos* cause the natural motion of simple bodies? As Helen Lang points out,[39] Aristotle does not refer to *topos* as a cause of the motion of inanimate bodies when he is considering this subject in *Physics* VIII.4. This is because Aristotle already assumes that the cosmos is determinate in relation to *topos*, which is to say that heavy things move down to their proper place and light things move up to their proper place. Within *Physics* VIII.4 Aristotle is merely concerned to identify the actuality which draws the potential to its proper actuality through motion: potency and act remain the framework within which he considers the subject, for he already assumes his account of *topos*. In *Physics* VIII.4 Aristotle identifies the generator as the cause of the motion of inanimate bodies in that the generator, for example a fire begetting another fire, gives all that is required for a particular natural motion. However, one might conjecture that *topos* is a mover of inanimate natural bodies in a general and secondary sense in being the actuality towards which a particular potentiality tends.[40] Bodies desire their actuality and are consequently able to be moved to their proper *topos*. Returning to the motor-causality principle, one can see that an object and its proper *topos* are not in constant spatial conjunction. Rather, they are 'together' through the potency–act relation, a kind of ontological kinship in which one is the *telos* or fulfilment of the other.

Given this account of inanimate natural motion, how does Aristotle understand the distinction between those inanimate beings in nature which are merely moved by others and those which appear to move themselves? Does the motor-causality principle reject the possibility of genuine self-motion by denying that some one thing can be the mover and the moved with respect to a single motion? Clearly not, for Aristotle is anxious to maintain the distinction between the living and the non-living in relation to the motion of which they are capable.[41] In outlining this distinction, he considers the movement of a living body composed of parts.[42] When a body is apparently moving itself (for example, when I walk across a room) it appears that one could say that my body is both the mover and the moved. However, in this example one can describe both a mover and something that is moved, for my body is divisible into parts; my legs are the mover and the remainder of me is the thing moved. Eventually, one will arrive at an unmoved mover which, in the case of living beings, is the soul.[43] The key fact for Aristotle is that no motion involving something moved is utterly simple, for every apparent self-motion in which the mover and the thing moved appear to be one and the same can be divided into the portions possessed by the agent (the mover) and the patient (the thing moved). It is not possible for a single simple entity to be both the mover and the moved in respect of the same motion, because these are utterly distinct processes. To postulate the coincidence of mover and moved in a single undifferentiated entity would be to break the law of non-contradiction.

However, Aristotle is now faced with the possibility of a 'flattened' universe of discrete ensouled beings which spontaneously initiate their own motion independently of one another. He rejects this possibility by maintaining that, although living beings are self-movers, their motion is not self-explanatory. Even animal movements are explained with some reference to external causes, and in order to explain this it is necessary to return to teleology. In *De Anima*, he states that

> these two, desire and practical thought, seem reasonably considered as the producers of movement; for the object of desire produces movement, and therefore thought produces movement, because the object of desire is its beginning. Imagination, too, when it starts movement, never does so without desire.
>
> (Aristotle, *De Anima*, III.10.433a)

Therefore, in a sense Aristotle describes the external object of desire as a mover of the animal. However, this motion remains as a genuine self-motion because of the deliberation or sensation which takes place in the intellective portion of the soul in addition to the action of the object of desire on the appetitive portion of the soul. An animal perceives an object of desire either through bare sensation (or, in the case of rational animals, sensation coupled with *nous*) and this provokes a desire in the soul which then initiates a motion for the attainment of the object which is the animal's goal. What distinguishes the motion of living things from that of inanimate beings is that the former motion is not fully determined in advance. Whereas a heavy or light object will always fall or rise in the absence of any impediment, an animal will not be moved by an object unless that object is perceived or understood by the soul as having significance for the animal in question. As David Furley has pointed out,

> An animal is correctly described as a self-mover, because when it moves, its soul moves its body, and the external cause of its motion is a cause of motion only because it is 'seen' as such by a faculty of the soul.[44]

It is both the external object of desire and the intellection of the soul concerning the significance of that object for the animal that for Aristotle provides the 'why' of the motion. This is to say that the animal must perceive the object as its appropriate goal and end before initiating motion.[45]

With regard to the foregoing distinctions, it is now possible to see that Aristotle does not regard the universe as a flattened series of autonomous and self-explanatory motions. There is a hierarchy of interdependent motions. Violent motions are in a sense the lowest form of motion, reminiscent of those in Plato's *khora* before the ordering work of the Demiurge begins: these are akin to 'wandering' or an imperfect *stasis* by the intimation of conflict rather than co-operation between the mover and the moved. Such motions prevent rather than provide fulfilment in a *telos*.[46] Those motions of the natural kind in which beings have within themselves the principle of receiving certain kinds of characteristic motion are the next within the hierarchy. These provide the means of the fulfilment of inanimate beings in a specified formal *telos* that is integral to their being. So too with natural motions in which the being possesses the active principle of its motion. This includes all animal motion. However, within the sphere of animal motions Aristotle makes further distinctions between those which are the result merely of sensation and the motion of rational human beings which also results from deliberation. With regard to animals which are self-moved, those capable of calculation are said to be more properly self-moved than those only capable of sensation. Aristotle states that

> in so far as the living creature is capable of desire, it is also capable of self-movement; but it is not capable of appetite without imagination, and all imagination involves either calculation or sensation. This latter all other living creatures share besides man.
>
> (Aristotle, *De Anima*, III.10.433b)

There is thus a clear distinction between animals whose self-motion results from mere sensation and those rational beings who deliberate concerning the ends at which their motion might aim. As will be seen below, this differentiation of the types of motion of which animals are capable is the basis of the identification of human beings as moral agents; that is, those whose motions are self-determined in relation to the ends they choose as significant to their proper being and its fulfilment.

Present throughout this hierarchy of motion and applicable at every level is Aristotle's principle that 'whatever is moved is moved by another'. This principle entails that the motion of inanimate bodies is never explicable with reference to the body alone; they are not self-movers because, in being moved, they are necessarily moved by another. Therefore, for Aristotle the universe is not a series of autonomous bodies whose motions require no explanation outside themselves; there is a crucial exteriority as the motion originates in another. In the case of the inanimate motions of natural bodies, this motion originates in the generator who gives to the body its nature and therefore its characteristic motion. In the case of animal motions, the motor-causality principle still ensures a differentiation not in terms of an exterior mover, but now in terms of an interior differentiation whereby the unmoved soul moves the animal in question. Moreover, in the case of animal self-movers, there are degrees of self-motion in which the more an animal's motion is explicable in relation to itself rather than another mover, the more truly it is said to be self-moved. However, even in the case of rational human beings, motion is not explained as a wholly self-explanatory and interior phenomenon. An external end is required as the object of desire which provides the 'why' (that is, the *telos*) of any motion. Therefore, by placing difference at the heart of his understanding of motion – both the difference between mover and moved and also the difference between the animal mover and the exterior *telos* of their motion – Aristotle describes a universe of relations in which any particular motion is part of the motion of another and, ultimately, a component in the teleological order of nature as a whole.

It can now be seen that, for Aristotle, any motion is not wholly self-explanatory and interior to the being concerned. In the next section, I will clarify the importance of the exterior vision of motion in relation to the distinction between *energeia* and *kinesis*. In being always related to an end or goal which is not yet contained within the moving being, motion will be seen to be the means by which any given nature can exceed its own already established being as it passes from potency to an ever intensifying actuality in the Good. This is to say that motion for Aristotle is ecstatic.

Ecstasy and intensification

Energeia *and* kinesis[47]

In *Metaphysics* Book 9, Aristotle differentiates between 'activity' or 'operation' (*energeia*) and motion proper (*kinesis*). These are distinguished by their relation to a *telos*. The former has no end or goal outside itself and is therefore its own end. It is, in a sense, perfect, self-sufficient, actual and always fully realised as itself. There is no 'limit' or determinate end to *energeia* and it may therefore be regarded not only as 'activity' but also as 'actuality'. Aristotle gives the activities of seeing, understanding and thinking as examples of actuality, for 'at the same time we see and have seen, understand and have understood, think and have thought'.[48] On the other hand, *kinesis* is directed to an end outside itself. It is for the sake of something else and never for its own sake. The movements of thinning, walking, learning and building are given as examples of *kinesis*. Thus Aristotle states that,

> Now of these processes we should call the one type *kineseis*, and the other *energeias*. Every *kinesis* is incomplete – the processes of thinning,[49] learning, walking, building. ... But [with regard to *energeia*] the same thing at the same time is seeing and has seen, is thinking and has thought. The latter kind of process, then, is what I mean by *energeia*, and the former what I mean by *kinesis*.
>
> (Aristotle, *Metaphysics*, IX.6.1048b)

In the case of *kinesis*, the *telos* is exterior and yet to be attained; in the case of *energeia*, the *telos* is immanent and 'all at once'. This is a distinction between activities which may be continued indefinitely (such as thinking or living well) and those which seek to come to an end (such as learning or building).

Motion now has a strange ontological status. It is distinguished from actuality or 'operation', and yet it is not in the strict sense a potency; motion hovers between potency and act.[50] One might ask what kind of a thing Aristotle supposes motion to be. To grasp his understanding of motion, it is necessary to see that he is not operating with a dualism of potency and act, but rather with a spectrum of degrees of potentiality and actuality. This notion of what might be called 'degrees of being' is expressed in the example of language-speaking presented earlier. To this we now briefly return.[51]

A person who cannot speak a particular language is in potentiality to speaking that language in a radical sense. This might be termed 'first potentiality'.[52] On the other hand, a person who can speak a particular language but is not in fact speaking that language at the present moment is in potentiality to speaking the language, but in a different and less radical sense. This might be termed 'second potentiality'. The state of being able to speak a language without exercising that ability might also be termed a kind of actuality (the actuality of being able to speak the language), or 'first actuality'. A person who can speak a language and is exercising that ability might be considered as being in 'second actuality' to speaking the language.

Aristotle wishes to identify motion with the constituting of a being in second potentiality to a *telos*. This is why he defines motion as 'the actualisation of what potentially is, *as such*'. One might say that before a person begins to learn a language, that person only 'potentially potentially' speaks the language. The motion of teaching a language is therefore the realisation of the student as a potential speaker of the language. This may be explained further with reference to a journey. If I am resident in Oxford, I am only potentially in Cambridge, or, one might say, only *potentially* potentially in Cambridge. When I leave Oxford on the bus to Cambridge, one might say that I am now actually a potential resident of Cambridge. In other words, in travelling on the bus from Oxford I realise my potential of being in Cambridge. Therefore, my motion on the bus towards Cambridge constitutes my potentially being in Cambridge, *qua* potentiality. It is this sense of motion hovering between act and potency that Aristotle is attempting to convey.

Once again, it is necessary to see how crucial is the identification of a *telos* in Aristotle's understanding of motion. If motion is the actualisation of something's potential *qua* potential, it must be the actualisation of a potential *to some particular thing*. This is to say that motion is the constituting of the being of the patient in relation to another external actuality, namely an actuality already possessed in some way by the agent of the motion. There is thus a crucial exteriority as the mover establishes the moved to another form which it does not yet possess.[53]

However, if motion is the constituting of a being in second potentiality to a given *telos*, how might one describe the transition from first to second actuality? In terms of a person who is capable of speaking a language but is not at present exercising that ability, is the subsequent exercising of that ability a 'motion'? Importantly, Aristotle replies in the negative.[54] The transition from first to second actuality is not an instance of motion but of what one might call 'energization'. The difference is that between motion (*kinesis*) and actuality (*energeia*). The former is directed towards an end outside itself whereas the latter has everything required within itself for the manifestation of its full 'actuality'; it is complete in a way that motion is not. Thus Aristotle states that,

> Since no action which has a limit is an end, but only a means to the end, as, for example, the process of thinning; and since the parts of the body themselves, when one is thinning them, are in motion in the sense that they are not already that which it is the object of the motion to make them, this process is not an action, or at least not a complete one, since it is not an end.
>
> (Aristotle, *Metaphysics*, IX.6.1048b)

Because motion is characterised as that which has an end outside itself, it is possible to describe motion as 'ecstatic'.[55] The being of something which is in motion is at every moment constituted in relation to that which is outside itself in the form of the *telos* of the motion. At every moment, the being of that which is in motion exceeds itself as it takes on a new form. Aristotle comments,

> For it is not the same thing which at the same time is walking and has walked, or is building and has built, or is becoming and has become, or is being moved and has been moved, but two different things.[56]

Thus although there is a constant subject which undergoes motion (frequently identified with underlying matter), nevertheless Aristotle characterises motion as an *ecstasis* in which a being may receive a new form. Because nature is identified more particularly with form rather than matter, Aristotelian motion is a genuine transformation whereby something may take on a new nature. By contrast, the being of what is actual is self-contained, it is within its own limits and, unlike that which is in motion, it is at every moment self-identical.

Because motion, in contrast to actuality, has an end outside itself, its goal or purpose is something other than motion. The essential purpose of motion is to cease, and thus it has been characterised by L. Kosman and J. Protevi not only as ecstatic, but also as 'auto-subversive', 'tragic' or 'suicidal'.[57] Motion seeks its own demise and is therefore deathly. To characterise motion as 'auto-subversive' is to draw a fundamental division between this category, on the one hand, which seeks its own disintegration, and, on the other hand, that which is replete, namely actuality (*energeia*). However, although it is clear that motion tends towards its own cessation, it is not clear that Aristotle himself views motion as 'tragic' or 'self-annihilating'. He states that

> the indefinite character attributed to motion is due to our inability to place it frankly either amongst the potentialities or amongst the realised functions of any kind of 'thing' as such. ... And the fact turns out to be that movement is a realization, but an uncompleted one; because a potentiality, as long as it is such, is by its nature uncompleted, and therefore its actual functioning – which motion is – must stop short of the completion: on the attainment of the end, the motion towards it no longer exists, but is merged in the reality. ... It remains then, as has been said, to define it as a kind of realization or attainment, but difficult to pin down.
>
> (Aristotle, *Physics*, III.2.201b)

It appears that, because motion is 'difficult to pin down' and hovers between potency and act, presence and absence, Aristotle does not regard motion as a kind of 'thing' in itself, but only as that which improperly borrows another's being, namely the being of the actuality (*energeia*) towards which it tends. It might be said that, rather than being deathly, motion is the intensifying of being as it approaches completion in actuality and partakes more fully in the being of the actuality towards which it tends. Therefore, rather than characterising motion as auto-subversive, Aristotle shows a tendency to characterise motion as an increasing plenitude or intensity of being in such a way that motion becomes 'merged' into a new actuality. Thus what binds motion and actuality together is the 'being' of actuality, for, in being the goal and the reason for motion, this guarantees motion's continuity and intelligibility. It is a notion of 'degrees of being' which prevents the rise of any dualisms between potency and act or motion and actuality. Instead, motion hovers between potency and act as it is the intensifying of the being of the potential towards the actual.

However, a contrast can be drawn here between Plato and Aristotle. For both, motion is a relation between mover and moved. For Plato, such relationality is present *within* the highest ontological level because, although the Forms are beyond motion and rest, nevertheless they participate in the being of one another and are finally unified in the simplicity of the Good.[58] By contrast, for Aristotle there exists merely one unmoved mover at the highest actuality which is not constituted by its relation to anything other than finite movers. Because Aristotle believes in the eternal nature of the moving cosmos,[59] the first unmoved mover is subject to a critique to be put later by the Neoplatonists: that is, its being is not simple and wholly one, being in some sense constituted in relation to a changing finitude.[60] Even if one considers the being of the unmoved mover in isolation from a cosmos in motion, there remains a disjunction between the relational nature of finite being, exemplified most particularly in motion, and that of the eternal one. It will be argued below that Aquinas resolves this difficulty by analogically referring all motion to the relations of the eternal Trinity.[61]

However, for the moment an important question arises concerning the 'direction' of motion. The above analysis associates motion with an intensification of being. There is an implicit ethical reference in such a specification: motion is regarded as the perfection of things in actuality as they become more truly themselves. Could motion not also be the means of a diminution of being? In other words, could motion in a sense be 'backwards' and the means of the disintegration of being? How, then, does one determine the 'direction' of motion?

The ethics of motion: place, limit and God

Initially, it is necessary to remember that for any particular being the direction of its motion is determined by its own nature. Any motion for any particular being is either natural (an intensification or perfecting of being) or unnatural (a diminution or imperfecting of being). This provides the framework for a determination of proper and improper motions. However, it was argued above that these particular motions are gathered into the motion of the whole cosmos as natural bodies share in each others' motions. Is there any sense in which Aristotle believes not just in the determination of direction of the particular motions of natural bodies, but also in the determination of the whole cosmos in relation to being? To answer this question, one may begin with a consideration of the determination of the direction of the first and most noble of motions, locomotion.

Aristotle states that 'if there must always be motion there must always be local movement, which takes precedence amongst all kinds of motion'.[62] This is not because all motion is reducible to locomotion, for Aristotle believes in the primacy not of atomised and locally material bodies, but of form within nature. Rather, locomotion is the first motion because it is capable of being the most perfect, closest in character to the eternal and thereby of the most intense variety. Within locomotion, it is circular motion in place which Aristotle believes is the first and most perfect motion, and this is associated most particularly with the motion of the first heaven.[63] This motion is most perfect because, in being circular, it has no beginning or end; unlike other motions, it has no contrary.

This primacy of circular local motion can be understood in relation to the distinction between motion and actuality. The former was said to have an end outside itself and to be thereby ecstatic. The latter, actuality, was said to be its own end and to be 'all at once'. Actuality (*energeia*) is therefore eternal, not in the sense that it is exercised unceasingly over an infinite time series, but in the sense that actuality is realised at every moment of its occurrence and that every moment is the same as any other moment. Thus the 'time' of any actuality becomes undifferentiable and perfectly one. In the case of motion, time is its divisible measure because every moment is different from the preceding and subsequent moments.[64] However, in the case of circular local motion, its end is as immanent as possible for a circle is complete. It is impossible, says Aristotle, for rectilinear motion to be so characterised. In the first place, such local motion could not be continuous by consisting in the motion of a body over an infinite straight line, for such a motion could never be complete; its end would be forever outside itself. Moreover, a body traversing a finite straight line will at some time cease and return upon itself in such a way that the movement of the body is composite and not singular. By contrast, circular local motion has no definite beginning, middle or end. No one point within the circle can be identified as a 'limit' towards which the motion tends. Any body in circular local motion is in potency only in the weak sense of being in potential to another place which is as equally the beginning, middle or end of the motion as any other place.[65] In being complete, circular local motion is nearest to actuality because it is not between contraries but is symmetrical. It is thereby the most perfect of motions in featuring an end which is realised at every moment in the form of the arrival at successive places which are equally a beginning, middle and end of the motion. Thus in being closest to actuality and in featuring a most faint potency, circular local motion can be everlasting and closest to the eternal. As the 'time' of actuality is an undifferentiable and a complete 'one', so circular local motion is 'one' and complete. Within the ambit of locomotion, rectilinear motion is therefore a faint reflection of a more perfect circular motion. Both varieties share a common factor: their motion is determined in relation to *topos*. Let us now turn briefly to a consideration of *topos* as that which determines the direction of local motion.

Initially, before considering Aristotle's definition of *topos* further, it is necessary to note that he considers *topos* in two senses.[66] First, it may be understood as the local or particular limit of any particular body. Such *topos*, says Aristotle, might be associated with the form of a body. Secondly, one might consider *topos* in a more universal or cosmic sense that includes the proper *topoi* of all things. This would be to identify the *topos* of any particular body by relating it to the ultimate limits of the whole cosmos rather than its own particular spatial limits. The distinction between these two modes of considering *topos* might be clarified as follows: it is said that I am immediately in the earth and this is my *topos*, whereas the earth is in the air and the air is in the heaven. Things are in universal *topos* in virtue of being in another, whereas they are in their immediate or 'local' *topos* by virtue only of themselves. In *Physics* IV, the main text under consideration here, it is this latter sense of 'universal *topos*' that Aristotle is primarily considering as it is common to all things. Therefore, his account of *topos* is intended to determine the 'where' of all things within the cosmos.

Aristotle identifies four characteristics of *topos*.[67] First, it is no part or factor of the body which is in *topos*, but is rather that which first surrounds the body. Secondly, *topos* is neither less nor greater than the body contained. Thirdly, a body may leave a *topos* and it is therefore separable from that body. For Aristotle, *topos* is therefore not the shape or form or surface of a body, nor the matter of a body, nor a spatial interval between container and contained.[68] Rather, *topos* is defined as the three-dimensional 'first motionless limit of that which contains'.[69] *Topos* is therefore distinguished and separable from that which is in *topos* in such a way that *topos* is not itself 'limited' (it is rather a limit), it is contiguous with that which is in *topos*, it is unmoved and therefore may 'hold' motion.[70] Finally, every *topos* can be described as 'above' or 'below' and all elemental substances have a tendency to move to their own natural *topos* in such a way that any movement or rest is characterised as 'upward' towards the outer limit or 'downward' towards the centre. Any rest is likewise identified as 'above' or 'below'. This is to say that *topos* is the limit of the first containing body because it determines motion in one of the six rectilinear directions: 'being in a *topos* means being "somewhere", and this means either above or below or in some other of the six directions; but each of these is a limit'.[71] Therefore, *topos* is a cause and principle of the determination of motion in respect of direction; and because *topos* determines the direction of motion, *topos* makes motion possible for without *topos* motion would be a nondescript and unintelligible 'wandering'. All natural elements will move to their proper *topoi* (fire up, earth down and water and air in their intermediate *topoi*) and thus *topos* for Aristotle is identified as a determining principle of the motion of bodies.[72] The relation of any body to *topos* is therefore understood within the framework of potency and act, for any body will either be in potency to its proper *topos* and attain that *topos* by motion, or be in actuality and gain its form in resting in its proper *topos*.

Local motion is therefore constituted by *topos*, for *topos* as 'the first unmoved limit of that which contains' renders all local motion determinate in respect of rectilinear direction. *Topos* therefore identifies the limits or goals of the local motion of bodies which are heavy or light. For a heavy body to fall and rest in its proper *topos*, for example, involves the attainment of actuality as a heavy body and thus the realisation of its form and goal. *Topos* determines the 'where' of both local and quantitative movement and constitutes, and therefore causes, the intensification of the being of bodies which are variously heavy or light. *Topos* can therefore determine the direction of motion in relation to potency and act, for the motion of a heavy body to a low *topos* near the centre of the cosmos will be a motion towards actuality, a motion to a higher *topos* will be a motion away from actuality, and likewise for any body in relation to its proper *topos*. It is therefore clear that *topos* identifies local motion as ecstatic in the sense that *topos* is an exterior limit not yet attained by a body in motion towards that *topos*. However, with regard to the 'direction' of motion, is there a 'limit', equivalent to *topos* in relation to local motion, which might likewise determine the direction of all varieties of motion as the intensification or diminution of being within the cosmos? To answer this question, it is necessary to turn away from the local motion of bodies to the motion of all things towards the Good, the first unmoved mover.

At the beginning of his *Nichomachean Ethics*,[73] Aristotle states that

> among the ends at which our actions aim there be one which we wish for its own sake ... and if we do not choose everything for the sake of something else, it is clear that this one ultimate End must be the Good, and indeed the supreme Good.
>
> (Aristotle, *Nichomachean Ethics*, I.2.1)

The Good to which human beings tend is identified as *eudaimonia*, namely well-being or pleasure.[74] In a much later discussion of the nature and quality of the more general notion of pleasure, Aristotle once again resorts to the distinction between *energeia* and *kinesis*, or that which has a goal or limit outside itself and that which contains its own goal and is 'all at once'. He states that

> the act of sight appears to be perfect at any moment of its duration; it does not require anything to supervene later in order to perfect its specific quality. But pleasure also appears to be a thing of this nature. For it is a whole, and one cannot at any moment put one's hand on a pleasure which will only exhibit its specific quality perfectly if its duration be prolonged. It follows also that pleasure is not a form of motion.
>
> (Aristotle, *Nichomachean Ethics*, X.4.1)

Aristotle goes on to state that pleasure is part of the perfection of an activity.[75] It is clear that pleasure is part of certain activities which are ends in themselves. Therefore, the goals of certain human 'motions', for example learning, are certain activities which contain 'pleasure' which is not itself a motion but is its own end and therefore akin to actuality.

How, then, might Aristotle's ethics be more broadly understood in analogous relation to motion? Aristotle understands training in the virtues teleologically as having an end outside itself, namely the attainment and exercise of virtue. However, the attainment and exercise of the virtues, like pleasure, are not means to a further end, but ends in themselves. The exercise of the virtues is a form of actuality and such a life is no longer in potency to anything further. The human life lived in exercise of the virtues is therefore a fully actual human life: it is a life which is worthy of being its own end and it therefore contains everything within itself. This is life lived within an ultimate end or *telos*, namely the good life.

What determines the motion of the human subject in relation to its actuality in a life worthy of being its own end? For Aristotle, what can determine such motion, and therefore what constitutes the exterior limit of all human motion – and indeed the ultimate limit of all motion in the cosmos – is the Good, which is seen to be the first unmoved mover and origin of all motion. There must be some being which is fully actual which acts as the source and goal of all motion in the cosmos and which therefore, in a fashion analogous to that of *topos* in relation to locomotion, constitutes the determining limit of all motion. For Aristotle, there must be a fully actual source of all motion because, within the order of being, actuality must precede potentiality as that which reduces potency to act.[76] In *Metaphysics* XII, Aristotle considers that fully actual being as the Good in a fashion analogous to the more specific relation of *topos* to local motion. Just as *topos* is regarded as substance and yet constitutes the ordering of the local motion of natural substances in the form of material bodies,[77] so more generally the Good is both a substance separate from the cosmos while at the same time being the ordering and determining principle of motion within the cosmos.[78] It is therefore that which is most fully actual, namely the first unmoved mover, which determines the 'direction' of all motion in relation to potency and act. The actuality of anything is its own proper good; in the case of human beings, this it the virtuous life which is worthy of being its own end; in the case of inanimate objects, this is rest in its proper *topos*. Actuality is therefore equated with the Good and likewise with the first unmoved mover such that all motion within the universe, as that which is ecstatic with an end, limit and final cause outside itself, is determined towards that one ultimate goal which is eternal perfection.

Motion can therefore be most generally understood as determined towards a unified end in the Good. As all things, by their natural motion, seek to attain their own actuality, they thereby seek their own particular good which partakes in the perfection of the whole cosmos. It is the Good which constitutes the ultimate and most general final cause of the intensification of being through potency to actuality in ecstatic motion.

In this chapter, it has been argued that Aristotle identifies motion at the heart of nature as the passage from potency to act. Such motion is explained not in terms of mechanistic chance and necessity, but rather in terms of teleology and the order of formal and final causes which render the cosmos intelligible. As for Plato, this ordered motion was seen to be the embodied means and precondition of the possibility of cosmological enquiry. Moreover, the motion of individual beings in nature was seen to partake in the motion of other beings and ultimately in the motion of the whole cosmos.

Motion was also seen to be ecstatic for Aristotle. Entities in motion transcend their own already achieved degree of actuality in being drawn into an ever intensifying being. It was argued that there is no formal disjunction between actuality and motion for Aristotle, but rather that the former is the intensification and perfection of the latter. Finally, it was suggested that all motion in the cosmos is determined in relation to the Good: motion towards the Good constitutes the intensification and perfecting of being, whereas motion away from the Good constitutes the disintegration of being. Once again, motion is teleologically ordered and determined.

So Aristotle describes a cosmology which seeks to explain the observed patterns of nature. The metaphysical and physical framework of teleology, potency and act, *energeia* and *kinesis*, substance and accident, matter and form, natural and violent, were to be the tools deployed by natural philosophers until the late Middle Ages as they attempted to explain the nature of phenomena such as motion. This therefore establishes the background for a study of the medieval theological investigations of motion. Yet as well as an anti-immanentist and anti-mechanistic understanding of nature, a crucial aspect of the Aristotelian approach to natural philosophy which distinguishes his thought from early modern science – a characteristic that will be continually revisited in the coming chapters – is the analogical application of the basic principles of a physics of moving bodies to other areas of enquiry, most notably in the present chapter that of ethics. One of the earliest attempts to study and extend an Aristotelian empirical, dialectical and analogical approach to nature and motion arises in the late twelfth century in the work of Robert Grosseteste, who was also one of Aristotle's first Latin commentators. Yet Grosseteste also worked within a prevailing Neoplatonism. I now turn to consider the work of Grosseteste, as the first to attempt a synthesis of the two traditions, and as a natural philosopher who is often thought to have anticipated the modern experimental method.

3   Light, motion and *scientia experimentalis*

Throughout the examination of the cosmologies of Plato and Aristotle in the first two chapters of this essay, a hierarchical understanding of the different sciences has been emphasised. At the base level is physics which has as its subject matter the motions of the realm of nature or becoming. This realm apparently offers little certainty for the knower. For Plato, we are confined to mere *doxa* in our investigation of nature. Yet it was seen that Plato wrote a cosmology which advocates a study of the realm of becoming which itself must take on the most regular circular motions of the cosmos. Through a particular quality of motion, the realm of becoming participates in the perfect, eternal and motionless realm of the Forms. Likewise, in adopting a more perfect motion of contemplation, cosmology can participate in the motionless certainty of the higher sciences of mathematics and dialectics. Within this hierarchy of sciences in which the lower participate in the higher, mathematics plays a crucial mediating role between cosmology and dialectics. Rather than express this in the modern parlance of reducing cosmology to mathematics, in which mathematics becomes a formal descriptive language, one might identify this as a mathematical investigation of nature which is made possible by nature's participation in the being of the realm of mathematics. For Plato, the incorporation of mathematics into cosmology elevates cosmology to participate in a higher level of being, certainty and beauty. Yet this is not to say that the cosmos is coterminous with the mathematical world. It is to say that the cosmos participates in the stability of the unchanging mathematical realm, and in turn the realm of the Forms and the Form of the Good, in its adoption of ever more perfect motions.

For Aristotle, the hierarchy and separation of the sciences is more intricate and clearly delineated. The sciences are productive, practical or theoretical according to whether their goal consists in the making of artefacts, the human performance of actions or the attainment of principles. For example, Aristotle describes the productive sciences of poetics and rhetoric, the practical sciences of politics and ethics, and the theoretical sciences of natural philosophy, mathematics and metaphysics. Within this hierarchy of sciences one finds a hierarchy of motion, yet motion or its absence is not, as one might think, that which constitutes the fundamental division between the sciences in any straightforward fashion. As was argued in Chapter 2, motion is intensified between *kinesis* and *energeia* in such a way that there are no dualisms for Aristotle between potency and act, motion and actuality, physics and metaphysics. Issues concerning motion appear at every level of the hierarchy of sciences, including the practical science of ethics. Nevertheless, in natural philosophy questions of motion are present most literally and properly. Yet these motions are related analogically to the energic actuality of the subject matter of 'first philosophy' or 'theology', all within a fundamentally teleological vision of the natural.

Having investigated the place of motion within the cosmologies of Plato and Aristotle, it is the task of the next two chapters to examine motion within the theological cosmologies and natural philosophies of the twelfth and thirteenth centuries. This was a time in which the works of Aristotle were being absorbed ever more completely into the body of learning within the universities, all within the context of a prevailing Neoplatonism. Was it the case that the new Aristotelian learning arrived in the universities of the twelfth and thirteenth centuries wholly to supplant a more traditional Neoplatonism, or was this latter tradition accommodated within the growing prominence of the Stagirite's natural philosophy?

Following the general availability of Aristotle's works in translation by the end of the twelfth century (works which were later the subject of multiple censures but which were taught within the university curricula at both Oxford and Paris by 1255),[1] it is perhaps tempting to think that natural philosophy throughout the later medieval period was predominantly Aristotelian.[2] Yet, as we will see, medieval understandings of nature and motion received prominent and central Platonic and Neoplatonic influences. Furthermore, it should be noted not only that later medieval natural philosophy is not exclusively Aristotelian, but also that many commentators on the advent of modern science frequently point both to the demise of the Aristotelian scholastic tradition and to the victory of an alternative Platonic approach and method, always prominent in some medieval natural philosophy, in the work of modern science's most prominent founders.[3] Alexandre Koyré wrote that 'the mathematisation of physics is Platonism ... from our vantage point we can see that the advent of classical science was a return to Plato'.[4] Although many now contest the so-called 'continuity thesis' first propounded by Pierre Duhem (the thesis that modern science flows from developments consonant with the natural philosophical frameworks of the late Middle Ages),[5] nevertheless it can be seen that traces of the Platonic tradition within which the likes of Galileo, Kepler[6] and Newton understood themselves to be working may be found in some form in medieval natural philosophy. Of particular significance is the work of Robert Grosseteste (*c*.1170–1253) and Roger Bacon (*c*.1220–*c*.1292). Their approach to natural philosophy bears resemblance to that of the early founders of modern science in two important ways, the first being distinctively Platonic: the belief in the importance and utility of mathematics in enquiry into nature, and the practice of some kind of experimental observation (which Bacon termed *scientia experimentalis*) in the confirmation or falsification of theory.[7]

From where does this emphasis on mathematics and the *experimentum* emerge? It has its conceptual origins in the Neoplatonic image of light as a 'formative power and form'[8] of nature and a means of knowledge by illumination. This Hellenistic tradition is the source of vivid light imagery deployed throughout early medieval Christian, Muslim and Jewish theology, for example in the works of St Augustine, St Basil, al-Farabi, Avicenna and Avicebron.[9] With scriptural precedent, light is associated with the life of God, emanation from divine being in the act of creation and the form of truth. Given that Grosseteste and Bacon were steeped in this tradition, particularly through the works of St Augustine and the mystical theology of the Franciscans,[10] it is unsurprising to find light as a central and unifying theme in their writings on natural philosophy, metaphysics and theology.[11] An emphasis on the nature and meaning of light forges a bond between observation, natural philosophy, mathematics, metaphysics and theology for a number of reasons. Initially, four of these merit particular mention. First, light was implicated in many of the most fascinating and mysterious natural phenomena: the rainbow, the halo surrounding the atmosphere and light's presence in the uncorrupted and perfectly moving celestial bodies. Secondly, it is light itself which is the form of truth and which makes all things both visible and knowable.[12] Thus observation, the *experimentum*, is intimately linked to the attainment of truth through the mediation of light, both spiritual and visible. Thirdly, through a long tradition of investigation into the behaviour of light (*perspectiva*), exemplified in the works of Euclid and Ptolemy's treatises on optics, it was known that visible light acts according to the strict patterns of a yet more real and abstract mathematical geometry. In true Platonic fashion, mathematics could then mediate between the Supreme Light or Highest Truth, and the weaker light reflected in created nature which is nevertheless an emanation from that Highest Truth. At the beginning of his treatise *De Lineis, Angulis, et Figuris*, Grosseteste writes a much quoted exhortation to the use of mathematics in natural philosophy: 'For all causes of natural effects can be discovered by lines, angles and figures, and in no other way can the reason for their action possibly by known'.[13] Finally, and perhaps most importantly, the Christian scriptures describe God as light, and Christ as the light of the world.[14] On these grounds writers such as Pseudo-Dionysius understood God to be the uncreated Light, and visible light to be God in action.[15] For Grosseteste and Bacon, to study light was to study God and all things in relation to God.

The present chapter will investigate the place of motion within this wider metaphysics and physics of light. With initial reference to the central theme of light, I will examine the role of the *experimentum* and mathematics in the natural philosophy and theology of Grosseteste, for here some have identified the faint beginnings of modern scientific procedures.[16] Beginning with his short tracts on light and truth, *De Luce* and *De Veritate*, it will be argued that a consideration of creation and light yields a distinct view of motion as a proportioned emanation from the Good in which truth is understood as illumination within the divine light. It will be seen that questions of motion are present at every level of Grosseteste's cosmological hierarchy of light. This will lead to an examination of the importance of observation within Grosseteste's natural philosophical method. However, *pace* Alistair Crombie, it will be argued that the practice of experiment is not introduced into natural philosophy to mitigate an inductive scepticism. Instead, it will be seen that for Grosseteste there could never be a 'problem of induction' as there was for the early modern scientists. Because of his Neoplatonic light metaphysics in which everything is ultimately illuminated by the Trinity, there is no dark, unillumined logical gap between the observation of singulars and the postulation of universal first principles of nature which must be traversed by a baseless intuition.

Having discussed Grosseteste's view, I will consider briefly Bacon's discussion of mathematics, *perspectiva* and *scientia experimentalis* in parts 4, 5 and 6 of his *Opus Majus*. This will lead beyond the immediate purview of the nature of motion itself towards a wider understanding of the medieval context which gave rise to the Newtonian approach to nature and, more particularly, motion. An analysis of Bacon's thought in this area will reveal intimations of the wider priorities and motivations of the early modern science of experiment. Three factors in Bacon's contribution will be seen as important: first, an emphasis on quantity rather than quality in mathematics and observation; secondly, a stress on efficient causality and force in relation to *perspectiva* and the doctrine known as 'the multiplication of species'; and finally a promotion of the technological application of *scientia experimentalis* coupled with the deployment of this science in displaying marvels within nature and thereby displaying God's existence to the sceptical and unbelievers. We begin with Grosseteste's cosmogony of light.

Robert Grosseteste: the science of light and the light of truth

De Luce

'The first corporeal form,' writes Grosseteste at the very beginning of his treatise *De Luce*, 'which some call corporeity is in my opinion light (*lux*)'.[17] In this first section, Grosseteste is attempting at once to demonstrate his opening proposition and account for the three-dimensional diffusion of matter into the universe. He begins with an observation concerning the behaviour of light: of its own accord, light spreads itself instantly from a single point in every direction (unless obstructed) and so forms a sphere. Light is dynamic and possesses an instantaneous self-motion and self-propagation.[18] Continuing with a consideration of the nature of corporeity and matter, Grosseteste presents two propositions:

corporeity is that which necessarily accompanies the extension of matter into three dimensions, yet in themselves matter and corporeity are simple substances lacking dimension. However, a consideration of matter on its own could only be conceptual: it cannot be separate from form, and the form of matter we observe in the universe is diffused into three dimensions. That which of its own accord diffuses itself in this fashion is light. Therefore, concludes Grosseteste, light is the first bodily form – which some call corporeity; it necessarily accompanies and enables the diffusion of matter into three dimensions.[19] As a conclusion to this section, Grosseteste remarks on the excellence of light. Because it is the first bodily form, it is therefore the most noble and 'comparable to forms which exist separately' such as intelligences. Within the cosmological hierarchy, light as the first corporeal form stands at the hinge between the physical and separate substances, sharing more intensely in the nobility and greater being of the higher realms. Through the mediation of light as its first form, matter is thereby exalted to share in a greater being.

Having outlined the primacy and excellence of light, Grosseteste deploys the mathematics of relative infinities to explain the finite extension of the universe from a simple point. How could a finite corporeal universe emerge from a simple point lacking dimension? Referring to the authority of Aristotle, Grosseteste states that the 'quantity' of the cosmos could not be the result of a merely finite multiplication of a simple thing such as light because the ratio between something simple and something finite is itself infinite.[20] Because a finite thing exceeds a simple thing infinitely, the primordial light must be multiplied an infinity of times in order to extend matter and produce a finite corporeal universe.

However, does this mean that the cosmos will be one spatially homogeneous and undifferentiated mass? No, because the infinities by which light is multiplied may vary. For example, Grosseteste states that the sum of all numbers is infinite and yet is greater than the sum of all even numbers even though this latter is also infinite. Importantly, there can be proportions between infinities: the sum of numbers doubled from one to infinity is an infinite, as is the sum of half of all these doubled numbers, yet 'the sum of these halves must of necessity be half the sum of their doubles'.[21] Grosseteste's final proposition relating to infinities is that the infinite sum of all doubled numbers is not related by a rational proportion to the infinite sum of corresponding halves from which has been subtracted a finite number (for example, one). The proportion which remains will only be expressible as an irrational number.[22]

From these propositions concerning the mathematical proportions of various infinities, Grosseteste claims that light extends matter into larger or smaller dimensions according to the proportionate infinities by which it is multiplied, 'for if light through the infinite multiplication of itself extends matter into a dimension of two cubits, by the doubling of this same infinite multiplication it extends it into a dimension of four cubits'.[23] Thus the extension of matter in increasing sparsity through the universe is explained by the different infinities by which the point of primordial light may propagate itself, although matter is extended in such a way that there is no void.

It is possible to see in this section of *De Luce* a Platonic understanding of mathematics. Grosseteste begins with a primordial light which is wholly single and one. Considered mathematically, unity or the one is not the first number, but the principle of all numbers. Multiple entities participate in unity, for they are multiplications of a unity which is their conceptual and ontological basis. In an analogous fashion, light (*lux*), as the first bodily form, is not merely the first body: it is the basis, conceptual and ontological, of all material extension. This material extension into multiple proportions participates in the single unity of the first bodily form which is light. This is to say that the advent of the material realm which is Grosseteste's subject matter in *De Luce* is a participation in the mathematics of the one and the many, where mathematics forms a mediatory bridge between metaphysics, which is concerned with the higher being of unchanging simplicity (into which would fall *lux*), and physics, which is concerned with the multiplicity of differentiated, complex, moving beings in the cosmos (into which would fall *lumen*). The one and the many, *lux* and *lumen*, are not in dualistic opposition: the latter participate in the former.[24]

Returning to the cosmogony of light, Grosseteste comments that light extends matter into a spherical form in such a way that matter at the extremities is most rarefied. The limits of extension are reached when matter cannot be rarefied any further. Therefore, the full potentiality of matter for extension is reached at the extremity of the cosmos and this forms the 'first body' or firmament which surrounds or contains the multiple and more dense bodies of the cosmos. The firmament itself contains nothing but primary body and primary form and is in full actuality, possessing no further potentiality to change.[25] Once the firmament has been completed, it spreads light from itself into the centre of everything for

> since light (*lux*) is the perfection of the first body and naturally multiplies itself from the first body, it is necessarily diffused to the centre of the universe. And since this light (*lux*) is a form entirely inseparable from matter in its diffusion from the first body, it extends along with itself the spirituality of the matter of the first body. Thus there proceeds from the first body light (*lumen*), which is a spiritual body, or if you prefer, a bodily spirit.
>
> (Grosseteste, *De Luce*, pp. 54–55)

This is to say that the firmament is a body which is so rarefied that it resembles spirit more than corporeity in such a way that light, as its form, is able to propagate itself more fully and completely, not being hindered by a more dense and less actual matter. This light, which Grosseteste calls a 'bodily spirit', its motion being one of instant self-propagation in straight lines towards the centre of the cosmos,[26] extends with itself the spiritual quality of the primary body. In its motion towards the centre, this light constitutes the matter found in the regions below – and contiguous with – the firmament. By a continuous motion of light towards the centre of the cosmos in which the light becomes progressively less able to maintain the rarefaction of matter, a hierarchy of nine celestial spheres is gradually established, each more dense and less 'spiritual' than the last, until the 'thickened mass' of the matter of the four elements is established below the ninth sphere of the moon. Eventually, postulates Grosseteste, the light first reflected from the firmament lacks the power to rarefy fully the elemental matter of the next region as it descends through the hierarchy. Thus matter is formed which still has potential for further condensation and rarefaction – that is, rectilinear motion up and down. This begins with fire, the most rarefied of the elements, and descends through air and water to earth.

> In this way ... the thirteen spheres of this sensible world were brought into being. Nine celestial, the heavenly spheres, are not subject to change, increase, generation or corruption because they are completely actualized. The other four spheres have the opposite mode of being ... because they are not completely actualized.
>
> (Grosseteste, *De Luce*, p. 56)

Throughout this hierarchy, light is the principle of unity and motionless perfection as well as of multiplicity and moving potentiality. Every level of the hierarchical cosmos is analogically related to the next level through the mediation of light, for the lower spheres participate in the light of the higher.

From this cosmogony, Grosseteste derives the principles of the motions of the cosmos through the metaphysics of participation.[27] He states that the inferior bodies participate both in the light of the higher and thereby also the motion of the higher.[28] It is the incorporeal power of intelligence or soul which moves the first sphere with a diurnal rotation and, through the lower's participation in the higher, all other spheres. However, just as the lower receives a weaker light which is less capable of rarefying matter, so too motion is received with increasing weakness as one descends the cosmological hierarchy. At the level of the elements, Grosseteste states that motion is no longer received in diurnal form because, although the elements participate in the primary light of the first heaven, they receive that light in such a weakened form that they adopt a less perfect rectilinear motion. They possess 'a density of matter which is the principle of resistance and non-conformity'.[29] Therefore, it is clear that for Grosseteste the primary motion is the circular or diurnal rotation of the heavenly realms, and that this rotation is itself caused by the 'glance' of 'soul or intelligence': the circular rotation of the heavenly bodies is the corporeal form of the motion of soul or intellect. By contrast, the light which is in the elements 'inclines them away from the centre ... or towards the centre' in such a way that they are 'naturally capable of being moved in an upward or downward motion'.[30]

To conclude his treatise, Grosseteste adduces the unity of the cosmos through certain mathematical proportions which are, once again, characteristically Platonic. The highest body contains four constituent parts: form, matter, composition and the composite. In other words, the form, being totally simple and devoid of composition, is akin to mathematical unity. In contrast to this fundamental unity, matter constitutes the dyad on account of a two-fold potency: a receptivity to impressions received from without, and divisibility. However, there is also composition, this being akin to the number three, for in composition we find informed matter, materialised form and the order between these two. Finally, the quaternary is 'the composite proper, over and above these three constituents'.[31] The cosmos therefore constitutes a unity because it possesses all these principles which together are sufficient and necessary for completeness: 'something ... corresponding to form and unity, and something corresponding to matter and duality, something corresponding to composition and trinity, and something corresponding to the composite and quaternity'.[32] There is a fundamental unity in the multiplicity of being within the cosmos, seen most particularly in the multiplicity of motions which emerge from the simple circular motion of the first heaven. Moreover, this whole is one of five harmonious proportions found in the first four numbers (the numbers themselves plus the whole which they constitute), these producing harmony in 'musical melodies, in bodily movements, and in rhythmic measures'.[33]

Thus Grosseteste is able to formulate a cosmogony based on light with mathematics providing not just a conceptual hinge between metaphysics and cosmology, but also the ontological mediation between the simple, motionless singularity of the first bodily form and the moving multiplicity of an extended, material creation. Mathematics is more than a convenient or formal language for describing the cosmos, because number is integral to the being of the materially extended, moving and harmonically unified creation.

It is clear from other writings, however, that for Grosseteste light does relate not merely to the first corporeal form or to visible light in creation, but supremely to God himself. In his *Hexaëmeron*, a meditation on the six days of creation, Grosseteste relates light analogically to the life of the Trinity in a fashion reminiscent of the Neoplatonic emanationist doctrine of creation. He writes that,

> From the fact that God is a Trinity of persons, it follows that God is light: not bodily light but non-bodily light. Or rather ... beyond either. Every light has by nature and essence this characteristic, that it begets splendour from itself. The light that begets and the splendour that is begotten necessarily are locked in a mutual embrace, and breathe out their mutual warmth.
>
> (Grosseteste, *Hexaëmeron*, VIII.3.1, p. 224)[34]

Thus his cosmology and cosmogony are linked with the doctrine of God through analogical participation in the supreme light of the Trinity.

However, is Grosseteste's deployment of light in describing the life of God and the formation and motions of the cosmos merely a convenient metaphor? Or is light the basis of a wider metaphysics which also encompasses a theological understanding of truth and science? To answer these questions, we turn initially to the treatise *De Veritate*, before considering Grosseteste's commentary on Aristotle's *Posterior Analytics* – the first Latin commentary on this work – and his advocacy of the *experimentum*.

Light and truth

Grosseteste begins *De Veritate* with no less than seven arguments in favour of the proposition that there is a truth other than the supreme truth. Having adduced five arguments to the contrary, he marshals the authority of St Augustine in postulating that 'everything which is known to be true is observed to be true in the light of the supreme truth'.[35] However, does the light of this supreme truth obliterate all other truth, just as the light of the sun is able to eradicate the power of other illuminaries? Just as in the case of *De Luce*, Grosseteste is posing the problem of unity and multiplicity: how is the unity of the supreme truth related to the possibility of genuinely other, multiple truth? In order to answer this question, Grosseteste first enquires into the nature of truth.

Immediately, Grosseteste outlines truth as inhering in the eternal speech of God. Rather than being a mere adequation of speech and thing, truth is found fundamentally and eternally in the 'interior' speech of God, namely the emanation of the Son from the Father.[36] Not only is the speech of the Father adequated in the highest manner to the thing of which it speaks, it is that very adequation of itself to the thing it states. This speech forms the 'exemplars' of all things in such a way that 'the conformity of things to this eternal speaking is the rightness of them and the obligation to be what they are'.[37] For Grosseteste (following the tradition of the convertibility of the transcendentals) truth, goodness and being are convertible in that something is true in so far as it is what it should be according to its idea in the divine mind which emanates from the divine being in the eternal speech of the Father.[38] 'A tree,' Grosseteste claims, 'is a true tree when it has the plenitude of being tree and lacks the deficiency of being tree, and what is this plenitude of being except conformity to the reason of tree in the eternal Word?'[39]

If truth consists in the conformity of each thing to its reason or idea in the eternal Word, this implies that in order to obtain truth one must be able to observe both the created object *and* its eternal exemplar. For Grosseteste, therefore, created truth is attainable 'in so far as the light of eternal reason is present to the person observing'.[40] Therefore, created truth is attained in so far as the light (*lux*) of the divine reason is present to the observer. Just as a body cannot be seen to be coloured without the presence of an extrinsic light, so too something cannot be known within its created truth alone.

At this point in his treatise, Grosseteste makes a subtle alteration to a basic simile concerning knowledge and illumination in order to show how created truth is not rendered redundant by divine illumination, but is instead made possible by a participation in eternal truth.[41] The familiar simile likens the divine light to the light of the sun which makes created objects visible. However, the sun obliterates and renders invisible all other sources of illumination. Grosseteste claims that a more appropriate comparison might be drawn. The highest, eternal truth is not to the other created truths as the sun is to other luminaries in the sky. Rather, the highest truth is to created truth as the sun is to colour. The sun illuminates colour which, by 'participating' in this light, reveals the body. Importantly, it is not a deficiency of the light of the sun which makes colour necessary to the illumination of a body, but a hierarchy of illumination ensures that created bodies are drawn to reveal themselves: the sun draws the colour to be colour and reveal itself as such, while the colour in turn, because it is integral to the being of the body and not a mere 'secondary quality', reveals that body as, say, a strawberry. While it is true that no truth is perceived except in the light of the supreme truth, nevertheless created being participates in this truth and therefore is said to reveal itself in a secondary but real sense. This is reminiscent of a familiar account of causation: the created light is the secondary but immediate 'cause' of created truth, while the supreme light is the primary and most potent 'cause' of truth, being and goodness. Meanwhile, a final addition to this simile of the supreme truth and the sun brings further clarity to Grosseteste's notion of truth. Just as the weak eye is not able to see colour except in the light of the sun, but cannot look directly upon the sun, so the created mind can only see created truth in the light of the supreme truth, but cannot look directly upon the light of the supreme truth. The supreme truth is always mediated to created being.

This familiar and Platonic simile of the vision of the sun and the vision of the supreme truth allows Grosseteste to make a brief speculation on the relationship between the knowledge and truth attainable by the 'impure' and the knowledge and truth attainable by the 'pure in heart'.[42] This distinction, expressed here in terms of the *immundi* and *mundicordes*, might also be the distinction between pagan philosophy and the truth which can be seen in the light of Christian revelation. However, it is clear from Grosseteste's writing that pagan learning (by which we may assume that he is thinking particularly of the newly circulated writings of Aristotle upon which Grosseteste will later compose commentaries) has a measure of truth, albeit thoroughly mediated through reflection in created being; that is, at one remove from the truly real. And yet, of course, this 'one remove' is by no means off the mark: 'there is no one ... who knows any truth, who does not also know in some manner, knowingly or ignorantly, the supreme truth itself'.[43] The pagan learning of the philosophers has its own value and is, in a more indirect fashion, a knowledge of the supreme truth. However, there is a genuine difference between the illumination attained by Christian revelation and that attained by pagan learning. It is a difference which is in turn based upon an ontological difference between the light of the supreme truth (*lux*), which is eternal, unitary and simple, and that light reflected in creatures (*lumen*) which is created, multiple and thereby less certain. The light of the supreme truth is different from the created light, and yet the latter is not autonomous; it is dependent upon the former.[44] In a similar fashion, for Grosseteste, it appears that the sciences other than Christian theology are *different*, but not understood as *autonomous*, any more than created being is autonomous from the supreme truth, or *lumen* from *lux*.

A crucial issue now arises which relates to the motion of human knowing. To what extent is Grosseteste advocating what was later to be termed 'ontologism', namely the view that the mind directly intuits knowledge of God in all its acts of knowledge, before reaching heaven? In other words, is human knowledge the result of a direct illumination of the human intellective soul by the light of the divine? If this were the case, knowledge would be akin to God's motionless knowledge because we would have a vision of the eternal and unchanging divine ideas themselves. Or, is motion an integral part of the attainment of truth for Grosseteste, in which we come to knowledge in mediated form through time? In what follows, I will argue that Grosseteste does not rule out intuited knowledge of God even within the temporal life, yet he acknowledges that, because of our weakened post-lapsarian state, we require 'motion' in order to come to truth, and this necessitates the observation of corporeal being which eventually suggests the importance of the motion of the repeated observations of an *experimentum*.

Illumination, the senses and the motion of knowing

In his discussion of Grosseteste's understanding of truth, James McEvoy is anxious to exonerate his subject of the charge of ontologism because this theory of illumination renders the difference between the knowledge attained by the blessed and the knowledge attained by the *viatores* 'a matter merely of degree'.[45] With reference to the *Commentary on the Celestial Hierarchy*, McEvoy outlines Grosseteste's belief that the direct intuition of God in this life has been attained by only a very few privileged mystics (for example, Moses and St Paul)[46] who may then be referred to as the *mundicordes* or *perfecte purgate*. This is a momentary and anticipatory sharing in the beatific vision. By contrast, knowledge that we ordinarily attain, whether of God or creatures, is of a different order. Thus the attainment of truth is understood as a hierarchy extending from the weakness of pre-scientific opinion to the direct vision of God. How might we identify more precisely the difference between knowledge attained through the mediation of created light and the knowledge of the blessed in the direct vision of God? To answer this question, it is necessary first to consider Grosseteste's understanding of universals before progressing to the importance of sensation and their link with the hierarchy of the elements.

In his commentary on Aristotle's *Posterior Analytics*, Grosseteste makes a distinction between four kinds of universal.[47] The first kind are the ideas in the mind of God which are contemplated in the supreme light by intellects separated from phantasms. These are the principles of knowledge and being. The second kind of universal resides in the angelic *intelligentia* or *luce creata* whose knowledge of all subsequent creatures is derived from a prior and direct contemplation of the divine ideas. These universals exist within the thought of the *intelligentia* and illumination of the human mind, which is not able to contemplate immediately the supreme light, comes from the light of this *intelligentia*. The third variety of universal resides in the celestial bodies. A mind which is not capable of the contemplation of the supreme light or the angelic *luce creata* may find in the light of the celestial bodies the principles of the sublunary realm which is subject to motion. The fourth variety of universal is to be found in form which, in its turn, is able to illuminate the material in which it resides and with which it constitutes a composite. It is here that one might find the immediate formal cause of things and their universal principles. Finally, in addition to the four varieties of universal, Grosseteste comments on the very lowest form of 'knowledge' attainable by the *intellectus debilis*. The 'knowledge' attained here does not concentrate upon universals at all, but is arrived at through the observation of accidents.[48] The link between knowledge and being is undone in such a way that the principles of this 'knowledge' are merely the accidents of things and not their essence (*principia essendi*). Such 'knowledge' is thereby wholly uncertain and might be more properly termed 'opinion'.

Throughout this hierarchy of universals, the latter are dependent on the former, and each mediates light to the lower levels of created being. This is a hierarchy of knowledge attainable by human beings which is also a hierarchy of being. In extending Grosseteste's immediate descriptive terms we might also understand this as a hierarchy of the 'motion' or 'change' that is involved in the attainment of truth. At the highest level of contemplation – which is the beatific vision and the highest level of knowledge – we find the universals in the actuality or 'motionless motion' of the eternal emanation of the Word of God.[49] The contemplation of these universals or 'divine ideas', which the blessed share with the angelic *intelligentia*, takes the form of immediate intuition. This might even be understood as akin to Aristotelian *energeia*: this knowledge is fully actual and 'all at once', being replete and contained within its own limits for it seeks nothing beyond itself. The principles of this knowledge are 'the uncreated reasons of things'[50] and the pure and intuitive intellect can grasp these with complete certainty without the mediation of time. The next level of contemplation is through (rather than with) the angelic *intelligentia*. This *luce creata* is a reflection of the supreme light of the divine mind, yet it contains within itself the first intimations of motion proper, for this knowledge is not necessary and has proceeded into being.[51] In his *Hexaëmeron*, while describing the nature of the movement between morning, evening and the return to morning in the Genesis narrative, Grosseteste describes the 'motion' of the knowledge of the angels as a movement between the light of their contemplation of the divine light itself and the 'evening' (that is, the relative darkness) of their contemplation of their own natures through themselves.[52] In the 'evening' of their contemplation, they know that they are not God and are thereby returned once again to the 'morning' of the contemplation the divine nature. He writes,

> Therefore, this circling round from morning to morning in the knowledge of the angels is by nature a day. But these days have no temporal succession, and the seven days which are here recorded did not follow one another in time, but existed all together in the knowledge of the angels.
>
> (Grosseteste, *Hexaëmeron*, II.7.1)

Moreover, we have already learned from *De Luce* that the *intelligentia*, which is the first reflection of the supreme light, is the incorporeal moving power which moves the celestial spheres with a diurnal motion.[53] So at the next level of the universal we find the celestial spheres which contain within themselves the light or principles of everything that occurs below within the cosmic hierarchy. These bodies possess a diurnal rotation which is a motion most akin to the created actuality of the angelic *intelligentia* and the 'motionless motion' of the supreme light. As was seen in the discussion of circular motion in Aristotle, this diurnal rotation can be regarded as the most complete and self-contained local motion for, unlike rectilinear motion, it does not seek an end outside its own self-delineated limits.[54] Finally, the universals of the celestial sphere impart motion to the lowest level of universal, namely the forms of created beings. This final motion may be rectilinear to a greater or lesser extent depending on a created being's particular susceptibility to condensation and rarefaction.[55]

In addition to this hierarchy of universals and motion, there is also an important hierarchy of the elements and the senses. The cosmic hierarchy of elements begins at its height with the serenity of the heavens and the light of the 'fire' of the celestial bodies, and descends through the air to water and earth.[56] Coupled to the elements is the hierarchy of sense. Grosseteste quotes Augustine to the effect that in sensation we do not find simply the action of an object on the passive senses of a subject. The encounter between a sentient creature and the creature which is perceived is one in which the action of one upon the other is reciprocal. The sensitive soul, we are told,

> *acts* in the eyes through the pure shining fire 'when its heat is suppressed and made into its pure light. But in hearing the fire penetrates by its heat into the more liquid air. In smell the pure air comes through and reaches the moist exhalation. … In taste it comes through

and reaches the moisture that is more bodily: and it goes into this and
through this when it reaches the heaviness of the earth, and makes the
last sense, that of touch'.

(Grosseteste, *Hexaëmeron*, VI.1.3)[57]

Whereas this passage puts forward the notion that the sensitive soul acts
through the senses on the bodies sensed, elsewhere Grosseteste outlines what
was later to be known as the doctrine of 'the multiplication of species':
everything in the sublunary world acts on its surroundings through an
emanation of a likeness of itself.[58] Grosseteste states that,

> A natural agent multiplies its power from itself to the recipient,
> whether it acts on sense or matter. ... But the effects are diversified by
> the diversity of the recipient, for when its power is received by the
> senses, it produces an effect that is somehow spiritual and noble; by
> contrast, when it is received by matter, it produces a material effect.
>
> (Grosseteste, *De lineis, angulis, et figuris*, p. 60)

Within this exchange between sentient creature and that which is sensed,
one finds fidelity to the principle that 'like knows like': sight perceives 'fire'
or the rarefied *lumen*, hearing perceives the relatively rarefied air, smell
perceives the thicker air, taste perceives liquid or water, and touch perceives
earth. However, as we might expect, all sensation is linked analogically to
light, for 'light is the instrument of the soul in sensing through the senses of
the body'.[59] Again, Augustine is quoted to the effect that light is that by
which the soul acts in all the senses. Thus sensation is integrated into a
general light metaphysics, for just as each of the elements is a more or less
dense form of light, so too the sensitive souls of animals in their activity of
sensing use a form of light which is akin to the element which is being
sensed. Both the sentient creature and that which is perceived by the senses
are constituted and related in the activity of light. Moreover, this emanation
of species from all things, in being a form of light, acts in accordance with
the principles of geometrical mathematics. Thus Grosseteste can claim that
'all causes of natural effects can be discovered by lines, angles and figures'.

However, it is important to note that, just as the lower universals are
analogously related to the higher by participation, so too are the senses. For
Grosseteste, the lowest sense is that of touch because this is concerned with
inert earth which, amongst the elements, is the most dense and least actual
form of light. The sense of touch is analogously related to taste, then smell,
hearing and finally sight. Yet sight itself, the highest of the bodily senses,
is analogously related to the vision of the inner mind possessed by those
who contemplate the angelic *lux creata* and the divine light itself. At the
highest level of knowledge and 'sight' in the beatific vision, we 'see' the
light of the supreme truth with the purified and unencumbered eye of the
mind. This vision of God is the culmination by grace of the restoration of

humanity in the image and likeness of God.[60] Because this knowledge is fully replete, actual and unmediated, it involves no change or motion. This beatific vision is an activity of the soul which no longer requires the mediations of bodily sensation.

A link is established between, on the one hand, the universals and motion, and, on the other hand, the elements and the senses. Before coming to a conclusion about how we might draw a distinction between knowledge attained through the mediation of created light and knowledge in the beatific vision, it is first necessary to clarify briefly one more connection between motion and the particular knowledge achieved by corporeal sensation. Initially, one can note that sensation is the result of the motion of the species or likeness of a creature to reside in the sensitive soul of an animal. Sensation is, therefore, already a kind of abstraction, because 'the outward sense is a power of receiving and grasping sensible likenesses without matter'.[61] The various species which arrive in the sense organs are then collated in the 'common sense' (*sensus communis*) to form a less fragmented and more integrated sense impression. From here, the united 'likenesses' of the perceived object are passed into the memory. Properly speaking, it is the imagination which retains the forms which have been sensed, while the memory proper includes the estimative power of judging the forms received. The motions from the senses to the memory are shared by all rational animals. The final motion involves the excitement of reason by many memories that are held in the soul. These motions involved in sensing are described in the final chapter of Grosseteste's commentary on Aristotle's *Posterior Analytics*.

> In those who have this sense as well as retention, there is a gathering of one memory from many sensations and this is common to brute animals and rational beings; but in rational beings it is already the case that from many memories, once reason is excited, an experience is formed; this is not the case with non-rational animals. Therefore, from sense comes memory, from the repetition of many memories an *experimentum* and from the *experimentum* the universal which is apart from the particular, yet not separate from the particular but is the same as them, namely as the principle of both art and science.
>
> (Grosseteste, *Commentarius in Posteriorum Analyticorum Libros*, II.6.33–40)

This whole systematic procedure of sensation, which is already a form of abstraction made possible by the emanation of species due to the dynamism of light which constitutes the more or less rarefied sensible elements of fire, air, water and earth, was to yield universal principles of nature, what Grosseteste called *universalia complexa experimentalia* (complex experimental universals).

The hierarchies of universals, elements and sensations in which the higher mediate light to the lower also feature ever more perfect motions as one ascends towards the 'motionless motion' of the supreme light which is the source of all goodness, being, truth and knowledge. The distinction between the knowledge attained in the mediation of the created light and that attained in the beatific vision might therefore be made in terms of motion. The latter features an immediate and intuitive grasp of truth which requires no motion. As was claimed above, we might even consider this akin to Aristotelian *energeia*, for this knowledge is replete and not mediated by time. Yet as a result of the Fall, we do not share in the beatific vision. The sin which results from the Fall is understood as a disordered desire in which the human soul seeks first after corporeal things rather than the higher light of God himself. Grosseteste argues that if the intelligence were not weighed down by the body it would receive complete irradiation, and therefore knowledge and fully actual being, from the supreme light. Because this purity is obscured by corrupt desires,

> the powers of this rational soul born in man are laid hold of by the mass of the body and cannot act and so in a way are asleep. Accordingly, when *in the process of time* the senses act through many interactions of sense with sensible things, the reasoning is awakened and mixed with these very sensible things and is borne along in the senses to sensible things as in a ship.
>
> (Grosseteste, *Commentarius in Posteriorum Analyticorum Libros*, I.14.235 ff.)[62]

Therefore, in a fashion strikingly reminiscent of the Platonic doctrine of recollection, Grosseteste states that our souls require awakening by the repeated 'motions' in time of our corporeal senses.[63] Within this context, the order of being and intelligibility is the reverse of *our* order of knowing, for fallen humanity must begin with knowledge of the less intelligible created and corporeal being by means of the motions of sensation before achieving the vision of the supremely intelligible light of the divine ideas.[64] Thus the lower one descends towards corporeal sensation, the greater the 'motion' involved in knowing.

Is there, then, a dualism between the knowledge attained in the beatific vision and that attained by fallen humanity? Just as there is no dualism between Platonic being and becoming and Aristotelian *energeia* and *kinesis*, so too there is no dualism between Grosseteste's beatific vision and the knowledge that is attainable in this life, because amidst all these distinctions the former participate in the latter. In the case of Grosseteste, all levels of knowledge are analogically related in light. Yet distinctions can be made between the different sciences which deliver different kinds of knowledge, and these distinctions may be delineated by the need to be awakened from potency to act. This is to say that distinctions may be drawn with reference to motion.

In Grosseteste's commentary on Aristotle's *Posterior Analytics*, directly after his description of the need for sensation to awaken the soul to the light of truth, the author describes what many commentators cite as the description of an experimental procedure. Having descended from the supreme light of God to consider the place of corporeal sensation in Grosseteste's metaphysics, we now turn to examine in greater detail the place of the *experimentum* in the methodology of his natural philosophy.

The *experimentum*

Commenting on Aristotle's methodology in natural science, Grosseteste describes how scientific reasoning is found in the syllogistic demonstration of the links between causes and effects, premises and conclusions. There is a distinction between, on the one hand, the knowledge of facts ('that' or *quia*) and, on the other hand, knowledge of the cause or reason for the fact ('why' or *propter quid*). Expounding this distinction, Grosseteste writes that

> science which is acquired by demonstration is acquired through a cause of the thing, which may be either a proximate cause or a non-proximate cause. What is acquired through the proximate cause is called science *propter quid* and this is most appropriately called science, and demonstration by which this science is acquired is most properly called demonstration. But that which is acquired through any cause but the proximate is said to be science and it is said to be science *per posterius*, and the demonstration by which this is acquired is *per posterius* demonstration. But science *propter quid* and science *quia* acquired of the same subject differ in two ways, for science *quia* is acquired either not through a cause or through a non-proximate cause.
> (Grosseteste, *Commentarius in Posteriorum Analyticorum Libros*, I.12.23 ff.)[65]

There is therefore a distinction between science which accumulates facts (*scientia quia*) and that which is more properly called 'science' and gives a more genuine knowledge, namely the enquiry which seeks explanation and demonstration with respect to causes (*scientia propter quid*), those causes being the four Aristotelian varieties. As Crombie indicates,[66] Aristotle uses the following syllogism as an example of *scientia quia*: what does not twinkle is near, planets do not twinkle, therefore they are near. This does not produce the reason for the fact, but only the fact. Planets are not near because they do not twinkle, but because they are near they do not twinkle. We can arrive at *scientia proper quid* by rearranging the syllogism so that the cause, nearness, constitutes the middle term: the planets are near, therefore they do not twinkle. This provides the proximate cause (nearness) and therefore the reason for the fact that planets do not twinkle. For Grosseteste, therefore, science was a search for abstracted universal or principle causes, or reasons, for effects, that knowledge coming through demonstration.

How was knowledge of these principle causes to be achieved? Aristotle maintained that the attainment of the knowledge of these principles was through an inductive (or abstractive) and then deductive process.[67] The enquirer was to begin with what was first in the order of our knowledge, namely facts (*quia*) which were observed in experience. From this, inductive inferences were to be made to more general principles or forms of causes which were removed from the experience of singulars. However, these more general principles are the causes of particular experiences and therefore they are prior in the order of nature. These principles, being prior in the order of nature, could then be the basis for a deductive inference to the explanation and demonstration of observed facts (*propter quid*). On the role and importance of sense perception within this scheme, Aristotle comments that 'demonstration proceeds from universals and induction from particulars; but it is impossible to gain a view of universals except through induction … and we cannot employ induction if we lack sense-perception, because it is sense perception which apprehends particulars'.[68]

Coupled with this process of induction and deduction is one of 'resolution' and 'composition'.[69] As an example of this method at work, Crombie describes how Grosseteste's successors, amongst them Roger Bacon, sought to describe the 'common nature' of the colours of the spectrum.[70] Beginning with the enumeration of 'composite' phenomena in which the colours of the spectrum were observed (for example, rainbows, water spray, lenses, iridescent feathers, and so on), these were 'resolved' into the simpler set of attributes common to them all. Thus the 'common nature' of rainbows, sprays and transparent containers of water producing the colours of the spectrum was 'colours produced by differential refraction'. Meanwhile, the more general common nature of the various phenomena producing the colours of the spectrum was found to be 'colours produced by the weakening of white light'.[71] This process of resolution is essentially a process of simplification. By contrast, beginning with this 'resolved hypothesis', the process of 'composition' forms a more complex hypothesis. For example, in the case of the colour of the spectrum, the rainbow was a member of that most general class which produces the colours by the weakening of white light in differential refraction. It is now possible to be more specific and add that in the rainbow this differential refraction occurs through spherical drops of water, and moreover that these drops occur in very large numbers through rain. This process continues until 'the aggregate of *differentiae* become convertible with the rainbow itself', and one arrives at an appropriate causal definition.[72]

However, Crombie claims that Grosseteste was well aware that within the process of resolution, broadly coterminous with inductive reasoning, there are two related assumptions or intuitive leaps at work.[73] First, there is the assumption that when one phenomenon is observed to precede and be connected to a second phenomenon, the former is in truth the cause of the latter. Secondly, there is the assumption that general principles (for example, that the appearance of the colours of the spectrum is the weakening of white light) apply in all particular instances (for example, the appearance of this particular rainbow).[74] With reference to the difficulty raised by this first variety of inductive leap, Grosseteste writes as follows:

Can the cause be arrived at from knowledge of the effect with the same certainty as the effect can be shown to follow from its cause? Is it possible for one effect to have many causes? If one determinate cause cannot be reached from the effect, since there is no effect which has not some cause, it follows that an effect, when it has one cause, may have another, and so there may be several causes of it.

(Grosseteste, *Commentarius in Posteriorum Analyticorum Libros*, II.5.9–14)[75]

Crombie claims that Grosseteste introduces experimental verification and falsification to mitigate these difficulties. Having begun with 'facts' given by the senses, having reasoned about those facts through resolution and then composition to arrive at the reason for the facts, it is still necessary to return once again to observation to eliminate false causes or confirm true causes. Aristotle had himself admitted the need for the verification of theory when he wrote that 'credit must be given rather to observation than to theories, and to theories only if what they affirm agrees with the observed facts'.[76] Crombie claims that Grosseteste adds to Aristotle's thought in advocating falsification as well as verification, and in the development of the method of verification and falsification into 'a systematic experimental procedure' which assumes, first, that the forms of nature always act in a uniform way so that the same cause will always produce the same effect, and, secondly, that the principle of economy in seeking the *propter quid* of natural phenomena is a real principle of nature itself.[77] This method was developed and utilised in investigations into the nature of stars and comets, optics and astronomy.[78]

Alongside this experimental procedure, Grosseteste provided some considered reflections on the nature of measurement.[79] As has been seen, the use of mathematical geometry was suggested by the Neoplatonic view that light propagated itself in straight lines and measurable angles, and closely associated with such use of number was the possibility of measurement. Grosseteste, in referring to Aristotle's understanding of time as the number of movement with respect to before and after, suggested that the rates of the local motions of bodies could be measured just as their extended magnitudes could be measured and compared. However, for Grosseteste, all measurement was in some sense a matter of mere convention and included an element of inaccuracy.[80]

It is, however, within the theological context of the implications of the Fall for human knowledge that Grosseteste, in his commentary on the *Posterior Analytics*, advocates what appears to be an experimental method.[81] He argues that an abstraction from many singulars must be made before one can arrive at a universal concept. For example,

when someone many times sees the eating of scammony accompanied by the discharge of red bile and he does not see that scammony attracts and draws out red bile, then from the frequent perception of these two visible things [he/she] begins to form a notion of the third, invisible element, that is [in this case] that scammony is the cause that draws out red bile.

(Grosseteste, *Commentarius in Posteriorum Analyticorum Libros*, I.14.254 ff.)

Once the reason is awakened, the memory leads the reason to conduct an experiment by giving someone scammony to eat 'after all other causes of red bile have been isolated and removed. ... And this is the way by which one proceeds from sensation to an experimental universal principle'.[82]

Grosseteste here outlines an experimental procedure which requires both the exclusion of possible causes not included in a central hypothesis, and repeated observation. Although this example of experimental procedure which refers to scammony was derived from Avicenna, Crombie sees in Grosseteste's method an attempt to overcome a 'logical hiatus' in scientific procedure which is highly reminiscent of modern Humean concerns. Crombie claims that Grosseteste knew that a gap had to be traversed between, on the one hand, the assertion of a formal definition or a regularly occurring series of events, and, on the other hand, the assertion of a theory stating a universal and causal connection. He writes that

to leap this gap in the logical process of induction he envisaged an act of intuition or scientific imagination, corresponding to Aristotle's *nous*, by which the mind reflecting on the classification of facts produced by induction suddenly grasped a universal or principle or theory explaining the connection between them.

(A. C. Crombie, *Robert Grosseteste and the Origins of Experimental Science 1100–1700*, p. 71)

However, as Eileen Serene has argued, there are some significant problems in Crombie's analysis of Grosseteste's thought concerning experiment and induction.[83] In outlining these difficulties, it will be seen both that Grosseteste is faithful to the doctrine of Neoplatonic illumination in his scientific methodology, and that an experimental practice, when appropriately placed within the hierarchy of sciences, is highly conducive to his theological cosmology.

In her analysis of Crombie's reading of Grosseteste, Serene notes that there are two distinct ways of understanding induction within Aristotelian science.[84] The first she calls the orthodox view which holds that induction is a necessary but not sufficient condition for the apprehension of first principles. This is to say that something has to be 'added' to induction in order to arrive at first principles. Typically, those who read Aristotle as holding this view of induction claim that *nous* is that which is added to intuitive induction in order to arrive at first principles. However, a scepticism arises (anticipating that of Hume) because this leap of *nous* is not self-justified and its content is not clear. By contrast, Serene prefers to ascribe to Aristotle the 'empiricist' view of the inductive procedure which states that induction is a sufficient condition in the apprehension of first principles. On this understanding, induction requires nothing added to it, but is instead a *state* of the enquirer. Therefore, an alternative reading of Aristotle's view of this matter would not regard *nous* as something added to induction, but rather as a state of the knower. Serene quotes Barnes to the effect that '*nous*, the state or disposition, stands to induction as understanding (*episteme*) stands to demonstration. Understanding is not a means of acquiring knowledge. Nor, then, is *nous*'.[85]

Serene rightly suggests that Crombie ascribes the so-called orthodox view of induction to Grosseteste, in which something is added to induction in order to arrive at first principles.[86] In the orthodox interpretation of Aristotle, that which is added is understood as *nous*, and it is, according to Crombie, replaced in Grosseteste's thought by divine illumination.[87] Moreover, according to Crombie, Grosseteste makes a distinction between intuitive and enumerative induction.[88] The former relates to the supposition that actual causal connections cannot in practice be observed: all we observe are the conjunctions of phenomena, so in addition to our observations we must add a leap of intuition to form first universal principles relating cause to effect. The latter enumerative induction relates to the supposition that one can form a generalised universal principle about a genus or species from a limited set of observations. This form of induction therefore requires the assumption that nature is uniform. It seems that Grosseteste did indeed hold to this assumption of nature's uniformity.[89]

Because the assumption of the uniformity of nature mitigated the problem of enumerative induction, Crombie argues that for Grosseteste the problem of induction was focused principally on the intuitive variety where one could not directly observe the connections linking supposed causes with certain effects. Also, Crombie understands Grosseteste's notion of induction to consist in a 'leap' which requires something in addition to intuitive induction in order to arrive at first principles, this being divine illumination in Grosseteste's thought. Within this context, a very modern looking scepticism apparently arises concerning the validity of this leap and the conclusions which may be reached. Given that the action of cause on effect cannot be directly observed, the natural philosopher is faced with the need to distinguish between a number of causes which in theory could be linked to a given effect:

> Can the cause be reached from knowledge of the effect with the same certainty as the effect can be shown to follow from its cause? Is it possible for one effect to have many causes? If one determinate cause cannot be reached from the effect ... it follows that an effect, when it has one cause, may have another, and so there may be several causes of it.
> (Grosseteste, *Commentarius in Posteriorum Analyticorum Libros*, II.5.9–14)[90]

To mitigate this scepticism, Grosseteste apparently adopts two strategies. The first and most important is to include positively by verification *and* exclude by falsification some possible causes of given effects. This is the purpose of the *experimentum* and, according to Crombie, this is the significant move by Grosseteste towards a modern scientific approach to nature. However, a number of possible explanatory causes may survive the process of falsification, so a second strategy must be adopted in order to assuage the apparent inductive scepticism. This strategy is the invocation of the principle of parsimony whereby one always opts for the simplest explanatory cause available.[91] However, these strategies are not in themselves sufficient to achieve the Aristotelian goal of providing the uniquely necessitating causes of particular effects. Therefore, scientific conclusions are always revisable and probable, and experiment must forever continue in order to mitigate both intuitive and enumerative inductive scepticism.

Following the criticisms of Serene, it is now possible to examine some of the weaknesses of Crombie's account. First, and perhaps most importantly, if Crombie is correct in ascribing to Grosseteste the orthodox view of induction (or 'abstraction' – he uses the words interchangeably), then Grosseteste has abandoned a thoroughgoing theory of knowledge as divine illumination, and, at the same time, provided the means of separating natural philosophy from theology. This is because, according to the orthodox view of induction, something has to be *added* to the inductive inference in order to arrive at first principles. This establishes a dualism which separates observation by the senses and the induction in the human mind – which we might refer to as *scientia quia* – from the explanatory conclusions which may be attained – what we may refer to as *scientia propter quid*. So divine illumination is invoked as *an addition* from outside to an otherwise autonomous abstraction from observed phenomena. In other words, the theological doctrine of truth as irradiation in the divine light is apparently introduced to mitigate a weakness in a distinctive and autonomous form of knowing which otherwise makes no reference to divine illumination.

Having made this distinction between the intuitive or abstractive grasp of phenomena and that which is added, namely divine illumination, Serene points out that Crombie is able to make a further distinction between Grosseteste's supposed theoretical and practical responses to the problem of intuitive induction.[92] The theoretical response is to suppose that all certain knowledge is provided by divine illumination and this was the only source of metaphysical certainty as this light was added to our inductive grasp of phenomena. The practical response was to engage in an *experimentum* (such as that relating to scammony and red bile described above) which assumed the uniformity of nature and the principle of parsimony as well as seeking to verify and falsify explanatory hypotheses within controlled observations. Once again, this appears to constitute a tacit separation of natural philosophy from theological accounts of truth: the former adopts a 'practical' response and the latter a 'theoretical' response.

Could this be Grosseteste's view? Surely not. Serene is right to resist the ascription of the orthodox view of induction to Grosseteste. Divine illumination cannot be *added* to observation and inductive or abstractive knowledge of particulars, because the latter, as we have seen, is just as much

Light, motion and scientia experimentalis   73

the result of irradiation by the divine light. All knowledge and being are forms of light which have their origin in the supreme light – the ideas – of God. The knowledge which comes from the inductive or abstractive process is itself the effect of the species which emanate from every creature, each creature itself being a more or less rarefied form of light. The soul, into which sense perceptions enter to be gathered into the memory, is a form of dynamic, spiritual light. All of this is a more or less spiritual form of divine illumination, so divine illumination could not be 'added' to inductive or abstractive knowledge of particulars as if it were something juxtaposed. As Serene tacitly suggests, divine illumination could only be the state of knowledge which one has after a successful induction. To paraphrase Barnes, divine illumination stands to induction as understanding stands to demonstration. Understanding is not a means of acquiring knowledge. Nor, then, is divine illumination, for divine illumination *is* knowledge.

Serene makes some interesting distinctions between Aristotelian *nous* and divine illumination which might clarify Grosseteste's view further.[93] As has been seen, Crombie equates these two in his analysis. However, Aristotle merely claims that we are in a state of *nous* if we have a genuine grasp of a first principle, but not every such impression of a sure grasp of a first principle might be termed *nous*. In the case of divine illumination, all knowledge is somehow analogically related to the first and supreme light, so, as Serene claims, this admits of degrees in a way that *nous* does not. In the above analysis of the distinction between illumination by the supreme light in the beatific vision and illumination by the *luce creata*, the degrees of illumination or knowledge were described in terms of motion. Here there is no dualism, but a clear distinction, between knowledge and ignorance. This suggests something important about Grosseteste's conception of natural philosophy.[94] Because of this doctrine of illumination, for Grosseteste all knowledge will always be provisional and corrigible until we finally arrive at the beatific vision. This includes knowledge of nature because complete knowledge of creatures is found in contemplation of their exemplars in the divine ideas.[95] By contrast, Serene points out that Aristotelian *nous* is possible more immanently in such a way that our knowledge is only corrigible when it is mistaken. For Grosseteste, our knowledge is corrigible even, or especially, when it is correct, because this knowledge can always be further irradiated by the supreme light of God to which it is analogically related by participation.

What, then, is the place of the *experimentum* in Grosseteste's schema? If we are to understand the role of experiment, we must first remember the context in which Grosseteste produces the much quoted example of the eating of scammony and the production of red bile. In this passage, he has just considered the effects of human sin and weakness on knowledge and illumination. The soul requires awakening by the motions of sense perception. This suggests the importance of considered or controlled observation of phenomena. It is the motions of repeated observations which awaken the soul to form universal principles from observed singulars by the light of the divine. Repetition of experiment is not important because of the need to overcome the problem of enumerative induction, but simply because our souls are asleep and require a wake-up call. However, although this involves the motion from ignorance to knowledge, it is analogously related to the motionless motion of the intuitive grasp of all things in the beatific vision. The *experimentum* is not then a different kind of knowledge, but a different degree of knowledge which will pass away once the effects of the Fall are assuaged by divine grace. Experimental practice provides a knowledge which is still analogically related to illumination in the beatific vision in such a way that natural philosophy and theology are not separated. Moreover, the *experimentum* is not the criterion of truth for Grosseteste, but merely the first and very important step on the way to a fuller and more scintillating illumination. It provides a knowledge which, although correct and true (but, importantly, not merely probable), is corrigible and capable of being filled with ever greater light. Grosseteste rightly places knowledge from the senses in *experimenta* in an appropriate context: 'It is not in sensation that we know; but it is as a result of sensation that knowledge of the universal comes to us. This knowledge comes to us via the senses, but not from the senses'.[96]

Finally, it is important to note that the *experimentum* makes supreme sense within a theological doctrine of divine illumination. Under the so-called orthodox view of induction in which something must be added to an inductive inference and that process justified, a Humean scepticism will always remain concerning the legitimacy of this reasoning. This scepticism is due to an unbridged dualism between the inductive abstraction of universals from singulars (sense perception) and the knowledge that is gained from that perception. Under Grosseteste's doctrine of illumination, the inductive abstraction is analogically related to knowledge and being through the mediation of light. There is no 'chasm' of dark ignorance to be traversed between sensation and knowledge, for the whole is filled with light. Thus the practice of experimental observation is integrated into this theological vision, yet its appropriate place is maintained in the hierarchy of analogically related science. Its truth is a result of irradiation from a higher light, and yet it will pass away at the *eschaton*.

Such is the alliance between experimental verification and falsification, on the one hand, and a Neoplatonic doctrine of divine illumination, on the other, in the work of Grosseteste. The practice of experiment and the importance of mathematics, however, receive much more sustained consideration in the work of one of Grosseteste's greatest admirers and theological heirs, the Franciscan Roger Bacon. In Bacon's *Opus Majus*, written for Pope Clement IV in 1267, we find a work dedicated to persuading its intended recipient of the need for a revision of the curricula and renewal of true learning in the universities of the thirteenth century. Of the seven parts of this lengthy and often verbose work, we find sections dedicated to mathematics, the science of *perspectiva* (sometimes referred to as 'optics', although the medieval variant referred to matters beyond the behaviour of visible light) and another to *scientia experimentalis*. With particular reference to these portions of Bacon's central work, two questions will occupy the final section of this chapter: how did he build upon the tradition of Grosseteste in his understanding of experimental practice; and what might this suggest for the future investigation of the nature of motion? Initially, Bacon will be seen to share much in common with Grosseteste in the understanding of the relationship between theology and the other sciences. However, three significant moves arise which mark developments of Grosseteste's thought and anticipate later natural philosophy. First, he reduces mathematics and experience more exclusively to quantitative analyses rather than the more Platonic and Aristotelian emphasis on the qualitative. Secondly, there is an emphasis on efficient causality, and consequently on 'force', which will supplant the focus on final causality and motion *per se*. Thirdly, he places *scientia experimentalis*, which is developed so as to include particular technologies for the production of scientific facts, within a central position at every level of his hierarchy of the sciences, including theology itself.

Roger Bacon: truth and experiment

In *Opus Majus* part 2, Bacon addresses the relationship between the philosophical sciences and theology and follows Augustine in stating that 'the gold and silver of the philosophers did not originate with them but are dug out of certain mines as it were of divine providence'.[97] So, while it is the case that theology is the mistress of all other sciences, nevertheless 'we Christians ought to employ philosophy in divine things, and in matters pertaining to philosophy to assume many things belonging to theology, so that it is apparent that there is one wisdom shining in both'.[98] Within the philosophical sciences which may be put to use in the service of divine science, Bacon includes mathematics, *perspectiva*, *scientia experimentalis* and moral philosophy. The last of these is the closest to theology because it provides knowledge of the end and purpose of both the practical and speculative life, namely salvation. He states that,

> This active science is called the moral science and the civil science, which instructs man as to his relations to God, and to his neighbour, and to himself, and proves these relations. ... For this science is concerned with the salvation of man to be perfected through virtue and felicity; and this science aspires to that salvation as far as philosophy can. From these general statements it is evident that this science is nobler than all the other branches of philosophy.
> (Bacon, *Opus Majus*, part 7, first part, p. 635)

However, all the philosophical sciences are 'vitally necessary' for theology to reach its end.[99] How are these sciences related to one another? As in the case of Grosseteste, whose thought Bacon so admired and praised, light is an analogical mediator: 'although in some measure the truth may be said to belong to the philosophers, yet for possessing it the divine light first flowed into their minds, and illumined them from above'.[100] As well as claiming that the divine light illumined even the minds of pagan philosophers, Bacon also produces a detailed 'genealogy of the illuminated'[101] in which he argues that the passage of wisdom and learning is traceable from Noah's sons through the Chaldeans, Solomon and the Egyptians to the Greeks. Plato accordingly came to share the Hebrews' knowledge of the divine name – 'I am who I am' – while Aristotle was the most effective at ridding all previous philosophy of errors.[102]

However, in addition to the deployment of a genealogical defence of the value of philosophical learning, arguing that the light of scriptural writings has been shed upon the 'unbelieving philosophers', Bacon deviates from Grosseteste's cosmogony of light in referring more particularly to an Avicennian distinction between the passive and active intellect.[103] This distinction, which will be examined further in Chapter 5, is intended to account for the motion of the human mind from potency to the actuality of contemplation. The passive human intellect has the capacity to receive 'sciences and virtues' from another source, namely the active intellect. Is this a distinction within the soul? This is a problem similar to one discussed in the previous chapter, that is whether the soul can be said to be a complete and self-explanatory self-mover: are the passive and active intellects resident within the soul in a way such that the motion of human contemplation from potency to act is self-explanatory? Aristotle expressed this issue in terms of light, stating that 'Mind in the passive sense ... becomes all things, but mind has another aspect in that it makes all things; this is a kind of positive state like light; for in a sense light makes potential colours into actual colours'.[104] An active intellect is required to enter and illuminate the passive intellect just as light brings colours to actuality in illumination. Bacon follows the Avicennian tradition of the separation of the agent or active intellect, arguing that an actuality which is prior both temporally and onto-logically is required to account for the motion of the human intellect from potency to act in contemplation. This active intellect is not a part of the soul, but is an unmixed intellectual substance separated essentially from the passive intellect. The active intellect influences and illuminates the passive intellect for the recognition of truth. It is present in the soul not according to substance, but as a moving force is present in that which it moves. The active intellect, in being simple and incorruptible, is, for Bacon, God, who 'has illumined the minds of those men in perceiving the truths of philos-ophy' in such a way that 'it is evident that their labour is not opposed to divine wisdom'.[105] I will return to this separation of the active intellect below in discussing Bacon's understanding of knowledge.

Having established that the divine light can shine in philosophy when it is put to use in the service of theology,[106] how and why does Bacon deploy the particular philosophical sciences? This question will be answered with reference to the central sciences which compose parts 4, 5 and 6 of the *Opus Majus*.

Mathematics, perspectiva *and* scientia experimentalis

While his predecessors wrote succinctly about the importance and utility of mathematics, Bacon produces a very lengthy exhortation to the study and employment of this science.[107] He adduces many reasons for the importance and utility of mathematics, amongst them its simplicity (apparently, even the clergy are able to grasp its central tenets), certainty and its required use in the understanding of the other sciences.[108] In this respect, there is a Platonic note to Bacon's thought. However, whereas for Plato mathematics was concerned with beauty and proportion as well as quantity, Bacon is distinctive in emphasising an apparent reduction of physics and mathematics to quantity. Place and time are both reduced to questions of quantity.[109] Also, 'the greater part of the category of quality contains the attributes and properties of quantities'.[110] Although the relations of proportions as well as the geometrical, arithmetical and harmonic means are central to Bacon's understanding of mathematics, nevertheless 'whatever is worthy of consideration in the category of relation is the property of quantity' in such a way that 'all the categories depend on a knowledge of quantity'.[111] Moreover, an emphasis on mathematics and quantity suggests an intimate link with sciences which have sense perception as their particular medium, most notably (as will be seen below) the so-called *scientia experimentalis*. For Bacon, nothing can be perceived without quantity, this being especially a matter of sense perception.[112] The human mind is particularly orientated towards the acquisition of knowledge in this way because the forms of bodies, unlike the forms of incorporeal things, are present in our minds such that we are 'vigorously' occupied with bodies and quantities.[113] Therefore, there is an intimate link between the quantity which is the special occupation of mathematics on the one hand, and the sense perception of *scientia experimentalis*, which is also preoccupied with quantity, on the other. The measurement of quantities in nature suggests itself forcefully, so it comes as no surprise to find Bacon proposing the employment of measuring instruments and tables, all regulated by 'rules invented for the verification of these matters, to the end that a way may be prepared for the judgements that can be formed in accordance with the power of philosophy'.[114] This remark, stipulating 'rules' which must be followed in establishing or verifying facts about nature, is a very early intimation of the priorities of modern experimentation: the need for what historians of early modern science have labelled 'technologies' for the production of matters of fact.[115] These take the form of practical, social and literary regulations for the ordered investigation of nature. These rules and practices, for example the use of a measuring instrument or table in a particular and regulated way, provide the 'technology' which produces matters of fact concerning natural phenomena. Of course, Bacon's approach was extremely vague when compared with the practices of the experimental scientists of the seventeenth century, yet he foresaw the need not only for the bare observation of nature, but for the communally *regulated* observation of nature.

So there is a link between mathematics which deals especially with quantity and the observations of the senses which are also orientated towards the perception of quantity. Is there another link between mathematics and the senses? Two require mention. The first refers to the need for the confirmation of mathematical propositions by observation. Although mathematical propositions are free of error, this does not mean that they rid us of all doubt. A mathematical proof must appeal not only to the reason, but also to the senses: we must be able to *see* that such a proposition is true. Bacon writes that

> the mind of the one who has the most convincing proof in regard to the equilateral triangle will never cleave to the conclusion without experience, nor will he heed it, but will disregard it until experience is offered him by the intersection of two circles.
>
> (Bacon, *Opus Majus*, part 6, ch. 1, p. 583)

Here, the work of reason and experience, mathematics and *scientia experimentalis*, are complementary.

The second and perhaps more significant link between experience and mathematics relates to 'the multiplication of species'.[116] This is a doctrine of physical causation, including the cause of sensation, which is indebted to a Neoplatonic tradition extending from Plotinus, Augustine, al-Kindi and Avicebron to Grosseteste's physics and cosmogony of light.[117] A 'species' for both Grosseteste and Bacon is the force or power by which any object acts on its surroundings. It is 'the first effect of an agent; for all judge that through species all other effects are produced'.[118] A species is a likeness of that from which it emanates, and the medium through which any being acts. The species resembles the agent and they have the same specific nature, even though in some cases we refer to the agent and the species using the same term (the species of colour is colour), while in others we do not (we do not call the species of man 'man', for the being of the species is, according to Bacon, insufficiently complete to warrant this ascription, although it is univocal with the human agent). Perception occurs by the propagation of species, for everything emits a likeness of itself that is gathered in the senses. Both substance and singulars as well as universals produce species, although the common sensibles, for example motion, size and shape, do not.[119] How are species generated? As D. C. Lindberg points out, species are not, as it were, deposited in a recipient, neither do they flow into a recipient from outside.[120] Instead, species are generated by a 'true alteration and bringing forth out of the active potentiality of recipient matter'.[121] The picture Bacon has in mind is of an agent producing its species out of the active potentiality of an adjacent recipient. This adjacent recipient then elicits species from the next adjacent recipient, and so on. In this way, species are multiplied and transmitted. For Bacon, there is no unmediated action at a distance: one being acts on another through a medium which enables the transmission of species. The relevant passage should be quoted at length:

> But the species is not a body, nor is it changed as regards itself as a whole from one place to another, but that which is produced in the first part of the air is not separated from that part, since form cannot be separated from the matter in which it is, unless it be soul, but the species forms a likeness to itself in the second position of the air, and so on. Therefore, it is not a motion as regards place, but is a propagation multiplied through the different parts of the medium; nor is it a body which is there generated, but a corporeal form.
> (Bacon, *Opus Majus*, part 5, 9th distinction, ch. 4, p. 489)[122]

As Lindberg points out, this is essentially Bacon's understanding of the physics of the propagation of force.[123] Crucially, this multiplication of species, which has as a primary incidence the propagation of light, takes place rectilinearly in such a way that causality can be examined by geometry and the other divisions of mathematics to form the science known as *perspectiva*, or optics.[124] Therefore, it is by means of the multiplication of species, which takes place according to the principles of mathematics, that forces are propagated and efficient causes brought to bear on agents or recipients. The universe, for Bacon, is saturated with efficient causes. He states that

> everything in nature is brought into being by an efficient cause and the material on which it works. ... For the active cause by its own force moves and alters the matter, so that it becomes a thing. But the efficacy and power of the efficient cause and of the material cannot be known without the greater power of mathematics even as the effects produced cannot be known without it. There are then these three, the efficient cause, the matter, and the effect.
> (Bacon, *Opus Majus*, part 4, 2nd distinction, ch. 1, p. 129)[125]

As will be seen in Chapter 6, this emphasis on force and efficient causality (rather than Aristotle's material, formal and final causes), along with the emphasis on mathematical quantity, is hardly Aristotelian or Platonic, but strikingly reminiscent of the concerns of Isaac Newton in his investigations into the nature and properties of motion.

How does the multiplication of species relate to Bacon's understanding of knowledge? In a recent study, Olivier Boulnois has shown how Bacon is part of a Franciscan Neoplatonic tradition (congruent with the thought of Avicenna rather than Augustine)[126] which anticipates modern conceptions of knowledge as lying within a dualism between the knowing subject and the known object.[127] For Bacon, a species migrates from its source by propagating itself in the immediately adjacent medium. These species are purely material likenesses of the object by which they are produced and they proceed by efficient causation to the knowing subject. They also carry certain spiritual qualities and are therefore indifferent to material and spiritual nature. These species are not, as in the hylomorphism of Aristotle and Aquinas, the form of the object because, as was seen above, 'form cannot be separated from the matter in which it is, unless it be soul'. So the species produce a sensible effect or material impression in the knowing subject, once again by means of efficient causation. The mind must synthesise this material impression into a mental form, its own immaterial mode of thought. As Boulnois explains, one can now consider 'thought' from two quite distinct points of view: first, there is the object of thought, the thing in itself, and secondly there is that by which one thinks, the latter being a representation in the mind of the former, yet with 'diminished being'.[128] The thing that comes to reside in the knower, in not being the form of the exterior object itself, is rather a representation of the object imposed on the mind just as a seal is pressed into wax. Because of the disjunction between the representation of the object and the object in itself, it seems that there is an incommensurable distance between what we know of things and things in themselves. In other words, there seems to be reason for real scepticism about our perceptions of the world, for what is the real connection between the object in itself and the object of my thought?[129]

By contrast, as will be seen in greater detail in the next chapter, for Aquinas knowledge is a case of the form of an object, having already been synthesised, migrating to enter the mind. In other words, rather than knowledge being mere 'representation', the form of the object now *informs* the mind in such a way that the object really does, in an essential fashion, reside in the mind of the knower. As Pickstock remarks, on this view knowledge is ontological, because an object repeats itself, although non-identically in a different mode, by informing the mind of the knower in such a way that the soul really is 'in a manner all things'.[130] By contrast, the view represented by Bacon, which will attain its clearest expression a generation later in the work of Duns Scotus, views knowledge in terms of units of information imposing themselves by efficient causation upon the mind which is a merely passive recipient of atomic 'bits' of information. Moreover, as was seen earlier, Bacon also understands the agent or active intellect to be separate from the soul. Knowledge now emerges essentially from this separate agent intelligence and the material object from which a species emanates merely *occasions* knowledge in the mind. Later, for Duns Scotus and the nominalists, this introduces the possibility of knowledge being miraculously imprinted upon the human mind, even in the absence of material objects of knowledge. God may indeed circumvent material reality altogether and impose a species upon the mind which becomes an *a priori* subject of knowledge.[131]

Despite the scepticism concerning the veracity of experiential knowledge in the nominalist tradition of the later Middle Ages, Bacon nevertheless has also been thought to be a precursor of modern science principally because of his advocacy of *scientia experimentalis*. What is the nature of this science? Bacon distinguishes it from the knowledge which comes about by reason which, although making us grant the conclusion, does not make that conclusion certain *for us*. To reason must be added *scientia experimentalis*. This science and the knowledge it provides is further distinguished from divine illumination, yet importantly Bacon counts them both as forms of 'experience'. The senses alone are insufficient for the attainment of knowledge, yet neither are they dispensable because our minds are primarily orientated towards corporeality. As for Grosseteste, we require divine illumination and *scientia experimentalis*. So he claims that 'the grace of faith illuminates greatly, as do divine inspirations, not only in things spiritual, but in things corporeal and in the sciences of philosophy'.[132]

The *scientia experimentalis* has three principal characteristics or tasks.[133] First, it investigates the conclusions of other sciences by means of experiment. This science, therefore, not only has its own domain, but is also involved in verifying the claims of other sciences. It also distinguishes between true knowledge and 'the mad acts of magicians, that they may not be ratified but shunned'.[134] More specifically, Bacon mentions the conclusions reached by *perspectiva* concerning the nature of the rainbow. He goes on to provide a lengthy description of an experiment which may be conducted to calculate the elevation of the rainbow and aid in the understanding of its cause.[135] Meanwhile, the second task of *scientia experimentalis* is to produce new instruments and data which can be put to use in many areas. Three examples are provided: experiments with the spherical astrolabe, the prolongation of life by experimental medicine and the formulation of improved health regimes, and finally experiments in alchemy. The third 'prerogative' of this science relates to its ability to discover the secrets of nature. This power (which is apparently quite separate from the sciences of astronomy and astrology) involves attainment of the knowledge of the future, the past and the present. Bacon appears to be claiming that this science interprets history, reveals the present and, importantly, is predictive in character.

However, it is in Bacon's advocacy of the technological application of *scientia experimentalis* and his sense of a need for 'progress' where we find an early intimation of the concerns of modern experimental science. For example, he considers how experiment might give rise to weapons which can be deployed in warfare, thus reducing the need for direct hand-to-hand conflict. These inventions include explosive devices, malta (a kind of bitumen which 'burns up' an opponent), yellow petroleum (which cannot be extinguished by water) and poisonous gas.[136] As well as putting *scientia experimentalis* to use in this way, Bacon also sees that there is a theological purpose for this mode of enquiry: it can reveal to us the literal meaning of things, and therefore the literal meaning of scripture. He states that,

> I have showed above that the literal meaning consists in expressing the truth in regard to created things by means of their definitions and

descriptions, and I likewise showed that reasoning does not arrive at this truth, but that experiment does. Wherefore this science next to moral philosophy will present the literal truth of Scripture most effectively.

(Bacon, *Opus Majus*, part 6, ch. 12, 3rd prerogative, p. 631)

Going even further, Bacon claims that this science can be put to use in Christian apologetics, 'not by arguments but by works'.[137] What does Bacon have in mind? It seems he believes *scientia experimentalis* may be deployed to show 'the glorious truths of God' revealed in the marvels and wonders of nature. This can be seen particularly in his treatise *Epistola de secretis operibus artis et naturae*, written around the middle of his career.[138] This work includes discussion of 'perpetual lamps', distorting mirrors and mechanical devices. In an assessment of the nature of thirteenth *scientia experimentalis*, Lynn Thorndike remarks that this search for marvels in nature coupled with Bacon's general credulity and superstition betray not a precursor of early modern science but a thinker firmly rooted in the Middle Ages.[139] However, it is certainly the case that, as recent scholarship in the history of science shows, many of the early modern experimenters were not so much concerned with the technological employment of science as with the theatrical display of curious natural phenomena.[140] To choose just one example, in 1675 Leibniz attempted to found a privately funded 'academy of sciences' which would exhibit all manner of natural wonders under contrived experimental conditions. This would include the use of the telescope, the air-pump and calculating machines.[141] As Lorraine Daston and Katharine Park comment, 'Although this particular Leibnizian fantasy was never realised, it nonetheless captured the atmosphere of wonderstruck novelty that suffused natural philosophy and natural history throughout the seventeenth century'.[142] Bacon certainly appears to anticipate this search for wonders and marvels in nature which are brought to view in contrived situations. As regards his general credulity, it is true that Bacon held many fanciful beliefs concerning, for example, dragons. Yet his lack of scepticism itself does not render him at odds with the concerns of a later experimental science. There is no benefit in raw experimental data. One requires a theoretical framework within which to interpret one's findings, a framework about which one must be credulous.[143]

Such an emphasis on the power of *scientia experimentalis* to reveal the wonders of God's work in nature betrays a final important link between Bacon and his modern scientific successors. His *scientia experimentalis* anticipates what was later to become a form of natural theology. Bacon writes that

if we proceed to the conversion of unbelievers, it [*scientia experimentalis*] is evidently of service in two main ways with numerous subdivisions, since a plea for the faith can be effectively made through this science, not by arguments but by works, which is the more effective way.

(Bacon, *Opus Majus*, part 6, ch. 12, 3rd prerogative, p. 632)

What is being suggested is that *scientia experimentalis* leads us to believe in many things which we cannot explain (for example, the mutual attraction of magnets). If this is the case as regards 'the vilest of creatures', how much more should we be ready to believe the things of God, even though we cannot explain them? *Scientia experimentalis* points our senses to the marvels of nature which are also the marvels of God. In a strikingly similar fashion, the science of the early modern experimenters was understood to be useful to theology in revealing the wondrous order and divine design of nature. Robert Boyle, for example, writing in the seventeenth century, dubbed the scientific experimenter 'the priest of nature'. The arguments of experimental scientists could be traced back to Moses and they were to provide arguments for the existence of God. Boyle wrote, 'if the world would be a temple, man sure would be the priest, ordained (by being qualified) to celebrate divine service not only in it, but for it'.[144] As will be seen in Chapter 6, Newton himself understood his *Principia* as a work of theology useful for the conversion of non-believers.[145] Bacon clearly anticipates this role for a science which, as well as being employed in the guise of technology, can reveal marvels to the human senses which thereby serve to convince us of God's existence.

In this chapter I have attempted to show that many of the concerns and priorities of the early modern investigation of motion were already present in thirteenth-century natural philosophy, albeit in a sometimes hidden form. It is not the case that Grosseteste and Bacon were 'laboratory scientists'. They were certainly armchair natural philosophers, and the various strictures against regarding them as men of the seventeenth century confined to the thirteenth should certainly be heeded. However, even if Grosseteste and Bacon describe very few experiments and leave little evidence that they themselves made significant progress in understanding nature through experiment, nevertheless the framework and motivations for this science are clearly visible. Grosseteste developed Aristotle's emphasis on 'experience' to include the verification and falsification of hypotheses. He also deployed mathematics as an explanatory science within his cosmogony of light, from which emerges a Platonic understanding of the nature and hierarchy of motion. I have argued that the *experimentum* makes good sense within the Neoplatonic framework of Grosseteste's natural philosophy. The notion of truth as illumination even evades the so-called 'problem of induction' while describing truth, and all the sciences which lead us to the truth, in terms analogically related to the life and light of the Trinity.

Meanwhile, it was seen that Bacon developed the tradition of the *experimentum* still further into a distinct *scientia experimentalis* which can be deployed within any science, including theology. In acknowledging that 'experience' is necessary *in addition* to basic mathematical or syllogistic reasoning in order to rid our minds of doubt, Bacon is moving beyond Aristotelian *empeiria* towards a *cognitio per experientiam*. However, in maintaining a doctrine of the separate agent intellect and a notion of the migration of material species which, by purely efficient causality, impose themselves upon the senses, Bacon clearly anticipates an understanding of knowledge as representation. So *scientia experimentalis* is not really an exalting of the importance of the examination of the material moving cosmos. For Bacon, unlike Grosseteste who maintains a more Augustinian position, *scientia experimentalis* is part of a system of sciences which collectively are designed to mitigate an increasing scepticism and distrust of the world around us. Experience of the material world is now only an occasion for knowledge, and true reasoning is *a priori* and takes place by examining the mental object which arrives from the separate agent intellect.

Before turning to early modern science and the theological and philosophical context of its investigations of motion, some of which has been pre-empted in the subjects of the present chapter, further analysis is required of motion in the later medieval period. In the next chapter, an assessment will be undertaken of the nature and status of motion in the theology of St Thomas Aquinas, an immediate successor of Grosseteste and Bacon. Here we will find a more exacting synthesis of Platonic and Aristotelian themes to form a thoroughgoing theological account of motion within areas as diverse as the physics of bodies and the motions of grace within the Church.

4   St Thomas Aquinas
The God of motion

In the previous chapter I described two attempts in the late twelfth and early thirteenth centuries to accommodate the newly available Aristotelian physics and metaphysics within a prevailing Neoplatonism. I argued that Grosseteste analogically relates creation to its divine source through the mediation of light. Within his cosmogony and cosmological hierarchy one also finds that distinctions between types of being and the sciences which investigate them can be formed through the analogy of motion. Beyond this theological cosmogony and metaphysics one can find early intimations of the priorities of early modern science and its investigations of such phenomena as motion. For Grosseteste, the *experimentum* is included within a wider metaphysics of illumination in which the human mind requires the movements of sensation in order to attain the actuality of knowledge and illumination. However, I argued that there are not two guarantees of truth for Grosseteste, theology and experiment. Rather, the latter is incorporated within a wider understanding of knowledge as illumination in the light of the divine. Meanwhile, for Bacon, *scientia experimentalis* attains new priorities which are at once both technological and formative of a theological apologetics: this science can assist in the development of such things as weapons while also revealing the wondrous works of God to the sceptic or unbeliever. Thus natural philosophy, in adopting an immanent end somewhat diverted from that of theology itself, begins the process of separation which, as we shall see, leads to the removal of such concepts as motion from the purview of the theologian to confine them within an apparently separate physics.

Following the work of Grosseteste, and coterminous with Bacon's attempt to transform the curricula in the schools, one finds a more exacting theological synthesis of Platonism and Aristotelianism in the work of St Thomas Aquinas (*c.*1224–1275). Whereas Grosseteste had written commentaries on some of Aristotle's works and Bacon had extolled the value of natural philosophy within education and the work of the Church, Aquinas not only produced commentaries on the Aristotelian corpus and other treatises of Neoplatonic origin, but also put Greek learning to work in vigorous fashion to expound the Church's *sacra doctrina*. In the previous chapter, I suggested that Grosseteste regards all knowledge, whether of pagan or

Christian provenance, to be a more or less scintillating illumination in the divine light. So how does Aquinas understand the relationship between pagan reason, which taught so much about, for example, the physics of moving bodies or the motions of the intellect, and the higher science revealed by God and taught within the Church?

The higher science that Aquinas calls *sacra doctrina* receives its clearest and most exacting exegesis in his principal work, the *Summa Theologiae*.[1] Before investigating the place of pagan natural philosophy, such as Aristotle's physics of moving bodies, in relation to theology, one might first identify the nature of *sacra doctrina* in whose service Aquinas deploys natural philosophy. He provides his account in the first question of his *Summa Theologiae*. This teaching has the articles of faith as its premises and God as its end.[2] It is not a science like others, having a strictly delineated subject matter (for example, arithmetic deals with number, ethics with the principles of action); rather, *sacra doctrina* delivers *salus*, for it is 'anything that breeds, feeds, defends, and strengthens the saving faith which leads to true happiness'.[3] It looks at things differently to other sciences, namely in the light of divine revelation, and thus 'the theology of *sacra doctrina* differs in kind from that theology which is ranked as part of philosophy'.[4] It does not focus on God and creatures as if they were thereby unrelated or in opposition, 'but on God as principal and on creatures in relation to him, who is their origin and end'.[5] This early comment on the 'shape' of the vision of *sacra doctrina* already hints at the motif of circular motion: God is both principal, from which creation emanates, and end, to which creation returns.

Meanwhile, one might clarify a potential misconstrual of Aquinas's understanding of his subject matter: is *sacra doctrina* merely a matter of the identical repetition of propositions revealed by God in such a way that the discourses which rest on human reason, such as a natural philosophy of motion, are rendered wholly independent? Not at all. Human reasoning, which does not constitute a separate trajectory to the grace which comes through *sacra doctrina*, 'should assist faith as the natural loving bent of the will yields to charity'. This 'assistance' is not, however, due to any weakness in the science of *sacra doctrina* itself; rather, it arises because our understanding is wanting and is more conditioned to operate through natural reason. In this way, reason may be deployed 'for the greater clarification of the things it [*sacra doctrina*] conveys'.[6] So just as nature is at its most natural when perfected by grace, so human reason is *at its most rational* when it is engaged in clarifying and making manifest the implications of the message of *sacra doctrina*.[7] This is made explicit by Aquinas in his commentary on Boethius's *De Trinitate*:

> The gifts of grace are added to nature in such a way that they do not destroy it, but rather perfect it. So too the light of faith, which is imparted to us as a gift, does not do away with the light of natural reason given to us by God. ... Rather, since what is imperfect bears a

resemblance to what is perfect, what we know by natural reason has some likeness to what is taught to us by faith.

(Aquinas, *In librum Boethii De Trinitate expositio*, 2.3.responsio)[8]

Aquinas states that there can be no contradiction between reason and faith, and that reason can 'demonstrate the preambles of faith, which we must necessarily know in [the act of] faith'.[9] Furthermore, it is crucial, he argues, that philosophy is brought within the bounds of faith and not vice versa. In other words, philosophy does not perform the function of apologetics, but the service of clarifying our minds.

Yet *sacra doctrina* is not sufficient unto itself and does not provide satisfaction of the desire of human beings to know: it is based on a still higher and most certain science, namely God's own, which he shares with the blessed. So this *sacra doctrina* which he now gives to the Church 'is like an imprint on us of God's own knowledge, which is the single and simple vision of everything'.[10] It reorientates us towards an end that we could not know by reason alone, the end which is salvation and the discourse of the blessed.[11] Much of this first question of the *Summa* thereby seeks to describe not only the nature of *sacra doctrina* itself, but the drawing together of numerous sciences under the sway of *sacra doctrina* in such a way that they are directed to a further divine purpose which they could not discern of themselves. As Aquinas himself states, 'those who use philosophical texts in sacred teaching, by subjugating them to faith, do not mix water with wine, but turn water into wine'.[12]

So, just as for Grosseteste, there are not two guarantees of truth, one theological and one philosophical. Rather, the latter is deployed to clarify our minds concerning the teaching of the former. What, then, is the particular science of motion which Aquinas deploys in order to expound and clarify *sacra doctrina*? It will be the first task of this chapter to delineate his understanding of the natural philosophy of motion, particularly as this is found in the commentaries on Aristotle's *Physics*. However, I will show that Aquinas does not adopt Aristotle's physics uncritically; at certain crucial points he extends the Stagirite's understanding of motion in more Neoplatonic directions. For example, it will be seen that motion finds its highest created origin in the intellective motion of the angels around the divine. The ultimate origin of motion, however, is God himself; the divine reality is intimate to every motion, even that of inanimate bodies. It will be seen that motion for Aquinas is never a simple and self-explanatory category. It is not a purely physical displacement. Ultimately, all motion returns us to God.

Having described the hierarchy of motion allied to the cosmological hierarchy, I will return to the *Summa* to examine Aquinas's deployment of this physics of motion in his exposition of *sacra doctrina*. I will begin, therefore, with the introductory questions concerning God. We will see how Aquinas's initial contributions on the nature of God, which constitutes a *via negativa* at the beginning of the *Summa*, formulates a radical distinction between finite and infinite which can be expressed in terms of motion: on the one hand, God is simple and beyond all motion, lacking potency and in full actuality, while, on the other hand, created nature is 'the distinctive form or quality of such things as have within themselves a principle of motion'.[13] How is this ontological difference traversed? I will suggest not only that God is related to creation as the first unmoved mover, but that this understanding is further enhanced by Aquinas as motion is related through analogical participation to the immanent life of the divine. This leads to an examination of motion's relation to the immanent life of God in the emanation of the persons of the Trinity, the principle of all motion. I will argue that Aquinas's understanding of creation, understood as the 'motion' of the emanation of all things from God, is founded on a participation in the eternal Trinitarian life of God. In turn, this will lead to the third section and an examination of how God is not only the principle of motion, but also motion's term or *telos*. Therefore, the third section of this chapter will focus on teleology within nature and the concomitant ends of all things. This will be suggestive of the importance of motion within Aquinas's ethics which involves a crucial emphasis on God as the prime mover of creation towards its proper end through the work of grace. The mediation of that grace through Christ's passion and the Church is the subject of the fourth and final section of this chapter. Here we will find the means by which humanity is returned to participate most completely in the 'motionless motion' of God in the beatific vision. This return reaches its apogee in the Eucharist where one finds a fundamental unity in the midst of the diversity of creation: the principle and end of motion, namely God himself, 'incarnate' under the creaturely appearances of bread and wine in such a way that we receive motion's beginning, end and way into our very selves; thus we are reorientated by grace to return to God. I begin, however, with a consideration of the physics of motion.

At the limits of Aristotelian motion

Motion and self-motion

Aquinas wrote extensive commentaries on many aspects of Aristotle's thought and it seems that he takes much of his understanding of motion from this source. In the earlier discussion of Aristotle, it was seen that he settles on a definition of nature as 'the distinctive form or quality of such things as have within themselves a principle of motion, such form or characteristic property not being separable from the things themselves, save conceptually'.[14] In this definition of nature as an inherent principle of motion, Aristotle considers motion within the division of potency and act to be 'the progress of the realizing of a potentiality, *qua* potentiality, for example the actual progress of qualitative modification in any modifiable thing *qua* modifiable'.[15] Importantly, motion is never a nondescript 'wandering' without an identifiable origin and terminus. Motion takes place between contraries, for example, 'in substantive existence ... form and shortage of form; in quality, white and black; in quantity, the perfectly normal and an achievement short of perfection'.[16] Furthermore, for Aristotle all motion implies a *telos* towards which the motion tends. If there were no *telos*, there would be no motion, for the *telos* is the reason for the motion. Motion is never for its own sake, but for the attainment of some end or goal which lies outside the motion in question (in contrast to *energeia* or 'actu-ality', where a being's activity or operation is immanent and fulfilled at every moment).[17] Hence motion is seen to be 'ecstatic': a moving body constantly exceeds itself as it tends towards the full actuality which lies beyond in the *telos* of its motion. On this general view, heavy or light objects, for example, have a teleologically defined motion: light bodies tend to rise up to their proper place, heavy bodies tend to fall to their proper place. Nevertheless, it is seen that motion has a strange ontological status because it hovers between potency and act. In order to identify motion within the spectrum of potency and act, Aristotle formulates the example of someone learning to speak a language.[18] A person who cannot speak a particular language is in potency to speaking that language in a different and more radical sense than someone who can speak the language but is not at this moment exercising that ability. The person unable to speak the language is in what one might call 'first potentiality' to speaking the language. The person who is able to speak the language but is not doing so has all the skills necessary to converse and is in 'second potentiality' (or 'first actuality') to speaking the language. The person who is speaking the language is in 'second actuality'. Motion, for Aristotle, is the constituting of a being in second potentiality to a *telos*. Thus, as L. A. Kosman has pointed out, the motion of building is the actualisation of bricks and mortar as *potentially* a house.[19] So given these various factors, in his commentary on the *Physics*, Aquinas can follow Aristotle and insist that 'all motion is from something to something',[20] and

> it is clear that nature is nothing but a certain kind of art, i.e., the divine art, impressed upon things, by which these things are moved to a deter-minate end. It is as if the shipbuilder were able to give to timbers that by which they would move themselves to take the form of a ship.
> (Aquinas, *Commentary on Aristotle's Physics*, II.14.§268)[21]

Linked with the teleology of motion in Aristotle's thought is a distinction discussed in Chapter 2 of this essay: that between 'natural' and 'violent' motion.[22] The former is a motion for which a particular being has a natural receptivity, for example a heavy object moving downwards. By contrast, a violent motion is one for which a particular being has no intrinsic intention: it is a motion to which the particular being contributes nothing and against which it may resist, for example a heavy object rising. This distinction is developed further in order to formulate a principle at the heart of medieval Aristotelian physics, and central in Aquinas's thought, namely *omne quod movetur ab alio movetur*, literally rendered as 'everything which is moved is moved by another'.[23] This is the so-called 'motor-causality principle'. Aquinas discusses this principle in his commentary on the relevant chapter of Aristotle's *Physics*. Initially, it is clear that all instances of violent motion originate, by definition, from some source other than the body moved and therefore submit to the principle that 'everything that is moved is moved by another'. A violent motion is one for which a being has no intrinsic receptivity, and it therefore must originate from another. In these instances there are clearly movers and things moved, and these remain separate. Aside from the relatively unproblematic instance of violent motion, one can claim that there are those things which do not have within themselves the active principle of their own motion and are therefore clearly 'moved by another'. These possess only the passive principle of receiving certain motions. The movement of inanimate objects, for example, is always received from an extrinsic mover. This is clearly a central concern for Aristotle and Aquinas because they wish to maintain that the explanation of the motion of the fall of the heavy body is not found solely within the body itself. The distinction between animate and inanimate motion, between living and non-living creatures, could thereby remain clear.[24] However, when one observes the fall of a rock, is the rock the source of its own motion? Where might one identify the mover of the rock? As was seen in the earlier discussion of Aristotle, the mover in such a motion is not identified with its substantial nature, namely the form of an inanimate body (its 'heaviness'), because this is merely the *passive* principle of receptivity to certain motions. For Aristotle, the mover in any natural motion of an inanimate being is anything that removes an impediment to the motion (for example, a column holding a heavy object) or the generator of the inanimate body which 'gave to it the form which such an inclination [to a particular motion] follows'.[25] This is to say that the rock is already constituted in second potentiality (or first actuality) with regard to being in its natural place. The rock has already *been moved*, and yet while it is held in a high place it maintains a second potentiality, namely a potentiality to being lower.[26] It merely remains for the motion to be completed. Therefore, it seems that the motor-causality principle applies: the rock is moved by a separate mover, either its generator or one who removes an impediment to the motion.

However, Aquinas notes that there can be some doubt as to whether a mobile animate being which seems to possess the principle of its own motion is moved by another. In such instances of self-motion, it does not appear straightforward to identify a mover that is distinct from that which is moved. Animals appear to be entirely self-moved. Following Aristotle, Aquinas considers the movement of a body composed of parts, for example a person walking across a room.[27] In this instance, my body is divisible in such a way that my legs are the mover and the remainder of me is the thing moved. The key fact for Aquinas and Aristotle is that no motion involving something moved is utterly simple, for every apparent self-motion can be divided into the portions possessed by the agent (the mover) and the patient (the thing moved). In fact, in the case of what is homogeneous and without parts, such bodies cannot *be moved* other than by something else, for they cannot hold within themselves the differentiation between something moved and the mover.[28]

The principle *omne quod movetur ab alio movetur* means that whatever is moved requires a mover that is not also the thing moved. And why is it not also able to be the thing moved? Because to be moved and to move are utterly distinct processes, and in formulating this principle Aristotle and Aquinas wish to point out nothing more than this. A mover, which is in actuality, moves that which is in potency to act. Actuality must ontologically precede potentiality. This distinction between potency and act is preserved in the distinction between a mover and that which is moved. However, in maintaining a strict distinction between mover and moved in such a way that animal self-motion is analysed into the movement of one part of a body by another, are Aristotle and Aquinas not in danger of another equally problematic conclusion: the denial that there is anything that might be described wholly and properly as *self*-motion?[29] Clearly not, for this would be counter-intuitive and contrary to much of what they say on the matter elsewhere, particularly in relation to the difference between living and non-living things in nature.[30] So, with regard to, for example, my self-motion across a room, we have seen that this is executed by parts of my body. However, eventually one may arrive at a mover which is not itself moved, namely my soul.[31] There will be some mover in which no further division is possible, where no further differentiation can be made between mover and moved – one is simply left with a mover. Therefore, could one conclude from this that motion is utterly spontaneous within nature? Could one surmise that the soul is the only explanation which Aristotle and Aquinas require for the motion of animals? For Aquinas, this could not be the case because the soul has a most fundamental composition by which it is moved, namely of essence and existence. The soul is not a necessary being. There is one 'motion' which the soul cannot account for, namely its 'motion' into existence, its passage from pure potency in the mind of God to some degree of actuality and, crucially, its continuation in being. It is not utterly simple in the manner that, as we will see, God is utterly simple. It is the case that for Aquinas the soul is the mover of the human body.[32] However, the soul is 'moved' by its divine generation. Therefore, the divine is, for Aquinas, the first, most general and therefore most potent cause of any particular motion. Yet the *per se* mover of a living being in any particular motion is the soul of the living being. However, the soul is moved in respect of another motion, namely its passage from potency to act in its generation by God. Thus the soul is the *per se* mover of my body in this particular motion and God is the more general cause of this and every motion.[33] In this respect there is a subtle difference of emphasis between Aristotle and Aquinas. For the former, the generation of, for example, the soul is not an instance of 'motion', this category being strictly retained to local, quantitative or qualitative instances of change.[34] The soul is thus not subject to motion in the strict sense, but only to generation by the eternal mover, a generation which, as we will see in more detail shortly, is mediated through the celestial spheres. For Aristotle, it is therefore not wholly apt to refer to the soul as moved *per se*. In those instances in which Aquinas is considering motion according to Aristotle's strict definition, he maintains the same position.[35] For Aquinas this leaves open the possibility that one could postulate a finite immaterial substance as a more autonomous and distinct source of motion (defined only as local, quantitative and qualitative) within the universe. As we will see when considering the divine creative act, elsewhere he wishes to maintain the more Neoplatonic notion that all things, including motion, have their discernible origin in a principle 'motion' (or *emanatio*) from the divine being, and are sustained and gain their character by their participation in the emanation of all things from God.

Before progressing to consider further the hierarchical nature of Aquinas's cosmos, one final potential difficulty with the motor-causality principle and the associated proposal that motion is passage from potency to act requires clarification. The difficulty is most clear in the case of human contrivance, or 'art'. In a way that is analogous to the example of heat (in order to move from cold to hot, that which is cold must have a mover which is already in act with regard to being hot), is it the case that that which brings a building from potency to act must itself be a building? Is it the case that that which brings a painting into act is itself a painting? Must God be in some sense univocal with the creation he moves? Aquinas tackles this question by stating initially that the form of an agent is sometimes received in the effect according to the same mode of being that it has in the agent. For example, a fire generates a fire. However,

> in other cases the form of the agent is received in the effect according to *another* mode of being; the form of the house that exists in an intelligible manner in the builder's mind is received, in a material mode, in the house that exists outside the mind.
>
> (Aquinas, *Summa Contra Gentiles*, II.46)[36]

I will argue below that this motion of human contrivance is a clearer participation in the divine creativity which emerges from the emanation of the divine ideas in the mind of God.

Meanwhile, the realisation that even the self-motion of human beings and the motions involved in their creativity are not self-explanatory but point beyond themselves to an exterior source suggests a hierarchical pattern of motion within the cosmos which stretches to an wholly actual and unmoved origin. I now turn to examine the importance of this hierarchy in Aquinas's physics. At its crown will be found a celestial diurnal motion which is the principle and measure of all motion. However, as a first point of enquiry, I return to the dictum 'everything that is moved is moved by another' in order to see the interconnectedness of Aquinas's universe and the dependence of inanimate natural motion on the movement of higher beings within this cosmological hierarchy.

Cosmological hierarchy, the primacy of rotation and the touch of God

In his extensive discussion of the principle that 'everything that is moved is moved by another', James Weisheipl comments at length on the view that the substantial nature of an inanimate being is its mover, recognising that this interpretation has a notable provenance in the work of, amongst others, Averroës, Peter Olivi and Duns Scotus.[37] However, for Weisheipl, it is clear that in Aristotle's thought nature is not a mover but rather 'an active source of spontaneous activity'.[38] In the same way that a point is a principle (the necessary beginning) of a line without being the cause of the line, so nature is the intrinsic principle (but not a mover) in the growth of a plant, the beating of a heart or the rising of a light object. In the case of the motion of inanimate objects, we have seen that the moving cause is attributed *per accidens* to the remover of any obstacle to the natural motion in hand, and *per se* to the generator of the object in motion. Weisheipl notes that with regard to the motion of non-living beings, each entity has everything it requires from its generator in order to exhibit its characteristic natural motion: a body is in second potentiality (or first actuality) to being in its proper place. We do not have to search beyond this to find the mover. This allows Weisheipl to reject a common interpretation of the motor-causality principle, namely the view that 'whatever is in motion is *here and now* being moved by another'.[39] When applied to the downward fall of an inanimate object, this interpretation of Aristotle's principle would imply the need for the *constant* conjunction of a mover with that which is in motion, what Weisheipl refers to as 'the specter of *motor coniunctus*'.[40] He is anxious to maintain a fidelity to the passive sense of *movetur*, namely that when something is *moved* it requires a mover, but not when it is in motion. It does seem from the above discussion that Aristotle, and Aquinas in his commentary, do not require a constant conjunction between mover and moved in, for example, the fall of a heavy body. The generator of the body does not need to be in contact with the body when it falls: neither temporal nor spatial proximity are required. This natural motion downwards now takes place 'spontaneously' by what one might call 'energisation', unless there is an impediment. According to Weisheipl's interpretation of Aristotle and Aquinas, a heavy body which is held in an unnatural place has already been moved; its subsequent being in motion requires no explanation beyond the body's generation and the removal of the obstacle to its downward motion. There is no need to stipulate a constantly conjoined efficient cause of the motion.

Crucially, in rejecting the notion that Aristotle and Aquinas are committed to the view that every motion requires a constantly conjoined mover, Weisheipl is able to delineate a *consonance* between the medieval understanding of motion and that of early modern science.[41] In particular, he is keen to exonerate Aristotelian physics of the charge that its conception of projectile motion is too crude to account for and predict motion within nature. Aristotle's physics has been radically misconceived because he is not, in truth, committed to the notion that moving bodies require constantly conjoined movers.[42] This has apparently led to a failure to recognise the philosophical sophistication of medieval natural philosophy. However, while concurring with the general thrust of Weisheipl's position, David Twetten has suggested that, in a sense, Aquinas is committed to the idea of a 'constantly conjoined' mover in any motion.[43] To this possibility we now turn with an initial and brief excursus to consider once again Aristotle's definition of motion before examining the natural motion of inanimate bodies within the wider context of Aquinas's hierarchical cosmology. This will lead us to draw a stark distinction between Aquinas's understanding of the physics of motion and that of the early modern physicists.

It is undoubtedly the case that, with regard to projectile motion, which was violent and mysterious for the ancients (after the projectile has left the hand of the thrower, by what means does the projectile continue moving?), Aristotle is committed to a constantly conjoined mover.[44] He rejects a theory often attributed to Plato which suggests that the continual displacement of air around a projectile provides the requisite motion: air pushed out of place by the projectile returns at the rear of the projectile to move it forwards. Aristotle, by contrast, argues that 'the prime mover [for example, the thrower of a projectile] conveys to the air (or water, or other such intermediary as is naturally capable both of moving and conveying motion) a power of conveying motion'.[45] However, with regard to the *natural* motion of inanimate bodies, it seems that a source of motion which is coterminous with the moving body is not required. As we have seen, a heavy body which is held at a high place is in second potentiality to being at its lower, proper place. This second potentiality is also regarded as a 'first actuality'. This in turn is reduced to a second actuality. This dual description – second potentiality/first actuality – reflects the difficulty of defining motion, for motion is opposed to what is fixed or static, and definitions are 'unmoving' and therefore unable to capture motion. One is frequently led to define motion in terms which border on the tautologous: for example, motion is a 'passage' or 'transition'. This is why Aristotle, and Aquinas following him, identify motion as the constituting of a being in second potentiality, namely as the realising of a potentiality, *qua* potentiality.[46] Motion is what one might call an 'active potency', hovering between potency and act. So at this point, poised between potency and act, Aristotle and Aquinas are very anxious to stipulate that the constitution of a being in second actuality is not *another* motion which thereby requires another mover beyond the generator of the body. For example, if a heavy body begins to fall and is stopped by an impediment, the motion which commences when the impediment is removed is not *another* motion requiring *another* explanation. All downward motion (natural motion) of a heavy body is explained with reference to the body's generator and any remover of an impediment to its natural motion. There is only one motion which *intensifies* and becomes more immanent as it approaches second actuality (that is, its natural downward place). Hence for Aristotle, to constitute a body in second actuality from first actuality is not a question of the body being moved, but rather of 'energisation'.

However, if one examines Aquinas's hierarchical cosmology and his view of causation in relation to God, this matter can be seen differently. In understanding the motion of inanimate bodies, Aquinas is quite clear that the mover or generator is part of a pattern of more general causes which, in being more general, are more potent. He adopts the principle, familiar from the first proposition of the *Book of Causes*, that, 'Every primary cause infuses its effect more powerfully than does a secondary cause'.[47] This means that, in any motion, there are both particular and more general, or universal, causes. For example, an axe cuts wood, and yet the axe is in turn reliant upon the power of the woodcutter. The motion of the axe in cutting the wood participates in the motion of the woodcutter, and likewise the woodcutter participates in the power of more universal, fundamental and potent causes of motion.[48] These causes extend through a hierarchy to the heavenly bodies in such a way that all motion is in some sense caused by the motion of the heavenly spheres. In order to understand why this is the case, one must first see the double excellence in the celestial motion.

First, this motion of the celestial spheres is local, and such motion is the first of all motions. Why is local motion the primary motion? Because all other motions require a local motion for their occurrence, for the mover must be brought into the proximity of that which it is to move.[49] Moreover, Aristotle and Aquinas regard the heavenly bodies as beyond the other types of motion found in terrestrial bodies, namely qualitative and quantitative motion, and also the broader category of generation and corruption.[50] These bodies constitute a 'fifth' element lacking contraries such as heavy and light, hot and cold. Thus the celestial realm is not subject to the motion between these opposites.[51] The second excellence of heavenly motion lies in its position as the principle and measure of other motions which are posterior.[52] How does such motion become the measure? By virtue of its unity. In his commentary on Aristotle's *Metaphysics*, Aquinas writes that the unit, namely that which is singular, becomes the primary numerical measure by which every number is measured and known. Multiplicity emerges from, and presupposes, a prior unity, and the unity becomes the principle and measure of all that follows. However, the measure of things extends beyond pure number, because there are basic unitary principles in all classes, and from these emerge the diversity of things. For example, in the case of words the unit is the vowel or consonant, in music the lesser half tone, and in motion 'there is one first measure which measures the other motions, namely, the simplest and swiftest motion, which is the diurnal motion'.[53] The diurnal (circular) motion, seen most particularly in the heavenly spheres, becomes the principle and measure of other motions because, unlike rectilinear motion, it is singular and complete, lacking beginning and end and featuring no contrary. Unlike circular motion (to which rectilinear motion cannot be reduced), rectilinear motion can be divided into up and down, left and right, forward and backward. Thus in being the most unitary motion, lying within its own bounds, circular motion can become the principle motion by which all other motions are measured and known.[54] Also, this motion is primary with regard to time because it alone can be perpetual and come closest to eternity in being 'complete' at every moment, not having a discernible beginning or end.[55] Finally, because the circular motion of the heavens returns upon itself, its principle and end are one in such a way that it maintains a more perfect 'motionless' form. This celestial motion is therefore closest to actuality and donates to the cosmos a measure of permanence and stability. The importance of this circulation and its associated permanence is also explained in Aquinas's commentary on Aristotle's *De Anima*. In the generation of like from like – man from man, tree from tree – one finds a circular motion in which a species finds its permanence, stability and actuality.[56] Principle and origin are the same. In this way, 'it is natural for living things to make another of a sort like themselves, so that they always participate, as much as they can, in the divine and immortal'.[57] It is as if the circular motion of the sublunar species – the generating of like from like – is borrowed from the circular motion of the heavens themselves.

Therefore, because the circular motion of the heavens is the most actual in being complete and unitary, it is ontologically prior to other motions and therefore the more universal and potent cause of other motions. Thus, Aquinas concludes, 'the many and various movements of bodies on earth must be causally derived from motion in the heavenly bodies'.[58] Yet Aquinas does not confine himself to metaphysical speculation concerning motion; he also appeals to an observed hierarchy in the heavens.[59] First amongst these is the first heaven or first orb, namely the stars themselves. This motion is most perfectly circular and cannot, because of its uniform perfection, produce difference. So the coming-to-be and passing away of terrestrial bodies, that is the circular motion of generation and corruption, is achieved by the motion of the celestial body which incorporates the difference of 'near' and 'far'. That body is the sun, whose motion lies in the oblique or elliptical circle known as the zodiac. This circle, says Aquinas, falls away on either side of the equinoctial circle in such a way that it is at one time near and at another far. This causes, for example, the generation of plants in spring and their passing away in the winter, a circular motion which, as was seen in Plato's cosmology, encompasses in the differentiability of time that which is known completely in eternity.

Before progressing to consider how Aquinas's cosmological hierarchy informs an understanding of the motor-causality principle, one might pause to enquire concerning the extent to which it is possible to assert in any way the veracity of this thirteenth-century astronomy in the face of quite opposed contemporary understandings of, for example, the material nature of the planets.[60] Even at a most basic level, it is now known that the planets are not composed of a so-called fifth element, they are not eternal as the Aristotelians believed, and the sun does not move further or nearer to the earth – the converse is the case. So considering Aquinas's understanding of celestial motion in this way is not to suggest that his understanding of the nature of the celestial realm can simply challenge or supplant that of contemporary scientific astronomy or astrophysics in a fit of eccentric nostalgia. Yet Aquinas's cosmology – like that of Plato – is not of this kind. Unlike much contemporary science, he is not concerned to produce a purely physical, univocal description of apparently discrete objects in the universe. Rather, his priority lies in discerning relationality and the analogical *meaning* of things within a unified cosmos according to certain theological principles. In this regard, two primary aspects of Aquinas's vision remain important for our understanding of motion of all kinds. First, it is clear that in emphasising the crucial role of the planets in all motion, Aquinas can maintain that any sublunar motion is not discrete or self-explanatory. It makes no sense in the Thomist vision of the universe to understand motion through the examination of individual bodies outside a cosmic context, simply because there are no such bodies. Aquinas is not engaged in abstract thought experiments of the kind employed by, for example, Newton or Descartes, who consider a supposedly idealised motion of a single body in purely empty space.[61] For Aquinas it seems that a genuine plurality and hierarchy of causes of motion can be maintained in a fairly straightforward fashion through attention to, for example, the importance of the motions of the sun for the passage of the seasons, or the motions of the moon in the generation of tides. Although this seems quite consonant with our present understanding of, for example, the effects of the sun and moon on the earth, the emphasis is nevertheless at variance with much contemporary scientific procedure: Aquinas searches for universal efficient *and* final causes (the more general those causes, the more potent and extensive their reach) rather than the purely immediate, immanent and efficient causes of things. Adopting such a strategy does, of course, lead Aquinas to a motionless first – and most general – mover which is the origin of things; motion remains an ontological category (integral, not incidental, to the being of things) in such a way that metaphysics and physics are understood to be in intimate proximity. Secondly, and more specifically, one can insist on the primacy of circular motion as that which (like the eternal in relation to moving being) is able to measure all other motions through the mediation of time. Thus time is not arbitrary or separate from motion, but is part of the *measured* harmonic proportions which give rise to a *uni*-verse.[62] One may also assert that circular motion does lend stability to a moving cosmos through this motion's complete and unitary nature. This stability saves the cosmos from an aimless wandering by providing non-identically recurring goals for the motions of things. For example, by the constant repetition of the seasons, the goal of regeneration and flourishing is placed within the grasp of creatures. This is, then, part of a teleological view of nature which sees circular motion as mediating graspable ends to a cosmos in motion. So it is not mere eccentricity or an obfuscation which returns us to Aquinas's natural philosophy, but the possibility of a different, theological construal of motion's nature and meaning which also considers origin, context and purpose.

Continuing now to consider the motion of inanimate bodies and the motor-causality principle (*omne quod movetur ab alio movetur*), in what sense are heavenly bodies not only the movers of inanimate bodies in the terrestrial realm as a universal and potent cause of generation, but also 'constantly conjoined' movers? It seems obvious that the celestial realm is not a spatially or locatively proximate mover. Moreover, for Aquinas the celestial realm is separate and distanced from terrestrial motions not merely in a spatial sense. In being beyond all motion except the most perfect, the heavenly bodies are not knowable under the normal categories (quality, locus) by which one categorises sublunar bodies. Thus Aquinas comments that, as well as the local distance between the terrestrial and celestial realms which hinders our knowledge of the latter, 'the accidents of heavenly bodies have a different motion and are wholly disproportionate to the accidents of lower bodies'.[63] Although one might not wish to maintain with Aquinas that the planets are composed of some unique fifth element, nevertheless it is the case that what impinges upon our minds most vividly concerning celestial bodies is precisely the motion which they do possess most clearly, the circular motion. One should also recall that Aquinas is not making mere assertions about a supposed 'divine' celestial realm, for there is indeed a sense in which the planets are not in potency to the kinds of motion to which sublunar bodies are liable. For example, one would need to assert the existence of absolute space in order to claim that celestial bodies are liable to rectilinear motion of 'up' and 'down' in any straightforward fashion.

However, despite the apparent distance between sublunar bodies and the heavens, in another way Aquinas considers the celestial spheres and their motion to inhere most potently and deeply in the movement, and therefore the being, of all terrestrial entities. How is this the case? Aquinas is clear (following Aristotle) that if the celestial motion were to cease, terrestrial motion would also cease. He states that 'after the cessation of the celestial movement, action and passion will cease in this lower world' and 'if the first movement which is that of the heaven ceases, all subsequent movements will cease also'.[64] Aquinas goes on to claim that a body has a two-fold action. In the first instance, the action is by motion, and, in the second instance, the action is more akin to that of the separate substance whereby a body communicates a certain likeness of its form in the surrounding medium; for example, the air receiving the light of the sun. If the heavenly bodies ceased their perfect circular motion, the first type of action would cease, namely that relating to motion, which in turn gives rise to generation and corruption. In this way, all terrestrial motion requires the constant action of a higher motion.

There is a further way in which the heavenly motion maintains a constant 'proximity' to the motions of lesser bodies. For Aquinas, it is not wholly adequate to describe the heavenly bodies as 'above' the terrestrial. They are only above by their association with fire, the most spiritual of the elements which has a natural motion upwards. Yet being 'above' implies a contrary relationship to being 'below', and earlier we saw that the heavenly bodies do not contain these kinds of contraries which would place them in potency to a motion other than the perfectly circular. It is more appropriate to describe the heavenly motion as 'around', 'containing' or 'enveloping' other motions, for the perfect circulation is a principle and end of all lesser, rectilinear motions. In this way, the perfect contains and is not compromised by the imperfect; the latter, in being less perfect and contingent upon participation in the former, must maintain a 'proximity' to that which is more actual and real.[65] Thus it seems that Weisheipl's interpretation of 'whatever is moved is moved by another' is not wholly accurate: Aquinas, following an Aristotelian cosmology, finds terrestrial motions, of which inanimate natural motions are a variety, to be contained within a constantly active and more perfect celestial motion. Even inanimate motions, for Aquinas, are never wholly self-explanatory; they always refer beyond themselves.

However, if one progresses further in Aquinas's cosmological hierarchy, it is possible to make this interpretation of his understanding of motion clearer still in such a way that the distinctiveness of his view in relation to that of early modern physics becomes more apparent. The celestial spheres, although possessing a perfect circular motion, are nevertheless moved, whether by means of a distinction of parts within the spheres themselves (for example, they may be ensouled and therefore 'living' after the fashion of terrestrial creatures) or by a separate mover conjoined by contact of power.[66] What could be the mover of the heavens? Their circular motion does not aim at another motion beyond; it is its own end in being singular, continuous and never ceasing. So the motion cannot be caused by an intrinsic natural principle originating in a generator, that principle tending to something beyond itself, for the celestial spheres are beyond the contraries between which lesser motions hover, whether of quality, quantity or place (understood as 'up' or 'down'). Therefore, the celestial bodies must be moved by some kind of mover, a soul or 'separate substance'.

Now before proceeding further, it is important to note that in formulating his view concerning the mover of the heavens, Aquinas worked from the principle that the soul's union with the body takes place for the sake of the soul and not the body (just as matter exists for the sake of form). Therefore, the issue is: what kind of soul – nutritive, sentient or intellective

– might benefit from union with a celestial body? According to Aquinas, it cannot be a nutritive soul, for such a soul pertains to generation and corruption to which the heavens are not prone. Neither can it be a sensitive soul, for a celestial body does not possess sense organs. Also, Aquinas comments that the senses are composed of elements mixed in proper proportion and these elements, in being subject to contraries (up or down, hot or cold), are alien to the heavens.[67] So one should conclude that the celestial spheres are moved by intellective separate substances in order that they might participate in God's creative act by taking their place as instrumental causes within creation. This is to say that it is of the goodness of God that he confers being on all things and, because all things find their perfection in participating in the divine, they seek to become like God in being the causes of other things.[68] It is by imparting a perfect motion (local and circular) to the celestial spheres that the separate substances, the angels, communicate their own being and find their perfection as instrumental causes in God's creative act.

But what aspect of their being do angels communicate through the celestial bodies? As was intimated above, intellection. In his treatise *De Veritate*, Aquinas describes the kind of intellection possessed by separate substances.[69] Their knowledge is not discursive like that of embodied intellects. In other words, they do not know one thing from another; such knowledge requires motion between things, for example when I know the sun from its effects in generating plants. Angels do not receive their knowledge from bodies (as is the case with human beings) but by an inflow of intelligible forms from the mind of God.[70] Following Pseudo-Dionysius, Aquinas likens this intellectual activity to circular motion. He states that, in the act of knowing God, an angel's gaze does not pass from one thing to another, but is fixed upon God alone. In this way,

> it is said to be moved about God, as it were, in a circular motion, because he does not arrive at God as at the end of cognition that had its beginning from some principle of cognition, but [his knowledge is] like a circle, without a beginning or end.
>
> (Aquinas, *De Veritate*, 8.15.ad 3)

Aquinas goes on to quote Pseudo-Dionysius to the effect that 'angels are moved "in a circular motion which is simple, without beginnings, and rich with everlasting illuminations of the good and the beautiful"'.[71] God's self-knowledge is compared with the centre around which the circular motion of the angelic intellect turns. The motion of angelic intellects partakes of the unity of the centre without achieving its perfection. It is this circular motion of intellective knowledge which is imparted to, and thereby embodied in, the movement of the celestial realm. This is to say that cosmic motion is itself a participation in, and representation of, a higher vision of God.

Returning once again to the lowly nature of the motion of inanimate bodies, we can see that even this motion has as its more potent and general cause the hyper-angelic knowledge of God which is in turn embodied in the perfectly circular local motion of the celestial spheres. In this way, the separate substances, in being joined to the celestial bodies as a mover to that which is moved,[72] can participate in divine creativity by communicating the stability, actuality and completeness of the circular intellective motion of their own being to an embodied cosmos. Therefore, it is, as Aquinas states, for the purpose of imparting and embodying motion that separate substances are joined to the heavenly bodies.[73] Aquinas likens the separate substances moving the celestial bodies to the human intellectual agent bringing about the existence of an artefact.[74] The human intellect, in which resides the form of the artefact to be made, uses instruments (for example, tools) in bringing the proposed artefact from potency to act. Similarly, the separate substances use the celestial bodies as instruments in bringing into actuality the forms which reside in their intellects. The teleological order of the motions of nature indicates the intellective source of these motions.[75] Elsewhere, Aquinas likens the celestial bodies as movers to the way in which a king governs a political body. He states that,

> For the heaven, which aims at the universal preservation of things subject to generation and corruption, moves all inferior bodies, each of which aims at the preservation of its own species or of the individual. The king also, who aims at the common good of the whole kingdom, by his rule moves all the governors of cities, each of whom rules over his own particular city.
>
> (Aquinas, *Summa Theologiae*, 1a.82.4.responsio)

Here we find revealed in nature and politics a hierarchy of the causes of motion.

It is now possible to pause briefly in order to underline an important congruence between this view of motion and the Platonic understanding outlined in the first chapter of this essay and thereby point out that Aquinas's thought here lies at the very limits of Aristotelianism. In the *Timaeus*, it was seen that motion is the embodiment in time of that which is known in the eternity of the Forms and, particularly, the form of the Good. Here, in Aquinas's cosmology, we find that the celestial spheres are an embodiment of the angelic knowledge of the divine ideas and the means by which all that is known at once in the eternity of the Godhead can be embodied successively in the motion of time. Crucially, there is no dualism here between intellection and embodiment, or soul and body, because the motion of the heavenly bodies, which in turn is a general and proximate cause of even the most lowly and inanimate body, is analogically related to a supremely spiritual motion, namely the knowledge possessed by separate substances, those beings which we call angels.

However, for Aquinas neither the celestial spheres nor the separate substances are the most potent and general cause of motion, for they are not themselves wholly unmoved and self-sufficient. One must therefore search for a first unmoved mover which can be the source of all motion in the cosmos. This ultimate source we find in God, for he bestows that which is most general, namely being.[76] It is God's nature to exist, and therefore he causes existence in all things, from angels to inanimate objects. This causation, however, is not merely found when things begin to exist, for all things are continually maintained in existence by God. The divine is therefore present in things in a fashion that is in keeping with the way a particular thing possesses its existence. Because existence is most intimately and profoundly interior in things, 'so God must exist and exist intimately in everything'.[77] This may be related more specifically to motion. Aquinas maintains that motion is apart from the being of a thing. In other words, it has its ultimate source in another, because something in potency cannot reduce itself to act. Ultimately, all finite actuality, that is all created being, is, for Aquinas, participated being. Therefore, all motion, which is the reduction of potency to act, must have its primary origin in that which is supremely actual, namely being itself. This is God. 'Therefore,' claims Aquinas, 'it is impossible for the being of a thing to continue except through divine operation'.[78] Moreover, this distinguishes God's causation from all other causes. Created sources of motion, for example, the generator of an object or a woodcutter wielding an axe, merely give the form to a moving object or apply an instrument to action. By contrast, God not only donates form and applies things to action, he sustains form at every moment, 'just as the sun is called the cause of colours' appearing in that it gives *and maintains* the light whereby they do appear'.[79]

Yet how can God be said to 'touch' things in such a way that he may be described as a constantly conjoined mover of even the lowliest inanimate motions? Initially, one must note that there are two kinds of touching. First, there is physical contact, as when two bodies touch each other; this is locative and exclusive in restriction. Secondly, there is what Aquinas calls a 'contact of power', and the example he gives is of grief touching the grieved. It is in this second sense that God 'touches' even the lowliest bodies. His is a touch of power which sustains things in being and, both immediately and by the operation of intermediaries such as the heavenly bodies, moves things to their proper ends.[80] The use of 'touch', however, is crucial, for it indicates most clearly the radical intimacy between moving creation and God as its origin and sustainer. The basis of this notion of intimate touch in relation to causation is expressed in the detail of the well-known first proposition of the *Book of Causes* and Aquinas's discussion of the power of God. He states that 'The higher a cause the greater its scope and effect. The more efficacious a cause the more deeply does it penetrate into the effect' and 'God is the cause of the action of all things because he gives to everything the power to act and maintains it in being and applies it to action, and by his power every other power acts'.[81]

This Proclean notion of the intimacy of divine causation in all levels of the cosmological hierarchy has further implications for the link between intellect and touch.[82] For Aristotle, in his *De Anima*, and likewise for Aquinas in his commentary, touch is the most basic and necessary sense.[83] Yet this does not mean that the other senses are unrelated, for all sensation takes place 'by contact'.[84] What varies is the medium through which sensation takes place. For example, in the case of sight, the medium is light, in hearing the medium is air. It appears that, in the case of touch, sensation is most immediate. This is because, in the case of seeing, smelling and hearing, it is the medium which acts on the organ of sense. In the case of touch (or even of taste, which is a variety of touch), we perceive things at the same time as the medium. In other words, in the case of touch, the sensitive soul is moved *with* the medium and not simply *by* the medium. Whereas in the case of the senses of sight, hearing and smell there are a succession of more discrete motions (the object moves the medium which moves the sense), in the case of touch the motion is more unified: the object, the medium and the sensitive soul share a common motion. In touch, corporeality and intellect are joined in a common motion.

So what constitutes the medium in the case of touch? For Aristotle, the medium between the soul and that which is sensed is the body.[85] Yet flesh is definitely not some kind of passive 'stuff' through which we perceive objects. Rather, the body is a mean which must be in a harmonious proportion between the soul and that which is sensed.[86] What does Aristotle intend by this claim? He wishes to point out that in order to sense hot and cold, for example, the organ of touch must lie within a suitable proportionate mean between hot and cold. Aristotle is thinking of the familiar experience of one of our hands being particularly warm and the other cold so that our sense of touch is distorted: one hand may perceive a bucket of water as hot while the other perceives the same water as cold. So in order to sense by touch (or, indeed, any other sense), our bodies must be in potentiality to becoming qualitatively akin to that which is perceived and this involves having some appropriate harmonious relationship to the perceived object.[87]

Given the mediation which takes place in all sensation between the perceiver and that which is perceived, one can see that, as John Milbank and Catherine Pickstock point out, for Aristotle all sensation is triadic in structure.[88] There is mediation between *psychē* and *hulē*: the soul has contact with, or touches, that which is perceived via the corporeal elements. Going further along a Thomistic line, one can surmise that it is the divine 'touch' which maintains and exemplifies this triadic structure: created being, which has the two-fold composition of essence and existence, is sustained by its interweaving in the simplicity of *esse ipsum*. Because being is the most common and intimate first effect,[89] the divine touch – God's sustaining of the created being of things by means of his knowledge of all things – is the exemplar of the mediated and less perfect intimacy of all corporeal sensation by which *we* know things.[90] It might even be suggested that sensation (and therefore knowledge) consists in some way in a shared motion. In the case of our corporeal sensing of things, a motion must be shared by the perceived object, the medium and the soul. In the case of the divine 'touch', the creature which is known – namely, the creature which participates to a greater degree in *esse ipsum* – shares in the energic stability of the divine ideas. Ultimately, as we will see when considering God's immanent life and creative act, this triadic structure of our corporeal knowledge of others and God's knowledge of creatures finds its ground in the perfect triadic structure of the divine self-knowledge.

So having reached the most intimate and potent cause of motion in the divine 'touch', we are now in a position finally to revise Weisheipl's understanding of the principle central to Aristotelian physics, namely 'whatever is moved is moved by another'. It is the case that the motion of inanimate bodies can be explained initially by reference to the generator of the body or the remover of an impediment to the motion, and that the reduction from second potentiality (or first actuality) to second actuality is not another motion requiring explanation. However, for Aquinas, before one arrives at God there will always be too much potency and an explanation of anything will never be wholly sufficient. The motion of any body will always have a more potent and more general cause to which it is, in an ever more radical sense, conjoined. At all moments, all things are conjoined to God because it is his touch which donates and sustains being. As we will see in Chapter 5, this is a view of motion at variance with that of, for example, Newton. Motion will always, for Aquinas, be explained by the intimacy of God, for motion is, as for Plato, the embodiment of that which is known perfectly in eternity. Whereas we tend to imagine motion in terms of 'towards' or 'away' from something, for Aquinas it seems that motion is more fundamentally understood to take place 'within', or 'enveloped by', *esse ipsum*.

This brings to a conclusion the discussion of Aquinas's understanding of motion which is most clearly related to the principles of pagan natural philosophy. In the next section, I turn to examine in greater detail Aquinas's deployment of this physics within themes central to *sacra doctrina*. Whereas the previous discussion began with lowly and inanimate bodies and arrived at God as the intimate source of all motion, in what follows the natural philosophy acquired by Aquinas will be deployed throughout an examination of the *exitus* of all things from God and their *reditus* to him, which is to say that, whereas the physics of Aristotle follows a linear path to God, Aquinas follows the more perfect and complete circular motion of *sacra doctrina*.

Motion and God

God beyond motion

In the introduction to this chapter, we saw that Aquinas begins his exposition of *sacra doctrina* by delineating his subject matter. He is clear that natural philosophy, such as that described in the previous section, can be put to the service of this higher science. Having outlined the nature of this divine science, he continues in the opening questions of his *Summa* with an examination of what might be said of God and things in relation to God. Aquinas formulates what David Burrell calls 'the metalinguistic project of mapping out the grammar appropriate *in divinis*'.[91] He delineates some basic principles to be adopted if one is to speak appropriately of God (thus question 13, which discusses theological language more explicitly still, is perfectly in tune with the preceding questions),[92] beginning with a question that appears to be a piece of natural theology which one might even think is designed to ground the whole *Summa*: whether or not there is a God.

What is Aquinas's purpose?[93] Within the context of the opening of the *Summa*, this question, which includes the so-called *quinque viae* of 'proving the existence of God', must be read within the trajectory of the exposition of *sacra doctrina*. These 'proofs' deploy a physics of motion, and in turn a metaphysics of potency and act, in an attempt to conclude to God's existence. Is Aquinas's deployment of a physics of moving bodies designed to provide a foundation in reason for his further theological investigation? Certainly not, because the foundation of this enquiry has already been stipulated, namely God's own science, that knowledge of himself which he shares with the blessed. Rather, one might outline two general purposes of the second question of the *Prima Pars*. First, Aquinas is aware that, for us, 'God is' is not self-evident. Scripture tells us that 'the fool has said in his heart, "There is no God." '[94] Because it is the *fool* who denies that 'God is', Aquinas resorts to 'foolish measures' – looking on matters as the 'fool' might do, with an apparently bare metaphysics which is characteristic of the pagan mind – in order to show that 'God is' *can* be made evident, even in this way. In this context, it hardly seems surprising that Aquinas, unlike a modern natural theologian, does not allow himself to be delayed with this particular enterprise.

However, there is a second and perhaps more important purpose to question 2, one which refers not to the demands of the fool but to the demands of the theologian. It seems that Aquinas may be anxious to respond to those who are sceptical about the deployment of the new Aristotelian learning within an exposition of *sacra doctrina*. One must recall the context in which the *Summa* was written: much suspicion and controversy continued in the later 1260s and 1270s concerning the usefulness of this pagan metaphysics and physics.[95] Aquinas deploys both of these sciences in questions 2 to 12 of the *Prima Pars* (most notably, Aristotle's physics of motion), not primarily to prove what one already knows by faith, but just as importantly to display the efficacy of reason: that is, that metaphysics and physics are worthy of their appropriate place within an exposition of *sacra doctrina*. When properly deployed, Aquinas demonstrates that these sciences are at least consonant with the articles of faith. For example, in the case of the so-called *quinque viae*, physics and metaphysics can make evident *to us* what is self-evident *in itself*, namely that God is. Aquinas is not attempting to prove 'God is'; he is trying to demonstrate the value of other sciences in making clear to us the content and implications of *sacra doctrina*. He does so by displaying meta-physics and physics as sciences which participate in the truth given in *sacra doctrina*.[96] The purpose of question 2 is therefore wholly consonant with the themes of question 1: the nature and content of *sacra doctrina* and the value of other sciences in making evident to our often darkened, foolish minds the teaching that has been given by God to his Church. One can concur with Rudi te Velde's comment when writing about reason's place in the *Summa Contra Gentiles*:

> So the issue is not a defense of the 'reasonableness' of Christian faith before reason. Aquinas's objective is to confront natural reason with its own condition, to make reason aware of its limitations in order to prevent reason from unreflectively imposing its own limits on the search for truth.
>
> (R. te Velde, 'Natural Reason in the *Summa Contra Gentiles*', p. 58)[97]

Yet Aquinas is not so much concerned with the denigration of reason before revelation, but the establishment of the value of reason in its rationality. He is not so much trying to 'prove' God as 'prove' reason.

How, then, does Aquinas deploy this understanding of motion in question 2 in order to elucidate *sacra doctrina*? He addresses this issue most particularly in the first of the so-called *quinque viae*. In the introduction to this question, Aquinas states that he has shown that the purpose of *sacra doctrina* is 'to make God known, not only as he is in himself, but as the beginning and end of all things and of reasoning creatures especially'.[98] He is now to show God as the beginning and end of motion,[99] motion being a fundamental characteristic of created nature, for even if something is not itself in motion, it is a crucial aspect of that thing that it is in potency and therefore liable to motion. As was seen above, 'motion is nothing else than the reduction of something from potentiality to actuality. But nothing can be reduced from potentiality to actuality, except by something in a state of actuality'.[100] Why? Because something cannot be in potency and act at the same time with regard to the same thing: something cannot be both cold (in potency to being hot) and hot at the same time. Therefore, in order for something cold to become hot, something other is required which is already hot which can thereby bring about motion from cold to hot. 'It is therefore impossible that in the same respect and in the same way a thing should be both mover and moved. ... It is necessary, therefore, that everything that is moved is moved by another'.[101] This is the Aristotelian principle, familiar from our earlier discussion, that a distinction is always required between something that is moved (that which is in potency) and something that is the mover (that which is in act). If every-thing that is moved requires a mover, could this series continue to infinity? Both Aristotle and Aquinas answer firmly in the negative because the ulti-mate beginning and end of motion cannot itself be moved. Why not? Because when there is a series of movers in which each one moves the next, if there were not a first mover which is itself unmoved and therefore beyond motion, none of the other members of the series would be a mover or moved.[102] This leads Aquinas to propose that, from the reality of motion in nature, one must postulate that which is itself beyond motion in order to account for motion, namely something in full actuality which can be the ultimate source of the reduction of potency to act.[103]

Following Gilson, it is also important to note a more general reason why a regression of movers to infinity is absurd for Aquinas (and Aristotle alike).[104] In the earlier discussion of the natural philosophy of motion, it was seen that the causes of motion are ordered within a cosmological hierarchy. The movers within this hierarchy are superior to those beings which they move, simply because movers are in act rather than potency with regard to the motion in hand. Within a species, accounting for the coming-to-be and motion of individuals is straightforward – humans beget humans, stones move stones. But each individual moving cause within a species cannot account for the coming-to-be and motion of the species as a whole, because each individual's nature is defined by the species: were an individual member of a species purported to be the origin of that species' coming-to-be and motion, that individual would be self-causing, cause and effect, mover and moved. Therefore, one must search outside and above the species for the reason for the coming-to-be and motion of the individuals of the species. Whatever acts by virtue of a nature received from elsewhere is an *instrumental* cause of the motion or coming-to-be of others and is therefore related through higher causes to a first cause. For Aquinas, therefore, there is a hierarchical series of moving causes which, far from being infinite, is not even particularly numerous. Furthermore, one must remember that the hierarchy and series under consideration is not temporal: it is not a motion begun by a first mover in the past. Aquinas does not have in mind a first mover who acts only to set off a kind of 'domino effect'. He is seeking to describe a cosmology in which any motion at any time is unintelligible without a fully actual first mover who is the source of motion for all things in the present. This is to say that God is not at a greater distance from the universe now than at the moment of creation. He is equally the ultimate source of the motion of the tiniest stone now and the greatest bodies at the moment of creation; God is equally 'close' to all parts of the cosmological hierarchy. As we saw earlier, God touches even the lowliest motion.

Having stipulated that God is known in the depths of human reason as the source of motion, Aquinas continues outlining the 'logical grammar' of his work in pursuing a *via negativa* which leads to more speech about the divine, albeit for the moment in negative guise. He states that 'we cannot know what God is, but only what he is not'.[105] Therefore, all attributes characteristic of creaturely contingency which are thus inappropriate for the logical description of God are removed. The divine is described as simple (lacking composition), perfect (lacking any blemish), limitless (lacking any limit or qualification), unchangeable (lacking potentiality), eternal (lacking temporality) and one (lacking division). As one would expect in such a logical grammar, each description implies the others. For present purposes, I will focus briefly on three aspects of this grammar appropriate to God which relate most particularly to motion: simplicity, unchangeableness and eternity.

To say that God is simple is, for Aquinas, to say that God lacks any composition. As is apparent from the earlier discussion of the physics of motion, there is one composition which is crucial for motion: potency and act. God is pure, simple actuality, possessing no potency. There is nothing towards which God could move or be moved. Furthermore, this implies God's unchangeableness: that is, there are no contraries between which God lies.[106] Aquinas reiterates that, by unchangeableness, he means that God is beyond the motion which hovers between potency and act. There are, he remarks, 'operations' – such as willing and understanding – which the Platonists refer to as 'motion', but these do not involve potency and are akin to Aristotelian *energeia* – that which is unitary, 'all at once' and within its own limits.[107] We will see below that Aquinas analogically relates motion proper (which he excludes from God) to energic operation which the Platonists do refer to as a special kind of motion (and this Aquinas does include within God). For the moment, one further aspect of Aquinas's logical grammar concerning God requires mention, namely eternity. This has two characteristics: eternity is unending and it is an instantaneous whole, lacking succession.[108] Referring to God as eternal is to say that God, who is beyond motion, is not subject to time, which is the measure of motion. Following Boethius, Aquinas comments that 'eternity is an instantaneous whole whilst time is not, eternity measuring abiding existence and time measuring change'.[109] We saw above that the circular rotation of the heavens, in being the primary, unitary motion which is within its own limits, becomes the principle and measure of all other motions. Likewise, the eternity of God, in being perfectly unitary, measures being itself. When we refer to God's simplicity, we thereby refer to his motionless unity, which in turn implies that God's 'time', his eternity, is also complete and whole.

So God, in being simple, unchangeable and eternal, is wholly beyond motion. We have seen from the examination of Aristotle's physics that nature is a principle of motion and rest in those things in which it inheres primarily. How, then, can any created thing, whose being is divided by the distinction between act and potency between which motion hovers, have any knowledge of, or speak about, that which is wholly beyond motion and is perfectly actual? Are we confined merely to a *via negativa*, referring to God by negating aspects of created being, such as motion? In what follows, I will argue (in line with the first three chapters of this essay) that motion is not that which separates creation from its creator, but is the very means of their analogical relation. First, however, a word about analogical predication and our knowledge of God in Aquinas's thought.

Having discussed the negative approach to God in the opening questions of the *Prima Pars*, a logic appropriate to the divine reality, Aquinas, in the twelfth question, takes a step to mitigate what appears to be a thorough-going apophaticism. Through the first eleven articles he considers the possibility of a direct knowledge of God's essence, and concludes that God's essence cannot be 'comprehended', but that, by grace, a created mind might gaze upon God. This sight is not by sense or imagination, but comes in an illumination of the mind by divine grace. Just as for Grosseteste, this illu-mination by grace is not an autonomous alternative to the natural light of the mind, but is 'an increase (*augmentum*) in the power of understanding we call "illumination" of the mind, as also we speak of the intelligible form as "light" '.[110] This gaze belongs only to the blessed in heaven. Can anyone in this life see God's essence? Aquinas answers in the negative, because the way in which a thing knows depends on its being, and our souls are embodied, hence they cannot *by nature* know anything other than those things which have form in matter.[111] However, this does not mean that our knowledge of creatures (composites of matter and form) tells us nothing of God. Aquinas is clear that effects resemble their causes, and because creatures are effects of God's causal power, they will in some sense, however attenuated, resemble God, even though God will not resemble his creatures.[112] Things resemble God as possessing being, 'for precisely as things possessing being they resemble the primary and universal cause of all being'.[113] Therefore, each creature, in so far as it has being, is similar to its divine cause. Yet there remains a funda-mental difference between cause and effect, because God is his being, whereas creatures have being by participation (*per participationem*) and resemble God by analogy.[114] This is the framework which was later to become known as the *analogia entis* which is based upon the ontological difference between God and creatures.

This analogical participation is the basis of the resemblance, and therefore the intelligible connection, between God and his creatures. Furthermore, this ontological grammar, in which the convertible transcendentals of being, good-ness and truth find their origin and coincidence in the simplicity of God, is itself the basis of the apparently more purely linguistic grammar which Aquinas describes in question 13 of the *Prima Pars*. Despite the *via negativa* of the opening questions of the *Summa*, speech about God is not confined to negativity for Aquinas. To see why this is the case one must recall the relation-ship of resemblance between God and creatures. In so far as they exhibit their own perfection, creatures resemble their divine source, yet not as something of the same genus or species, but rather by a relation of causal dependence where the effects represent in divided and manifold forms the perfections that are contained wholly and simply in the source.[115] All perfections existing in crea-tures pre-exist in a united fashion in God.[116] Because perfections have their ontological source in God, so our perfection terms (good, wise and so on) in speech about the divine are predicated primarily of God, where they find the source of their meaning, and only secondarily of creatures. So speech about

God, the universal agent of perfections, cannot be *mere* equivocation (as, for example, when we speak about a 'river bank' and an 'investment bank') because this would deny any resemblance between God and creatures. Rather, the universal agent of perfections 'is to be called an analogical agent, as all univocal predications are reduced to one first non-univocal analogical predication, which is being'.[117] Thus, as John Milbank states, 'grammar here grounds itself in theology, not theology in grammar', for it seems that for Aquinas *all* speech (and likewise all resemblance) is grounded in the analogical participation of creatures in the transgeneric being that is God.[118] Moreover, Aquinas goes on to state that perfection terms are predicated analogously of God and creatures according to a certain proportion. Yet this is not the 'proper proportion' of a kind of mathematical ratio between God and creatures which purports to establish a determinate distance between the two,[119] but rather a relation of attribution which has the character of *convenientia*.[120] One might say that the proportion is one of an appropriate harmonious beauty between God and creatures – a kind of 'coming together' – which constitutes and indicates analogical resemblance.[121]

What, then, is the analogical role of motion in relating creatures to God who is apparently beyond motion? I now turn to address this issue by examining in more detail the content of that analogical relation, or *convenientia*, of all things with divine being, with particular reference to motion. Following Aquinas through the *Prima Pars*, I begin with a consideration of God's action in creation. Is this act itself any kind of 'motion', and how might motion be related more explicitly to God's knowledge and inner life?

Trinity, creation and motion

While it is the case that Aquinas is frequently reluctant to describe God's act of creation as any kind of 'motion', this is in order to avoid confusion and error. Given an Aristotelian definition of motion which has at its heart the passage from potency to act and the postulation of a subject which preceded the motion, there is always the danger of compromising the Christian doctrine of creation *ex nihilo*, that is creation from anything pre-existent other than the action of God himself. Thus Aquinas comments that 'creation is not a change (*mutatio*), except merely according to our way of understanding. For a change means that a constant is now otherwise than it was before'.[122] However, on other occasions Aquinas stretches his use of the term *motus* in such a way that it can be employed at least metaphorically, but without error, of the divine creative act and even of God's immanent and perfectly subsistent intellective life.[123] However, this requires some explanation, and it is necessary to begin with an examination of the character of emanation, for Aquinas refers to creation as 'the emanation of things from the first principle'.[124] I will examine the importance of emanation for an understanding of motion and its perfection in the eternal emanation of the Son from the Father.[125]

Emanation for Aquinas refers to the active self-expression of a nature in relation to others in the production of another self. No loss to the individual's being is implied, neither any gain. In the *Summa Contra Gentiles*,[126] Aquinas expounds his understanding of emanation and its hierarchical character. He begins by noting that 'one finds a diverse manner of emanation in things, and, the higher a nature is, the more intimate to the nature is that which flows from it'.[127] This is therefore a hierarchy of emanation based on the discreteness of emanation, or the ability to make a communication of oneself without losing that self. Inanimate bodies are said to have the lowest form of emanation. The only way in which they are able to communicate anything of their nature is through the external action of one upon another – for example, a fire acts on a combustible object and produces another fire. However, animate beings such as plants are better able to communicate their nature. As has been seen, they have a principle of movement within themselves. This is the mark of life in a plant, namely that 'that which is within them moves toward some form'.[128] However, although this implies some interiority, the plant finally loses itself through the emitting of a seed which eventually grows up and dies. Aquinas also comments that in truth there is very little interior subsistence to plant life because it must gain the means of its emanation from nutrients found externally. For Aquinas the highest form of emanation is not that which terminates externally from the being concerned (for example, the nature of a plant emanates and achieves the external form of a plant), but that which has an internal termination. Whereas the emanation of a plant has a partially internal beginning and an external end, the emanation of the sensitive soul has an external beginning but an internal end in sense perception. From the senses certain perceptions move to the inner sense and are gathered to reside in the memory. Thus these sense perceptions become part of the internal consciousness of an animal's life. However, this mode of emanation is not yet self-reflective because 'the principle and the term refer to different things'.[129] In contrast, a complete return to self is first found in the human intellect, for the intellect is capable of self-knowledge and understanding. Thus a human being is able to produce a communication of its nature, an emanation of another self, in such a way that reflection of the self on itself, namely self-reflection, is possible. Yet the human intellect is imperfect because it must take its first knowledge from without, namely through sense perception, before returning from the external object to arrive at knowledge of itself by its relation to the object in question.[130] This emanation is not entirely subsistent and without exteriority because sense perception is part of human intellectual activity.[131] The human intellect knows itself through others and not through itself. Therefore, next in the hierarchy of emanation are spiritual substances, or angels, who do not require sense perception but who know themselves through themselves. Their self-knowledge is not from anything exterior. Finally, perfect emanation is found in God whose intellect and act of understanding, unlike that of angels, is identical with his being. As Aquinas points out, God's being is his essence, and his understanding and intellect must be identical with his essence or they would be accidental. Therefore, God's being, intellect and understanding are one.[132] For the divine to know himself and express himself through that knowledge is the divine essence, the very divine life itself.

Within this same question in the *Summa Contra Gentiles*, Aquinas goes on to maintain that God's self-knowledge, although perfect, unitary and eternal, still maintains distinction. This distinction consists in the God who expresses his self-knowledge in himself and the God who is expressed or conceived, namely the Son who is the expression of the self-knowledge of the Father. The former is a perfect emanation of the latter in such a way that the being of both is identical and this emanation remains entirely immanent. Aquinas draws out the difference between divine emanation and self-understanding and its human form when he writes,

> in a man understanding himself, the word interiorly conceived is not a true man having the natural being of man, but is only 'man understood', a kind of likeness, as it were, of the true man which the intellect grasps. But the Word of God, precisely because he is God understood, is true God, having the divine being naturally, because the natural being of God is not one being and that of his understanding another.
>
> (Aquinas, *Summa Contra Gentiles*, IV.11.11)

As well as God's knowledge of himself through himself, elsewhere Aquinas outlines the sense in which ideas subsist in the divine mind and are therefore known by him.[133] He claims that these ideas are forms of things existing apart from things, and that the form of a thing can have two functions. Either it can be the exemplar or pattern of the thing whose form it is said to be, or it can be the means of knowing the thing whose form it is by its residing in the knower. Aquinas stipulates that in both these aspects ideas subsist in the mind of God. Yet as regards the latter, it can be seen that it is by God's interior self-knowledge, namely the emanation of the Son from the Father, that he knows other things by their proper ideas subsisting in him. In a sense, therefore, all things are known primarily and *per se* as they exist most perfectly in God's knowledge, and as they are therefore known in God's self-knowledge, in God's emanation. Importantly for Aquinas, this enables a consistency between the divine simplicity and the multiplicity of ideas in the mind of God. He explains this by the metaphysics of participation. God's knowledge is not the result of the multiplicity of creatures informing the divine intellect (for this would compromise divine simplicity), but rather the other way around. It must be remembered that God knows his essence not only as it is in itself, but also in the many ways in which creatures may participate in that essence. The divine ideas represent the myriad ways in which the one divine essence can be participated in to some degree by creatures. Therefore, 'in knowing his essence as imitable in this particular way by this particular creature, [God] knows his essence as the nature and Idea proper to that creature'.[134] These many ideas are the subjects of one, simple (self-)knowledge of the one divine essence.

However, is divine knowledge and the creative act only a matter of the difference between Father and Son, the God who expresses and the God who is expressed? What of the divine will and, more particularly, the Holy Spirit? As one would expect, Aquinas goes on to describe the place of the Spirit within the divine emanations and creative act, beginning with a lengthy outline of the scriptural evidence for the divinity and subsistence of the Spirit.[135] From this premise and teaching he seeks to make clear what we must understand of the Spirit with regard to God's immanent life and act of creation. Initially, Aquinas examines intellectual natures in general and states that there must be a will alongside intellect because such a nature must desire to know.[136] Crucially, intellects are not merely passive recipients of 'information' after the fashion of machines; all knowledge is at once *willed* or *desired* knowledge. Just as any natural thing has an inclination to its own proper operations, for 'it tends to what is fitting (*convenientia*) for itself', so too an intellectual nature has an inclination, which we call will, towards its own proper operation in knowledge.[137] Aquinas claims that, of all the acts which belong to the will, love (*amor*) is found to be a principle and common root. He describes this in terms of the 'affinity and correspondence' (*affinitatem et convenientiam*) between the principle of inclination in natural things (their form) and that to which they are moved. Once again, an analogy with physics is deployed: 'The heavy has such a relation with the lower place. Hence, also, every inclination of the will arises from this: by an intelligible form a thing is apprehended as suitable or affective (*conveniens vel afficiens*)'.[138] To be so affected towards something is to love that thing. Thus it seems that the form of *convenientia* between things, and the principle of the motion one to another, is love, and the motive for their coming together is a teleological desire for fulfilment.

What is loved, the beloved, is in the intellect by reason of a likeness of its species. For example, via sense experience a likeness of an object comes to reside in the intellect of a human soul and in this way it is known by the intellect. However, that which is loved also resides in the will. How? By reason of a proportion and affinity between the term or goal of a motion and the principle of that motion. For example, in the case of the element fire the upper place to which fire tends is 'in' the fire by reason of the lightness of the fire which gives it 'proportion and suitability' (*convenientiam et proportionem*) to such a place.[139] Likewise, in the case of the intellectual nature we might say that an object is in the intellect by means of species residing therein, while it is in the will by means of a 'proportion and suitability' with the object; this *convenientia* then becomes the principle behind the intellectual nature's self-motion towards knowledge of the object. The intellect loves the object and by the will desires to know.

However, in contrast to intellectual beings such as angels or humans, God is at one with his intellectual nature and, likewise, his will. The intellect and will are not accidents of the divine substance because God is simple and has no accidents. Moreover, will is in God always and only by way of act, never by potency or habit, and because every act of will is rooted in love, as we have just noted, in God there is love. The first and most appropriate object of the operation of the divine will is the divine goodness, and so God, because God loves himself and is beloved and lover, must be in his will as the beloved is in the lover.[140] How is the beloved in the will of the lover? As we have said, by means of a 'proportion and suitability' (what has been termed a *convenientia*) between the two. God has a most perfect proportion and suitability with himself because he is undifferentiated simplicity. Therefore, God is in his will with perfect simplicity. In addition, any act of will is, as Aquinas remarks, an act of love, but the act of the will is the divine being. So 'the being of God in his will by way of love is not an accidental one – as it is in us – but is essential being', hence the scriptural teaching that 'God is love'.[141]

Coupled with what has been said of God's self-knowledge in the emanation of the Son, we now have a two-fold picture of the divine life. On the one hand, God loves himself because, as we have seen, the 'proportionate and appropriate' end of God's operative will is himself and his own goodness. Yet this would not be loved if it were not known, and God knows himself through conceiving of himself in the eternal emanation of the Word. Yet it is not adequate to say that it is God's knowledge which is beloved, for God's knowledge is his essence. Therefore, coupled to the emanation of the Word must be a love whereby the lover dwells in the beloved, both in God's knowing and in that which is known. The love by which God is in the divine will as a lover in the beloved 'proceeds both from the Word of God and the God whose Word he is'.[142] It is as if the Father is the lover and the Son the beloved, but immediately and in eternity this is returned so the Son is the lover and the Father the beloved. This introduces a kind of circular dynamism to the inner divine life which Aquinas refers to as a kind of intellectual 'motion'.[143] He states that St Paul attributes to love and the Spirit a kind of impulse, and such 'motion' takes its name from its term which in this case is God. The things of God are holy, hence the love proceeding from the Father and Son is properly called the *Holy* Spirit.

With regard to God's self-knowledge and self-love in the persons of the Trinity, one can now understand how the universe has the divine nature as its cause. Aquinas states that,

> Effects proceed from an efficient cause because they pre-exist there; every agent enacts its like. Now effects pre-exist in a cause according to its mode of being. Since, then, God's being is his actual understanding, creatures pre-exist there as held in his mind, and so, as being comprehended, do they proceed from him.
>
> (Aquinas, *Summa Theologiae*, 1a.19.4.responsio)

This may be expressed through the familiar analogy of the artist. The idea of the painting pre-existent in the mind of the painter is in some way the cause of the painting itself. Thus 'God's knowledge stands to all created things as the artist's to his products'.[144] However, in addition to the knowledge of things, Aquinas also notes that an act of will is necessary in the act of creating: 'And similarly an intelligible form does not indicate a principle of activity, merely as it is in the knower, unless it is accompanied by an inclination towards producing an effect; this is supplied by the will'.[145] God is so inclined because his own subsistent goodness wills that other things be in such a way that 'by his will he produces things in being' and his self-love thereby becomes the cause of the creation of things.[146] In a similar fashion Aquinas elsewhere states that,

> It is ... from the fact that the Holy Spirit proceeds by way of love – and love has a kind of driving and moving force – that the movement which is from God in things seems properly to be attributed to the Holy Spirit.
> (Aquinas, *Summa Contra Gentiles*, IV.20.3)

It seems, therefore, that God's knowledge becomes the cause of creation and the ground of the continual subsistence of the cosmos, while the Holy Spirit, which proceeds from the Father and Son by way of love, is properly described as the principle of the motion of nature.[147] This means that what moves all things to their characteristic operation is love, namely a desire for fulfilment in the beloved. As we will see shortly, Aquinas even suggests that God is 'moved' to characteristic operation by the exchange of love which proceeds from the Father and Son, that which we call the Holy Spirit.

Therefore, it is through his knowledge of things and desire of his own goodness that God is the cause of the universe. Now this self-knowledge and self-love, the very life of God, his emanation and return to self, can be described as a kind of 'motion', or, as Wayne Hankey has described it, 'motionless motion'.[148] Hankey notes that, 'It is not possible for Thomas to affirm the *life* of God unless he find some sense in which motion belongs to the divine'.[149] It was seen above that both Aristotle and Aquinas are keen to maintain the distinction between the living and the non-living by means of the kind of motion of which they were capable: the hierarchy of life goes hand-in-hand with the hierarchy of motion. Indeed, both Aquinas and Aristotle define life as the ability to partake of self-motion and therefore 'we must note that life is attributed to certain things because they act of themselves and not as moved by other things; hence the more perfectly this is verified in a thing the more perfectly does it possess life'.[150] Following Aristotle, Aquinas also maintains that life is predicated primarily of the intellect, and that therefore 'the life of God is his intellect', that is his knowing, and what is known and the act of knowing are the same.[151] So in what sense can God be described as 'self-moving' and therefore as life in its perfection? Aquinas begins by stating that there are two kinds of action.[152] The first is that which passes to matter outside the agent concerned, for example locally moving another body or the heating of one body by another. The second is that which remains in the agent, for example understanding, sensing or willing. In the case of the first, the motion is completed not in the agent of the motion, but in another. In the second, the motion is the completion or perfection of the agent of the motion and not of another. Aquinas can now stipulate that, since motion involves the actualisation of the thing moved, this second type of action or motion, in being the actualising of the agent, is called movement in a proper sense. However, he does mention a text from Aristotle's *De Anima* which appears to state that this second class of motion, namely that which concludes in the actualising of the agent, is not really motion in the strict sense:

> Knowledge when actively operative is identical with its object. In the individual potential knowledge has priority in time, but generally it is not prior even in time; for everything comes out of that which actually is. And clearly the sensible object makes the sense faculty operative from being only potential; it is not acted upon, nor does it undergo change of state; and so, if it is motion, it is motion of a distinct kind; for motion, as we saw, is an activity of the imperfect, but activity in the absolute sense, that is activity of the perfected, is different.
>
> (Aristotle, *De Anima*, III.7.431a)

In other words, as Aquinas notes in his commentary on this passage, the motion of intellection (or sensing or willing) does not appear to be motion in the strict sense of the passage from contrary to contrary or the actualising of the potential. In Aristotelian terms, it may be regarded as *energeia*, a kind of constant similar to seeing.[153] Therefore, Aquinas concludes, this 'motion' (which is more commonly termed 'operation') is different from the strict Aristotelian definition of the *Physics*. However, elsewhere he does seem willing to assimilate the Aristotelian view with the self-moving soul of Plato when he writes,

> According to Plato ... that which moves itself is not a body. Plato understood by motion any given operation, so that to understand and to judge are a kind of motion. Aristotle likewise touches upon this manner of speaking in the *De Anima*. Plato accordingly said that the first mover moves himself because he knows himself and wills or loves himself. In a way, this is not opposed to the reasons of Aristotle. There is no difference between reaching a first being that moves himself, as understood by Plato, and reaching a first being that is absolutely unmoved, as understood by Aristotle.
>
> (Aquinas, *Summa Contra Gentiles*, I.13.10)

Again, Aquinas outlines a Platonic notion of the self-motion of God when he writes,

A will is set in motion by another when its main objective lies outside the person willing. The objective of God's will, however, is his own goodness, and this is his nature. And his will itself is also his nature. Therefore it is not moved by anything other, but by himself alone – if we may employ language which refers to understanding and willing as movements, as Plato did, when he spoke of the First Mover moving himself.

(Aquinas, *Summa Theologiae*, 1a.19.1.ad 3)

Moreover, these passages, in referring both to understanding (which in God is the begetting of the Son from the Father) and willing (which is the procession of the Spirit from the Son and the Father), indicate further the way in which knowledge is always willed or desired, for in God this understanding and willing are always simple and one. Elsewhere, Aquinas explicitly states that life is especially manifested in motion and specifically in self-motion and those things which put themselves into operation.[154] He states that if love, drive and motion are particularly suited to the Holy Spirit, as scripture suggests,[155] it is here that we find the dynamism of the divine life fully expressed: in the love that moves the sun and other stars.

Having sketched these aspects of Aquinas's reflections on motion, one can now see how his understanding of this subject is profoundly theological, having its ultimate origin in the doctrine of God. He begins with Aristotle and the distinction of motions which gives rise not only to proper and improper movement, but also to a hierarchy which is preserved through the possibility of maintaining a difference between the living and the non-living, those which are merely subject to motion and those which are self-moving. It has also been seen that Aquinas and Aristotle maintain an understanding of motion which saw this category not in terms of discrete moving objects, but in terms of participation in a wider and increasingly perfect motion through a hierarchical scale. Yet this motion is itself an analogical participation in the divine life, namely that which Aquinas is willing to refer to as divine self-motion. The hierarchy of motions can be understood as itself predicated on a hierarchy of emanation which is in turn a hierarchy of life and therefore a hierarchy of being. It was stated above that 'emanation' refers to the ability to make an actual communication of one's nature. This varies from the merely external, such as inanimate objects which make such communications only accidentally in so far as they are in spatial proximity, to the partially internal in the life of plants which are, to a minimal extent, self-moving. Next came animals and their ability to sense and store memories. However, it is in intellection, and the intellect of the human, that one first encounters a genuine emanation. A human being is able to make an imperfect communication of the self in the life of the intellect. That 'self-reflection' enables a more profound 'self-motion', for it permits the possibility of identifying the self in relation to the perfection towards which it tends, namely the 'true self', and initiating motion towards that true self through the desire of the will. However, a proper human self-motion still maintains a principle and term which are exterior. The principle is God himself, the term or *telos* is also God himself. Thus human life is not self-subsistent and its self-motion is imperfect. In terms of the example used above of the motion of a body by the soul, one saw that a human's self-motion is imperfect because it involves ultimate composition, namely a principle mover who is God, and also a *telos*, that to which the motion of the soul might ultimately pertain. It also remains as motion in the strict Aristotelian sense of passage from potency to act. It is only in God that principle and term coincide and that his self-knowledge and self-love are all one and properly interior in such a way that God never leaves himself. God can make an actual and perfect communication of his very self and therefore perfectly contemplate and know himself. This *exitus* from God, and his *reditus* to himself, is what Aquinas identifies as the emanation or procession of the persons of the Trinity, a perfect 'motionless motion'. In God's self-knowledge and love, which is his being and very life, all that subsists in the mind of God is desired and known. Aquinas states that 'God in his essence is the likeness of all things. Hence an Idea in God is simply the divine essence', and therefore in so far as God knows his essence, his very self, so too he knows all things. It is this knowledge of all things which is identified as the fundamental universal cause of a universe in motion. One can therefore see that the universe and its hierarchy of motions has as its origin the divine emanation, for in this emanation are all things known in God's eternity, a knowledge and love which is the cause of their flow from the divine being. From the perfect emanation of God from himself and to himself, one finds the form of the motion of the universe, its emanation from God and its return to him through the action of his will and intellect. Ultimately, it is God's perfect fullness of life in the form of his complete subsistence and self-comprehension that constitutes the crown of all 'motions'. From inanimate beings to spiritual substances, all motions are a participation in the eternal self-movement of the Godhead in that motion is the passage from imperfect to perfect, a self-actualising in different degrees which is achieved through varying degrees of self-knowledge and right desire.

Having outlined the content of the analogical relation between the motion of creation and divine being itself, one might now ask how humanity, lying for Aquinas within the moving material cosmos, might deepen its participation in God. In other words, if all motion is ultimately a participation in the emanation of the persons of the Trinity, how is humanity moved towards greater perfection in a more replete participation in that divine life? Is there any sense in which the practical science of human action might be expressed in terms of motion? A similar question was put to Aristotle in the second chapter of this essay where I enquired concerning the 'direction' of motion in relation to Aristotle's physics and ethics, asking whether motion might also be the means of the disintegration as well as the perfection of being. It was seen that all motion is orientated towards the Good as every being by nature seeks its fulfilment and actuality. For Aristotle, the Good is the first motionless determining limit (analogous to *topos* in his physics of locomotion) of all cosmological motion. It is the task of the next section of this essay to tackle a closely related question by following Aquinas to examine the nature of human virtue and natural law to see how humanity moves, and is moved, towards *beatitudo* – its perfection and end in God. I will argue that God is the first motionless limit of all human being, yet moving on from Aristotle, Aquinas proffers the motive power of grace which draws humanity to its proper salvific end. I begin, however, with the nature of human acts and dispositions.

Virtue, grace and motion

Principles of motion: habits and virtues

In the earlier discussion of Aquinas's appropriation of Aristotelian physics, it was noted that there are proper and improper motions for all natural bodies. These are defined by their *telos*. The natural or proper motion of, for example, a light body, is upwards. Such motion of inanimate bodies is determined and given by their generator. Further up the ontological scale, sentient animals which lack intellect and knowledge perceive objects as proper ends of desire and so move themselves to the attainment of these objects. Because such animals lack understanding, their motion remains substantially determined by their natures. For example, the sheep's fleeing from the wolf springs from a judgement that the sheep is not free to make or not make because it comes directly and spontaneously from the sheep's nature.[156] In the case of angels, they possess free will but in a more perfect sense than human beings because their choices emerge from an intuitive and energic apprehension of the truth. Because their apprehension is non-discursive and therefore devoid of motion, their position in relation to the Good has been decided from the moment of their creation. There is, as it were, no 'gap' between their intellect and will. Although their intellect is not their will (unlike the perfect simplicity of God), their intuitive apprehension of truth is coterminous with their willing of that truth. This is not to say that they are created in a state of beatitude, but they have always been in a state of grace, meriting eternal happiness by a singular and complete act of charity.[157]

The case of humanity, however, which lies at the centre of the cosmological hierarchy, is much more complicated. Whereas the motion of inanimate objects is determined solely by their nature and that of non-rational animals is the result of will uninformed by intellect, humanity possesses free will. The motions of the human being do not emerge only from a natural principle, but are a result of an intricate interweaving and interaction of intellect and will. To see why this is the case, one must again recall the principle that 'every agent, of necessity, acts for an end'.[158] Unlike the separate substances, man does not perceive his ultimate end in a non-discursive, immediate or intuitive way. This ultimate end, in being the universal Good, contains within itself all particular and contingent goods which are means to (that is, participate in) the ultimate end. The human will is moved by this end of necessity; it cannot fail to will the Good any more than sight can fail to see colour.[159] Just as in the case of Aristotle (and, one might add, Plato), the universal Good is therefore the determining limit of all motion, including the human, in such a way that motion is not qualitatively neutral. However, the will is presented not with the universal Good which, in conjunction with an intellect open to universal being, it wills of necessity, but with particular and contingent goods. The universal Good is therefore mediated to the human will through the particular, and these particulars are not, unlike the universal, good from every point of view.[160] Therefore, will must make an intellectual judgement about these particular goods, and it is therefore within its free power not to be moved by them. It is in the distance between the will and its object that the freedom of the will consists.

How, then, does the will move to its right object, its appropriate end? In order to answer this question, it is necessary to examine the interaction between will and intellect. Although the will is moved of necessity towards the universal Good which is its proper end (just as a heavy object is moved of necessity to its proper place), it is said to be self-moved in respect of the means to that end. The means are composed of things which, unlike the universal Good, are not good from every point of view, but only in some specific regard.[161] The will does, however, intend and will the means and the end in one complete motion, not in a series of partial movements.[162] However, voluntary activity also involves an element of choice as regards the means to an end. In the case of the willing of the ultimate end, the will in a sense borrows the motionless and universal stability of that end. Yet in willing the particular and contingent means to that end, we enter into a realm of uncertainty which requires discernment. The will is presented with choices and must pick its path to *beatitudo* through the thicket of constantly moving particular goods which are means to its ultimate end. Such choice is not undertaken by the will in isolation, but is preceded by judgement and counsel.[163] Aquinas tells us that the discernment about what is to be done is difficult and requires reason to institute an enquiry; this we call counsel. Yet the will and the intellect remain intertwined in their motions because, we are told, in the act of will one finds the order of reason which is choice, while in the act of reason known as counsel one finds the motive power of the will: it moves the intellect to counsel concerning what should be willed as a means.[164] Yet this intertwining is not a mixing of intellect and will: they remain distinct powers, and the act of choice is substantially an act of will because the will does not of necessity follow reason; rather, the will inclines to certain means as proposed by the intellective power.[165] This interaction of will and intellect reveals once again a crucial aspect of Aquinas's thought which might appear strange to modern eyes: as intellect and will are intimately intertwined in making any judgement, so also reason cannot purport to be an 'objective' realm divorced from desire and the good. As was seen when considering intellect and will in the Trinity, for Aquinas, all knowledge is willed knowledge. Put another way, there can be no separation between 'facts' delivered by the intellect and 'values' delivered by the will. One assents to something because it is good.

An important question now arises: are each of these interactions of will and intellect in the soul discrete? Does man, each time he seeks to move towards his ultimate end in *beatitudo*, embark on a fresh deliberation? Not at all. Aquinas has no sense of each human choice being a discrete 'calculation' or self-contained motion of the will after the fashion of a utilitarian or 'decision-based' ethic. Rather, human acts form a 'life' which is not a succession of discrete states: it is a continuous motion towards the universal end in being, truth and goodness. So, how do human individual deliberative acts cohere to form a narratable life of singular motion towards the good? The motions of man's soul may be collected to form what Aquinas calls 'habits'. A habit is a quality and disposition which modifies a being's substance. They differ from other qualities in determining the manner in which a being realises, or fails to realise, its own nature. Aquinas follows Aristotle and identifies a habit as a disposition whereby a being is disposed to good or evil. In other words, a habit determines the motion of a being towards fulfilment in its *telos* or away from such fulfilment towards the demise of its being: 'a habit implies a disposition in relation to a thing's nature, and to its operation or end, by reason of which disposition a thing is well or ill disposed towards its end'.[166] Therefore, a habit is good if it draws a being towards its form (that is, towards its goal) and bad if it draws a being away from its form. Habits are not, however, mere qualities or accidents of things; good habits are qualities and accidents of a special kind which lie at the heart of the nature of a being and therefore come closest to entering into the essence and definition of a being. There may be habits within all faculties of the soul, appetitive and intellectual.[167] Meanwhile, what determines the direction and place of the motion guided by habit is the universal good which is the motionless determining limit of all motion.

How are habits acquired? They are not the result of the accumulation of singular acts, but repetition, practice and training.[168] Aquinas implies that a habit – which is a quality or disposition to some kind of operation – comes about through an appropriate balance of the intellective and appetitive powers of the soul in such a way that a being (without changing its substance) takes that quality into its nature. In the case of a good habit, this leads to the greater actualisation of that nature. In a protracted and difficult passage, Aquinas goes on to explain the increase of habits through the Platonic metaphysics of participation.[169] In general, we might say that qualities which refer to the substance of a being, namely those which are integral to its definition, do not admit of degrees of 'more' or 'less'. For example, regarding shape, triangularity does not admit of degree; a figure has three sides or it does not. Such shapes or numbers include an essential and simple indivisibility in their definition which requires an indivisibility in those which participate therein.[170] However, many qualities involve a relation which is not quantitatively exact or simple and does admit of degrees of 'more' or 'less'. For example, knowledge may be greater or lesser according to its extent. Also, certain qualities which involve a complex relation of parts admit of degree. For example, health involves a balanced relation of humours in the body, and the proportions of these humours may vary to produce greater or lesser health. Ultimately, Aquinas claims that a quality such as a habit may be possessed to a greater or lesser degree by a subject depending on the intensity of its participation in the form towards which it tends: 'and equal knowledge or an equal degree of health can be more fully participated in by one than another, according to the aptitude which each derives from nature and custom'.[171] So for Aquinas a habit is increased not by quantitatively adding to the form (in the way that, for example, one might add to one's book collection by adding books), but by a more intense participation in the form on the part of the subject.

How does Aquinas understand this increase in intensity? By referring to motion. He states that an increase in a habit by increased participation in a form enables a kind of 'intensification' of motion towards a *telos*. Thus he states that

> motion increases the intensity as to participation in its subject, that is, in so far as the same movement can be executed with more or less speed or with more or less assurance. ... Yet a man's knowledge increases (as to the subject's participation in knowledge) in intensity in so far as one man is quicker and more sure than another in considering the same conclusions.
>
> (Aquinas, *Summa Theologiae*, 1a2ae.52.2.responsio)

Therefore, we might say that a good habit is the principle of an intensification of the motion of a being towards fulfilment, that is greater participation, in the form which is its *telos*. In giving assurance or readiness in the motions of the intellect and will, a good habit therefore provides spontaneity and delight in a being's operations.[172] A man who has the habit of love moves spontaneously and readily to operative loving just as fire spontaneously and readily moves upwards. In the case of knowledge, this habit is not a question of the mere accumulation of facts as if knowledge were quantitative, but a more ready or spontaneous application of knowledge in such a way that the knower has a more 'intense' knowledge. Thus good habits which are qualities of the human are a recompense for the anxious, plodding deliberations which might otherwise afflict the soul and which characterise so much of modernity's ponderous and calculative moral deliberations. Moreover, this is the ground on which Aquinas claims that the unity of a science is found in a stable habit of the intellect, a simple form or quality of the mind.[173]

In knowing the nature of habits, that they dispose us to move towards or away from the fulfilment of our natures, we are now in a position to consider those good habits which Aquinas calls virtues. These are habits and operative perfections of power which are always for the good.[174] They are, like habits, principles of motion towards the good.[175] Virtues are primarily of two kinds: intellectual and moral. The intellectual virtues are in turn divided between those of the speculative intellect (understanding, science and wisdom) and those of the practical intellect (art and prudence).[176] These intellectual virtues (which lead to the right use of reason and tend to action most particularly in prudence which is the habit of right counsel about what is to be done) are, according to Aquinas, only virtues in a secondary sense. A habit may become a virtue for two reasons: first, because it gives the ability to act well, and, secondly, because it ensures right performance.[177] It is only in the first sense that these habits of the intellective soul are virtues; that is, because they give the ability (rather than the actuality) to act well. Yet these intellectual virtues 'make us capable of good activity, namely to consider the truth, which is a good work for the intellect'.[178] By contrast, the moral virtues refer to practical action. They are 'a natural or quasi-natural inclination to do some particular action'.[179] So Aquinas is considering all powers of the soul, both intellectual and appetitive, when he claims that for people to act well they must be in possession of good intellectual habits but also those virtues which lead one to act well. The cardinal moral virtues are named as the actions of justice, temperance and courage.

If the habits known as virtues are principles of motion, what are these motions themselves? Aquinas calls them passions and he explains their nature very explicitly in terms of motion, drawing a clear analogy with the physics of moving inanimate bodies.[180] Passions are intense motions of the sensitive appetite within the soul which are, in themselves, neither good nor evil. Thus a passion, like any motion, always exists between contraries, for example love and hatred, desire and aversion, joy and sadness. Virtue establishes a mean between passions, which is to say that virtue directs the motions of the soul – the passions – to an appropriate rest in fulfilment. That appropriate rest will be in a mean between excess and deficiency. This characteristically Aristotelian doctrine is merely a statement of the requirement that 'the measure and rule of its [the appetitive soul] motions for objects of desire lies in the reason itself'.[181] Thus there is an appropriate or rational rest between contrary extremes of passion in any situation (for example, a mean between too much and too little love). Note that Aquinas is not suggesting a grey modicum of love in every situation. There may be situations in which a most passionate and intense love represents an appropriate mean. In its motion towards rest in this mean, the passions of the soul are guided by virtue.

Yet are the intellectual and moral virtues outlined above sufficient of themselves to guide humans to their ultimate end in *beatitudo*? Two further aspects of Aquinas's theological ethics require elucidation in this regard.

First, I will consider the nature and purpose of natural and eternal law to which man is subject; secondly, I will investigate the importance of the motive power of divine grace in moving man to his ultimate end.

Law and grace

Whereas the virtues are habits that come into intimate proximity with the nature of their possessor, in the case of law this is a directive power which, at least initially, works from without. Aquinas identifies a law as 'a kind of direction or measure (*regula et mensura*) for human activity through which a person is led to something or held back'.[182] More broadly, there are laws which govern and direct the activities of all things within the universe and in this sense law is, for Aquinas, a communal measure and rule which does not apply merely to individuals; all laws have as their ultimate end the universal good, hence those who legislate must have conformity to this common good always before them.[183] Are there different kinds of law? Aquinas describes a hierarchy of three, beginning with the eternal law.

The eternal law 'is nothing other than the exemplar of divine wisdom as directing the motions and acts of everything'.[184] This law is the supreme exemplar in conformity to which all things, including the other laws, find their fulfilment. Crucially, this law is not merely an act of divine fiat or an arbitrary decree of God's will. If this were the case, the divine intellect, for example, would fall under the absolute decree of the divine will. As we will see in the next chapter, the voluntarist view of law is characteristic of early modern science. By contrast, for Aquinas the eternal law is not superadded to anything by the divine, but is itself the divine essence.[185] Not only is the law not divorced from the law giver, but in the case of the eternal law, the law *is* the law giver. There is therefore nothing at all arbitrary about eternal law, for this law is the essence of the absolutely real, good and true. As such, it is not merely the higher beings in the cosmological hierarchy whose motions are guided by the eternal law; this law extends providentially throughout the cosmos to govern even the lowliest motion: 'Accordingly, God is said to command the whole of nature. ... That is why every motion and every act in the whole universe is subject to the Eternal Law'.[186] However, there are different ways of being subject to the eternal law. In the first, a being may be acted upon and receive from the eternal law an inner principle of motion. Such is the manner of inanimate bodies' subjection to the eternal law. In the case of rational creatures exercising will, they are subject to the eternal law in this first sense, but also by being 'a companion by way of knowledge', that is by according one's intellect, understanding and will to the eternal law which is the measure of truth, being and goodness.

Why does Aquinas insist on a notion of eternal law in addition to the divine ideas which are God's essence? It is important to recall that the notion of law is analogical. By describing the divine essence in terms of

eternal law, Aquinas is able to provide a notion of God's providential governance of the universe while relating all other laws, which are necessary for
the governance of human society, to the divine being. In this way, he ensures
that laws are never understood as the arbitrary dictates of any particular will
in such a way that they are divorced from an ontological basis. So, as an
example of another kind of law, Aquinas describes the natural law as nothing
other than a participation in the eternal law by intelligent creatures.[187] This
law directs us to our appropriate end in a fashion that is more immediate
and attainable to reason, for the natural law mediates to us the eternal law
by referring more particularly to our human nature as this is analogically
related to the divine nature.

It is important to note that by natural law Aquinas does not mean a list
of regulations, or even any particularly specific rules. A most typical
example of the natural law is the requirement (known by synderesis) to do
good and avoid evil. Neither is natural law a series of commands which
simply arrive and demand mere obedience and no further reflection. Rather,
the precepts of the natural law expressed in natural categories refer particularly to the guidance of human beings in motion towards their *telos*. Aquinas
refers to three stages, the third of which is the particular province of natural
law in relation to humans.[188] First, there is a natural tendency in people, in
common with all things, to preserve their natural being. Secondly, there is a
natural tendency, shared by non-rational and rational animals alike, towards
the pro-creation of each species. Thirdly, there is the natural tendency
particular to human beings which refers to their rational nature, that is to
live in accord with reason. Thus the natural law guides the appetite towards
the knowledge of truths about God and life within society; it is constituted
by those precepts which guide the motions of human beings towards such an
end. Such law is a basic framework which aids human souls – their appetitive and intellective powers – in discerning a motion towards the good.

At this point, one might even draw a clear analogy between spatial or
local motion and the ethical motions of the human soul. Place constitutes, as
it were, the limit and an aspect of the 'natural law' of the local motion of
bodies. For example, it is in accord with the natural law of heavy bodies that
they move to rest in a low place. The low place is part of the reason for the
motion – it helps to answer the question 'why?' With regard to human
beings, action in accordance with reason is the limit and a characteristic of
the 'natural law' of the motion of the human soul. For example, it is in
accord with the natural law of human beings that they move to rest in
rational operation. This mode of operation is the *telos* and therefore the
reason for the motion of the soul from irrationality or ignorance to rationality or knowledge. So one might say that the physical motion of bodies
and the ethical motion of human souls, which are subject to the different
sciences of physics and practical philosophy, are nevertheless held in analogical relation by their participation in eternal law which constitutes the
motionless limit and end of all motion, namely the universal good.

In this way, the exemplary pattern of the eternal law is providentially mediated to humanity through the natural law to provide some guidance in motion towards the good. The third position in the hierarchy of law, which in turn participates in the natural and eternal law, is human law. This variety of law is a means of educating through pressure and fear of punishment so as to instil the habits of virtue, particularly amongst the young and unformed.[189] Aquinas is quite clear that it is better that people act for the good by means of their own virtuous dispositions. However, human law is required to defend the peace of the community and maintain people in the basic virtues which refer to the common good.[190] It is a means of training people by providing essential precepts for their action.

The notion of law, however, leads us to question our ability to adhere to the law; or, to express this in a way analogous to physics, to move in accordance with the law's precepts. Even though in principle humanity's reason cannot be led astray as regards the precepts of natural law which are universally accepted, nevertheless by habitual sin humanity's ability to discern what ought to be done in particular instances in accordance with the law becomes compromised.[191] Because of this apparent inability, the authority of divine law is brought to bear in God's gift of the Old Law to the Jewish people. In a sense, the Old Law makes more apparent to the mind of humanity the precepts of natural law and the two are therefore coterminous. Yet it also adds certain ritual and judicial precepts by which God is to be appropriately worshipped.[192] What is the purpose of the Old Law? As it is the purpose of the divine law to establish friendship, and when applied to humanity it is to establish friendship with God and between people, the Old Law shares in this purpose by making humanity into a likeness with God in such a way that friendship can be established.[193] However, according to Aquinas, the more immanent end of the Old Law is to prepare a people to receive a New Law which is given in Christ. Thus the Old Law, in so far as its precepts are in accordance with right reason in the natural law, is binding on everyone; but in so far as it adds to the natural law to prepare a particular people at a particular time to receive the New Law in Christ, these precepts are binding only on the people to whom the law is given.[194] So the Old Law might be understood as a stage in a much wider process of drawing humanity to its final end. It makes clear the natural law while adding precepts which prepare for the receipt of the New Law which will have a universal power of moving people to *beatitudo*.

Before discussing the New Law and its power by grace to move humanity to *beatitudo*, one might ask why humanity is not, of its own accord, or even by governance of the Old Law, capable of attaining its ultimate end. Aquinas suggests that every form bestowed on created things by God has the power for a determined act which it is capable of bringing to fruition in proportion to its own proper endowment.[195] For example, humanity is capable of knowing things whose nature is proportionate to its own. Because no proportion can be fixed between *esse ipsum*, which is God, and *esse commune*, we cannot, by our own natural power, come to know God, despite it being the case that the vision of God is our end. Such is the *telos* of humanity because 'all men by nature desire to know' and ultimate knowledge can only be had in the first cause and end of all things.[196] However, this is not to say that we can know nothing of God, for we can have knowledge of the divine as this is mediated through those creatures who are proportionate to our nature. As has been argued in this and the previous chapter, there is no sphere of knowledge which lies outside of illumination by God. But neither are we to say that there is a sphere of supposed 'natural' ends, whether they be knowledge or goodwill, which are autonomous from the motive power of God. Such could not be the case for Aquinas. He explains that, although humanity is capable of self-motion, such motion is always ultimately a participation in the power of the first mover. Thus every formal perfection is from God as first act. We attain any perfection, whether it be knowledge or right action, by participating to some degree in God's moving of our nature towards intermediate or ultimate goals, for God is the ultimate source of all truth and goodness.[197]

It was the case that humanity was once capable by its own power of performing the good proportionate to its nature, such as the good of acquired virtue. Following the Fall, which fractured the natural powers of humanity issuing in the disintegration of virtue, the assistance of God is required to achieve even what previously was connatural to humanity. Certain natural goods do remain within humanity's power, such as forming friendships and communities. Aquinas uses an example from medicine: 'just as a sick man can of himself make some movements, yet he cannot be perfectly moved with the movements of one health, unless by the help of medicine he be cured'.[198] However, it is not the case that humanity was once capable of achieving its ultimate end of its own natural power; rather, even what humanity was once capable of achieving naturally now requires the infusion of divine grace. Therefore, we now turn to consider how God moves humanity by the power of grace towards intermediate ends and ultimately towards *beatitudo*.

According to Aquinas's understanding of grace, it is important to recognise that grace is not merely a force exerted on humanity by God to move people to the beatific vision. Such could not be the case, because a being which tends towards an end must have not only a natural appetite for the end and the requisite motion, but also a nature proportionate to that end.[199] If God were simply to move humanity to the beatific vision, this would constitute a violent motion because humanity would be destined for an end for which its nature was not prepared. Now the vision of God is something that is connatural only to God. Therefore, explains Aquinas, for humanity to achieve its ultimate end prepared by God, humanity requires not only the requisite motion and something which inclines the appetite towards that end, but also 'something by which man's very nature should be raised to a dignity which would make such an end suited to him'.[200] In addition to God moving humanity to its appropriate end, grace is also given as what Aquinas calls a 'habitual gift', namely a form or nature – a quality – by which humanity can move and be moved to the supernatural end appointed by God. Just as God provides for creatures not only by moving them to their appropriate ends but also in bestowing forms and powers by which they make that motion their own, so too God provides his grace by which humanity may make its motion to *beatitudo* its own. Importantly, with regard to natural bodies, 'the movements by which they are moved by God become connatural and easy to them'.[201] So too, by analogy, God 'infuse[s] supernatural forms or qualities into those whom he moves towards obtaining an eternal, supernatural good, whereby they may be moved by him sweetly and promptly towards obtaining the eternal good'.[202] Grace is therefore a particular love whereby God draws the rational creature beyond its nature to an ecstatic participation in the divine good, that motion being sweet and delightful.[203]

Grace can therefore be understood as a motion – God moving us to will and do the good – and also as a habitual gift or form implanted in us by God which can become a principle of motion.[204] Each of these aspects can be divided into what Aquinas calls 'operative' and 'co-operative' grace. On the one hand, when grace is understood as a motion, if God is the mover of an interior act of the will (for example, a will which desired evil begins to desire the good), then grace is called operative. In an act which does not remain immanently within the person but is external (for example, the will moves the whole person to an external act of generous giving) then this motion is properly ascribed both to the will and to God who confirms the will interiorly and provides the means for the exterior execution of the act. In this instance, grace is called co-operative because God co-operates with the motion of the person. On the other hand, when grace is understood as an habitual gift, then the effect is like any other formal cause; the first effect is being, the second is activity. Thus when habitual grace 'heals or justifies the soul' it is said to be operative, for God provides for humanity a form whereby it might move and be moved towards the beatific vision. When that same form is understood as the principle of the motion which proceeds also from the free choice of the individual, then grace is said to be co-operative.[205] In both kinds of grace, God is always the principle and universal cause, yet in co-operative grace humanity finds a more direct participation in its own motion towards *beatitudo*.

In addition to the distinction between operative and co-operative grace, Aquinas makes two further classifications. First, whereas the division into operative and co-operative grace refers to the effects of grace, one can also distinguish between prevenient and subsequent grace as regards cause or effect.[206] This distinction refers particularly to the motion which grace initiates. Aquinas claims that there are five effects of grace: the healing of the soul, the willing of the good, the efficacious performance of the good willed, perseverance in the good and the attainment of glory. Now when grace causes the first effect, namely the healing of the soul, it is said to be prevenient with respect to the second effect, namely the willing of the good. The second effect, the willing of the good, is said to be a subsequent grace. Similarly, the willing of the good is then a prevenient grace with respect to the efficacious performance of the good willed. So the distinction between prevenient and subsequent grace marks grace as a temporal effect in which one thing precedes another.[207] Thus Aquinas can describe the relationship between the stages in the motion towards the beatific vision. Meanwhile, a final distinction between different aspects of grace is that between grace freely bestowed and sanctifying grace.[208] The former is bestowed 'for the co-operation of one man with another' in such a way that grace may be mediated between people. Aquinas has in mind the grace which one person may possess to move another to God by, for example, teaching or prophecy. By contrast, sanctifying grace is that which orders humanity directly to its last end.

With regard most particularly to habitual grace, what form do these gracious habits take? There are three forms, and Aquinas calls them the theological virtues of faith, hope and charity. These are special sources of action by which humanity is moved towards *beatitudo*. They resemble those principles by which humanity is moved to its connatural end, yet these theological virtues are infused by God and made known to us by divine revelation contained in the teachings of sacred scripture.[209] These virtues endow intellect and will with the power to desire and know their appropriate end in God. First, there are certain truths which are not accessible to reason alone which are given in the virtue of faith, for example the articles given in the creed. Secondly, with reference to the will, hope and charity combine to form, on the one hand, the motion of wilful intention towards the beatific vision and, on the other hand, a particular cleaving of the will to the supernatural end, this latter being delivered by the virtue of charity.[210] Does any one of these particular virtues have precedence over the others? Although Aquinas explains that the theological virtues, as habits, 'are all infused together', nevertheless charity has a special dignity in being 'the mother and root of all the virtues, inasmuch as it is the root of them'.[211] For example, in the case of faith, which is a kind of mean between science and opinion, having its certainty in divine revelation while referring to things as yet unseen, although the intellect is brought to assent to certain truths through, for example, the freely bestowed grace of others, nevertheless charity is required as the form of faith in such a way that faith can be described as a kind of charity.[212] Aquinas explains that sometimes the mind is moved to assent by the will when, for example, the motive power of the object is insufficient to move the intellect. The will is thus moved to assent 'because it seems good or fitting (*bonum vel conveniens*) to assent on this side. And this is the state of one who believes'.[213] Thus the will must find itself in a relationship of charity with the object of faith in order to produce the assent of faith, that charity springing from a kind of *convenientia*. Such a desire is either caused in people through the sanctifying grace of others, or, chiefly, it is infused in humanity by God's grace.[214] The intellect is then brought to faith through the charitable assent of the will. This is not to say, however, that Aquinas has any kind of voluntaristic notion of faith. He is quite clear that the discursive thought of the intellect continues in parallel to the assent of the will, continually reflecting on the articles of faith to conform itself to the divine truth which is faith's object. Thus faith is not a matter of static certainty, but of the motion of continuous reflection on the truths of God.[215] Thus he can define faith as 'that habit of mind whereby eternal life begins in us and which brings the intellect to assent to things that appear not'.[216] The motive power is desire infused with charity.

How does charity relate to hope? Hope is principally a passion of the soul, but 'when we hope for anything as being possible to us by means of the divine assistance, our hope attains God himself, on whose help it depends'.[217] Therefore, hope becomes a theological virtue when it is infused by God to become a habit of the mind, making God its proper end. The efficient cause of the motion of humanity to its proper end is the divine assistance, and the final cause is the goal of such motion, namely eternal life. In these things rests the virtue of hope. This virtue precedes charity in the sense that hope is in God because of the expectation of the reward of eternal life. Put another way, hope is the love of God because of the good that one will thereby gain. Yet the more perfect one's hope, the closer it comes to genuine charity whereby one desires the end for itself. Thus hope leads to the love of God for himself in addition to the good that this brings to oneself.[218]

Aquinas is therefore quite clear that the development of humanity towards its end and goal can be expressed in terms which are analogous to those which describe the motion of inanimate bodies as described in natural philosophy. Thus the category of motion can be deployed analogously in both physics and the practical science of human action. I have argued that it is through the guidance of eternal, natural and human laws that humanity comes to acquire certain virtuous habits which can be the principles of motion towards a *telos* which is ultimately *beatitudo*, where humanity's natural desire to know is satiated. Such motion, in becoming part of humanity's nature, becomes delightful and swift, emerging naturally from human being just as downward motion emerges naturally from a heavy object.

A final crucial question remains to be answered. Aquinas writes frequently of God moving humanity to its final end, and of infusing virtues by operative and co-operative grace. How exactly is this motion executed by the divine and what is its form? In a previous section, I outlined the way in which the eternal divine processions of the 'motionless motion' of God are the basis of all other motion. In the concluding section of this chapter, I will examine how, for Aquinas, that eternal motionless motion is communicated within the moving universe. I will argue that this is effected primarily in the incarnation and continued through the sacraments of the Church. Here, one finds that motion becomes not that which sunders us from God, but the very context and means of our salvation.

Christ, the Eucharist and motion

In the previous section, I described the way in which, for Aquinas, the Old Law made apparent to humanity the natural law while also preparing a particular people for the receipt of the New Law. The natural law, mediated clearly through the Old Law in scripture, was thereby rendered effective in forming in humanity certain virtuous habits which are principles of motion towards the good. But what is the New Law and how does it differ from the Old Law? The New Law is described by Aquinas as 'the grace of the Holy Spirit, shown in faith working through love. Now men obtain this grace through the Son of God made man; grace first filled his humanity, and thence was brought to us'.[219] In the previous section it was stated that law is, at least initially, that which moves us externally to our appropriate end, or helps us to acquire certain virtuous habits by which we might move and be moved to our *telos*. By contrast, the New Law is internal, inscribed within the hearts of people. It becomes not just a rule or measure of the motion of humanity to its appropriate end, but the very principle of that motion as it reaches to the heart of human being. In other words, it becomes our nature. According to Aquinas's statement just quoted, it seems that this grace of the New Law is communicated through the Son and the Spirit. The question is: why and how?

Grace is principally the communication of divine goodness so that humanity may move and be moved to its appropriate end. As has been seen, this grace is something internalised within human being, either by infusion or mediation, in such a way that it dwells intimately with human nature, raising human nature ecstatically to partake of its final, supernatural end. One might therefore expect that if God is to infuse his grace into human nature, he would come so close to that nature as to join it to his own. This, for Aquinas, is what we find in the incarnation of the Son. Is the incarnation thereby a necessary means of God imparting salvific grace? As Milbank and Pickstock have argued, Aquinas understands the incarnation to hover between logical necessity, on the one hand, and pure caprice, on the other.[220] God could, by his *potentia absoluta*, have redeemed humankind without the need of the incarnation by a simple eradication of sins coupled with an act of reformation in the truth.[221] So why the incarnation? Because this means of redemption represents the supremely appropriate, fitting or convenient way of redeeming humanity.[222] This is an aesthetic construal of the incarnation; its reason lies in its intrinsic beauty, appropriateness or proportion – and therefore truth, for it was remarked earlier that *omnia in esse conveniunt*[223] – as regards its content and purpose. In communicating redemptive grace by which humanity might be moved to its ultimate end, Aquinas sees that it is most fitting that human nature should be joined to the divine nature fully in one person. Many reasons may be given for the fittingness or beauty of the incarnation.[224] Amongst them, Aquinas explains that humanity has, in the order of its nature, the divine for its *telos* and has been created for union with God in his intellect. An appropriate testimony to this is given in the union which takes place in one person visible to the senses, namely Christ. Nevertheless, explains Aquinas, the nature of God and the nature of humanity are preserved in the hypostatic union, the former suffering no loss and the latter no gain beyond the bounds of its species.[225] This is possible because of the natural *convenientia* between the divine and human, existing in the fact that God is both humanity's principle and term. Aquinas gives another reason for the incarnation when he states that humanity stands at the centre of the created cosmos and thus shares things in common with all levels of the hierarchy, from inanimate objects which are its tools to angels with which humanity shares an ultimate *telos*. Therefore, explains Aquinas, it seems most fitting that human nature, sharing most in common with all aspects of the cosmological hierarchy, should be joined to the universal cause of all things.[226] He explains this further in a way particularly appropriate to our present concerns:

> Lastly, man, since he is the term of creatures, presupposing, so to say, all other creatures in the natural order of generation, is suitably united to the first principle of things to finish a kind of cycle in the perfection of things.
>
> (Aquinas, *Summa Contra Gentiles*, IV.55.7)

This is to say that Christ is the embodiment of the full circular motion of the human soul. In one person we find creation's originating principle, for as was seen above, the emanation of Christ is God's self-knowledge by which he knows and thereby creates all things, while in that same person we also find a vision of the divine nature which is humanity's very goal. In Christ we find human nature joined to its beginning and end, and this gives it an almost 'motionless' quality. The whole circular motion of humanity is mysteriously revealed to us in a single visible person.

Aquinas is clear that Christ, in being both fully divine and human, has all that grace could ever effect. Because the closer a subject is to the inflowing cause, the more effective will be that action, it seems that, because Christ in his humanity is united most closely to God, his soul receives most fully and completely the grace of God.[227] Because the soul of Christ is not in itself divine, it is made so by sharing in the grace of the Godhead.[228] This grace of the Godhead may be communicated to humanity in a fashion reminiscent of freely bestowed grace. This is to say that Christ's teaching and example can instil in humanity the grace of virtue by which motion may take place towards God. For example, seeing human nature united to God and a body susceptible to suffering and death transformed into a resurrected body, humanity is stirred to the virtue of hope of eternal life.[229] Thus Aquinas writes that 'it was necessary for man to be solidly grounded in virtue to receive from God made human both teaching and the examples of virtue'.[230] However, is there any sense in which Christ mediates to us not just freely bestowed grace, but also sanctifying grace? Is Christ anything more than an excellent teacher and example whom we imitate? To answer this question, we first begin with justification and Christ's passion before considering the mediation of grace through the Church and the Eucharist.

When Aquinas talks of justification, he does so in terms of motion: 'the justification of the unrighteous is a motion in which the human mind is moved by God from the state of sin to the state of justice'.[231] Justice is a kind of rightness of order in humanity's internal disposition, whereby the highest powers of the soul are rightly ordered to God.[232] To arrive at justice, Aquinas states that there are four requirements: the infusion of grace; a motion of free choice directed towards God by faith; a movement of free choice directed towards sin; and the forgiveness of sin.[233] Thus the motion of justification takes place between the contraries of sin, or unrighteousness, and the forgiveness of sin. An analogy with the motion of inanimate bodies will assist in clarifying the role played by sin in this motion. By his power, God can infuse in people his grace so that their motion towards the good becomes 'delightful and natural'. Just as a heavy body's downward motion emerges from its natural principle (heaviness) given by a generator, likewise a human soul following an infusion of grace would renew its nature and make motion towards God its very own. However, as we have seen, there can be an impediment to the natural motion of an inanimate body, for example an obstacle which prevents a heavy weight from falling to its natural place. In a similar fashion, God may infuse grace so that humanity may move and be moved to its proper end, and yet there may still exist an obstacle to that motion. Aquinas refers to this obstacle as sin, and it is removed by forgiveness through the passion of Christ.

For Aquinas, Christ's passion renders satisfaction for the sins of humanity. To see why this is the case, we must return to the notion of *convenientia*. It is the case that God could, by decree, simply cancel the debt of humanity which is owed due to sin. In fact, we might say that, because this always remains within the divine power, God always was in himself eternally reconciled to humanity.[234] Thus it is humanity's reconciliation to God that is to be effected. For Aquinas, Christ's passion (namely his suffering and sacrificial death) are the most fitting and appropriate means of effecting the forgiveness of the sins of humanity.[235] First, Christ's passion is consonant with divine justice because Christ made satisfaction for the sin of the human race. Meanwhile, the sacrifice of a human would be unable to satisfy because a sin against God who is infinite likewise has a kind of 'infinite' character.[236] Therefore, secondly, Christ's passion is consonant with divine mercy because, humanity being unable to make satisfaction for its own sins, God gave his Son. Aquinas therefore states that, 'In so acting God manifested greater mercy than if he had forgiven sins without requiring satisfaction'.[237]

So Christ's passion removes the sins of humanity (in a way remotely analogous to my removal of an impediment to a heavy body's downward motion)
and enables the motion of human souls towards the beatific vision by means
of grace.[238] We are now in a position to see how that grace is mediated
through time and to all people in a way that is more than mere *mimesis*. For
Aquinas, grace comes through the Church, which is Christ's body of which
he is the head.[239] In his treatise *De Veritate* he states that, 'Christ and his
members are one mystical person. Consequently, the works of the head are in
some sense the works of the members'.[240] Thus it is that the head of the
body moves the other parts of the body as instruments and thereby
commands their motions. Thus Christ is head of his body the Church
according to order, perfection and power: order because the head is the first
part of the human body; perfection because it contains all the higher senses;
and power because the sensitive and motive power which rules in the head
thereby rules the motions of the remaining parts of the body.[241] Yet it is
according to the sacraments that Christ, the head of the Church, communicates not only his will but also his very substance.

In his discussion of the New Law, Aquinas comments that 'it is fitting
that the grace which overflows from the incarnate Word should be carried to
us by external perceptible realities'.[242] Those external perceptible realities are
the sacraments of the Church. A sacrament 'is a sign of a sacred reality inasmuch as it sanctifies people'.[243] This is to say that a sacrament as a sign also
effects something in humanity. The sacraments are visible, corporeal signs
which are fitting to our nature, because we first come to know the intelligible
from the sensible.[244] Yet it is not only because of our nature that sacraments
are corporeal. Aquinas claims that instruments must be proportioned to their
first cause, and that first cause is the incarnate Son. Because this latter is
corporeal, for the sacraments to be in harmony with their first and universal
cause, it is fitting that they bear a likeness to that cause. Therefore, the power
of the incarnation continues to reach us under appropriate corporeal and
visible signs.[245] Finally, it is important to note that Aquinas is clearly
anxious to maintain that, because the sin of humanity came by clinging to
visible, material things in a distorted way, it is appropriate that God use
these things as instruments for grace and salvation so that people be
reminded that, as created by God, material things are good by nature. They
can be the means of salvation when used in an ordered way.[246]

Amongst the seven sacraments of the New Law, Aquinas counts baptism
and Eucharist as primary. The former is a spiritual rebirth which signifies
and effects our entry into the mystical body of Christ. Thus Christ's grace is
present in baptism by his power.[247] However, in the case of the Eucharist,
Christ is substantially present. In conclusion, I now turn to consider the
grace of the Eucharist as the motionless point towards which all cosmological motion tends.

For Aquinas, the Eucharist is the sacrament of the Church's unity and,
whereas baptism is the beginning of the spiritual life, the Eucharist consti-

tutes the goal and consummation of that life.[248] How is this so? In the discussion above, the reasons for the incarnation and the grace received thereby were discussed. The Eucharist ensures the continuation of the communication of grace to humanity throughout history, for this sacrament makes Christ corporeally present in our midst in such a way that the infusion of grace takes place by our taking of Christ's resurrected body into our bodies. This is possible because the corporeal elements of bread and wine are mystically transformed into the substance of the body and blood of Christ. One may also note that Aquinas writes that 'after the consecration the substance of the bread and wine is neither under the sacramental appearances nor anywhere else. But it does not follow that it is annihilated; for it is changed into the body of Christ'.[249] However, the presence of Christ in the Eucharist is not local after the fashion of human corporeal presence because this would entail the accident of the dimension of the host having Christ's body as its subject. Christ's body cannot be the subject of accidents, for it is replete and, unlike our bodies which require the accidents of the dimensions of our limbs, requires no qualification. Moreover, the change of the bread and wine into the body and blood of Christ is not a 'change' (*conversio*) after the fashion of other changes. Aquinas writes, 'This conversion ... is not like any natural change, but it is entirely beyond the powers of nature and is brought about purely by God's own power'.[250] This change is not accidental or formal (as with conversions brought about within nature), but a mystical substantial change in the sense that God, by his power, transforms the substance of bread and wine into the substance of Christ's body and blood. Therefore, he refers to it by a proper name, 'transubstantiation'.

However, the integrity of the elements of bread and wine remain because their accidents are preserved. These accidents do not, as in nature, inhere in and qualify a substance, for Christ's body and blood are replete in themselves. Rather, these accidents are 'free-floating' and, contrary to Aristotelian metaphysics, have no subject.[251] These accidents of bread and wine inhere in *esse commune* for they maintain a composition of essence and existence. Therefore, the Eucharistic elements after the consecration do not participate in substance and exist by virtue of that substance which is thereby accidentally qualified. Instead, they exist, just as created things exist, by virtue of their direct participation in *esse ipsum*. This participation in *esse ipsum* is Aquinas's most fundamental ontological framework, for whereas substance stands in its own right accidentally qualified as this or that kind of creature, it is created being – substance or accident – which stands as this or that kind of being not in its own right but only by an improper borrowing of *esse ipsum*. Because of this most fundamental ontological framework, the participation of the natural accidents of bread and wine in a most intimate and mysterious way marks their *telos*. They are given the exalted privilege of showing the body and blood of Christ.

In taking this sacramental resurrected body of Christ into itself at the Eucharist, the Church is thereby made the body of Christ. This is not a feeding after the fashion of normal meals, but a feeding *par excellence*. Aquinas comments that, 'There is a difference between bodily and spiritual food. Bodily food is changed into the substance of the person who eats. ... But spiritual food changes man into itself'.[252] Thus humanity is transformed by receiving the grace of Christ not just through his teaching in the form of freely bestowed grace, but also substantially in the form of sanctifying grace. Thus Aquinas can refer to this sacrament as the end and goal of our highest spiritual life for we do inhere in the body of Christ itself, namely the Church which is made by the Eucharist.

However, the Eucharist has a crucial relation to the central topic of this essay, for in an unexpected way it displays to us in a single unity the completion of the multiplicity of cosmological motions. In an earlier discussion, I commented that in Christ's incarnation the whole human motion is mysteriously revealed in a single person. This is because Christ, as the second person of the Trinity, is the principle whereby all creation comes into being; in the emanation of the Son, God knows himself as – and this knowledge is – the foundation of creation. Yet Christ, as God, who is the universal good, is also the *telos* of creation, and the beatific vision is the *telos* of all intellectual creatures. In the appearance of the principle and *telos* of creation in the midst of creation, we find mysteriously revealed the whole circular motion of *esse commune* from its emanation to return in God. As Milbank and Pickstock comment, deploying an Augustinian motif, 'After expulsion from paradise, only the arrival of the goal in the midst of the way reveals again the way'.[253] One may therefore understand the Eucharist to be the continual arrival of the goal in the midst of the way which thereby incorporates us into the very means, or motion, of our salvation. We find in this sacrament a momentary still point in which the cosmological origin, *telos* and motion are mystically united and delivered into our very bodies.

The Eucharist unites in a further important respect. Aquinas remarks that the choice of bread and wine as the matter of this sacrament is by no means arbitrary. Bread and wine are used primarily because Christ himself used these elements, but also because they are the most common food of humankind and therefore prefigure that which is the essential nourishment for all, namely Christ himself.[254] In addition, the bread and wine, in being made of many grains and grapes, signify the gathering together of the Church community around the Eucharist. Moreover, they draw together the labours of many and various people into the event of the Eucharist. This gathering together also incorporates, by association, the motions of the heavenly bodies. In an earlier discussion, it was seen that, for Aquinas, the generation and corruption of sublunar creatures takes place because of the diurnal rotation of the heavens and the earth and the sun's varying proximity to one another. This very motion gives rise to the generation and corruption of the plants and animals which contribute to the growth of grain and grapes which are made into bread and wine. Therefore, gathered into this single sacrament are the motions of the cosmos which contribute towards the formation of bread and wine, the motions of human labour and culture, and the motion of humanity which finds its most appropriate motion emerging from the desire for Christ himself who is their principle and *telos*. So the Eucharist is the sacrament of unity in a radical sense, for it unifies all 'motion' by providing the still *telos* in the midst of the motion. It is as if the motions of the cosmos, at the moment of consecration, were mystically granted a glimpse of their final *telos* in the midst of the way and thus enabled to renew their motion by intimation of its eventual goal.

5 The isolation of physics

Throughout the first four chapters of this essay I have sought to elucidate an understanding of motion as an analogical and qualitative term which, amongst others, may relate the various practical and speculative sciences one to another, reaching even to God's own science and the divine life itself in the 'motionless motion' of the Trinity. For Aquinas the study of motion belongs first to physics, but is also investigated and the concept applied within ethics, metaphysics and *sacra doctrina*. For example, a physics of moving bodies is deployed in Aquinas's attempt to make evident to our clouded minds the existence of God. As has just been seen, motion reaches its apogee in the Eucharist where cosmic motion is ontologically unified with its origin and *telos* in being exalted to become the body and blood of Christ.

Before proceeding to consider a radically different understanding of motion in the theology and physics of the early modern scientist Isaac Newton, I now turn to consider first a view of natural philosophy which suggests a much narrower understanding of motion. More particularly, I will examine the influential work of the Persian Islamic philosopher Avicenna (Ibn Sīnā) (980–1037) and his isolation of physics from metaphysics. For Avicenna the proof of God's existence belongs exclusively within the realm of metaphysics rather than physics.[1] Physics continues with the sole task of studying particular principles and those things associated with natural, material bodies. Along with his Plotinian emanationist cosmogony, the notion of an agent intellect which is a substance separate from the human soul, and nature as an efficient cause of all motion, I will suggest that this leads Avicenna to isolate physics in a way that marks a sharp contrast with the Thomist tradition.

Following this study of Avicenna, I will examine briefly some developments in natural philosophy in the later Middle Ages which are suggestive of a quantitative and mathematical understanding of motion as a state. These changes in the understanding of motion also suggest, but in a different way, the isolation of physics from questions of ontology and theology. In particular, I will describe the impetus theory of projectile motion articulated by Jean Buridan (*c*.1300–*c*.1358). While this physics in many ways remains within the Aristotelian tradition, nevertheless it also points more clearly towards the revolution in natural philosophy inaugurated by Newton in the seventeenth century that will form the subject of the final chapter of this essay. I begin, however, with Avicenna's understanding of the nature of metaphysics and its relationship to the physics of moving bodies.

Avicenna on metaphysics and physics

At the beginning of Book IV of his *Metaphysics*, Aristotle states that there is a science which studies being *qua* being and the attributes which pertain to being by virtue of its nature.[2] Whereas physics studies beings which are subject to motion and allied to material nature, and mathematics studies its subject matter as immobile yet not separable from matter, metaphysics is the most universal science in not pertaining to particulars, but to being in general, its principles and causes.

In his *Metaphysica*, a portion of his most thorough philosophical work the *al-Shifā'* (*The Healing*), Avicenna follows the Aristotelian path in discussing the nature of the different sciences and in particular the science which studies being *qua* being.[3] Following Aristotle, he surmises that a science does not demonstrate the existence of its own subject matter or formulate its own principles, but rather seeks to enunciate the essential attributes of that which it studies. For example, physics does not demonstrate the existence of moving bodies or derive the principles which govern their motion, but accepts these things and, deploying the *principles* of motion derived from the higher science of metaphysics, outlines the essential characteristics of moving bodies.

So if, as Aristotle states, metaphysics has as its subject matter being *qua* being, what, for Avicenna, are the tasks of this science? Importantly, he is adamant that it belongs to metaphysics, not physics, to prove the existence of God as the first cause of the existence of the universe.[4] This is surprising because Aristotle, whom Avicenna seeks to follow so closely, seems to locate the proof of God's existence principally in the realm of physics. The deity is known as the first unmoved mover and, as was seen in the previous chapter, Aquinas similarly identifies God as the beginning and end of motion. Yet Avicenna argues that physics is concerned with moving corporeal beings and these characteristics do not in any way belong to God for he is simple, eternal and wholly beyond matter and movement. Therefore, for Avicenna it cannot belong to physics to prove God's existence because this science knows God only as that which is beyond its own subject matter. It is surely far more appropriate for metaphysics to undertake this task because it is concerned with being *qua* being abstracted from the characteristics of natural entities, for metaphysics studies being in general. So given that it belongs properly to metaphysics to prove God's existence, Avicenna can then surmise that God is not the proper subject of metaphysics because a science does not seek to prove the existence of its own subject matter. God cannot be at once the subject of metaphysical proof and the subject matter of that same science.[5] Rather, the subject matter of metaphysics is being *qua* being.

Whereas Aristotle's proof of God's existence from motion can only lead to the postulation of a first cause of motion, Avicenna sees that his purely metaphysical proof leads to the more fundamental first cause of existence itself. This is not to say that Avicenna's proof may not use aspects of created nature in its reasoning, but only that it does not employ any characteristic that is possessed by a created nature *qua* created nature. So metaphysics demonstrates that there are attributes possessed by existents *qua* existents and from these one can demonstrate a first and necessary cause of existence itself.

This already hints at the separation of metaphysics and physics. For Aquinas, God's existence can be made evident in both physics and metaphysics through the category of motion because this category is not confined exclusively and univocally to physical bodies but is analogically related to both corporeal and incorporeal beings. This means that, although God is indeed beyond matter and motion, certain aspects of creation such as motion can be applied analogically to make evident God's existence. The basis of this analogy is the realisation that motion is an ontological category and a means of deepening the participation in *esse ipsum*. In other words, when one speaks of a being's motion, one also speaks of the nature of that being's existence. So for Aquinas, there seems to be no fundamental division between a proof in physics and one in metaphysics. They might be mutually enhancing because the categories included within physics are not merely functional or nominal, but remain analogically part of being itself which is the subject matter of metaphysics. Avicenna, by contrast, takes a first step towards isolating physics from metaphysics by locating the proof of God's existence purely within metaphysics because this science apparently does not in any proper way include the concept of motion. Furthermore, following Herbert Davidson, one can see that Avicenna's restriction is not limited to proofs of the divine existence.[6] As Aristotle offers a proof of the celestial intelligences from the motions of the celestial spheres (and it was seen in detail above that Aquinas follows Aristotle), Avicenna offers a proof of the existence of the separate substances which takes as its starting point not the motion, but the existence, of the celestial bodies.[7] One might also infer the existence of the agent intellect (which will be discussed below) from the passage of human intellection between potency to act, yet Avicenna offers a more 'static' reasoning: the existence of the matter, forms and knowing human soul of the sublunar region imply the existence of the agent intellect from which they emanate. In this Avicennian vision, it now seems that motion is indeed that which divides God from creation and philosophical theology from physics.

In outlining this role for the science of metaphysics, Avicenna is faced with a particular difficulty regarding metaphysics' enquiry concerning God. As was stated above, for Aristotle a science cannot demonstrate its own principles but must rather accept these as originating from a higher science. Avicenna also believes that a science is not to demonstrate the causes of its subject matter.[8] Therefore, if the subject of metaphysics is being *qua* being, then this science cannot seek to demonstrate the principles of being. Rather, the purpose of metaphysics must be the discovery of the characteristics of being *qua* being. Surely, however, God is the principle of existence, so how can metaphysics seek to demonstrate God's existence, for in doing so it would surely seek to prove its own principle? Avicenna's answer to this difficulty is to claim that God cannot be the cause or principle of *all* of being *qua* being, for indeed he is not the cause or principle of himself. It is possible for God to be a principle of just part of the subject matter of metaphysics, namely created being, and for metaphysics thereby to maintain its role as the science which demonstrates God's existence. Therefore, the enquiry of metaphysics seeks to describe the attributes of being *qua* being, the demonstration of the principles of being whose existence is caused (that is, the demonstration of God who is the principle of caused existence), and finally the principles of lesser sciences. The crucial point for Avicenna is that metaphysics cannot seek to demonstrate the existence of all of being, but only of particular beings. God is included under this latter category, namely as a type of being, the kind which is uncaused. So Avicenna's answer to the question concerning the scope of metaphysics is to subsume God under the category of the subject matter of metaphysics, namely being *qua* being.[9]

How does this contrast with Aquinas's view? He addresses this issue in his commentary on the *De Trinitate* of Boethius.[10] Aquinas states that a science investigates a subject genus and, in so doing, must investigate the principles of that genus. There are, however, two kinds of principle. The first kind includes those principles which are complete natures in themselves as well as being principles of other things. As an example, one might think of the heavenly spheres which are both natures in themselves and the principles of the lower bodies. In one science we might study these as they are principles of other things, in another science we might study these as discrete natures. Meanwhile, there are those principles which are not complete natures in themselves, but are only known as principles of other things. Aquinas is here thinking of, for example, unity as the principle of number or form and matter as the principles of natural bodies.

Aquinas deploys the distinction between different kinds of principle to argue that the divine beings are the principles of all things (because they are actual, lacking potentiality) while also being complete natures in themselves. Therefore, they can be studied in two ways: first, in so far as they are the principles of other things; secondly, in so far as they are natures in themselves. Although these divine things are most intelligible in themselves (because they are actual), nevertheless their reality is too bright for our minds. Thus we approach them through a science which principally studies the effects of divine things and we make our approach through natural reason. This is the science pursued by philosophers and is called first philosophy, metaphysics or 'divine science' and it studies divine things only in so far as they are revealed as the principles of all other things. Divine things are tangentially the subject of the science which studies that which is most general and common to all existents, namely being *qua* being. There is, however, a science which studies these divine things not as they are revealed in being the principles of all other beings, but as they reveal themselves. This Aquinas calls the theology of *sacra scriptura* and it is based upon revelation. Therefore, while being *qua* being is the proper subject of metaphysics, Aquinas does not subsume the divine under this category. God is known to metaphysics only mediately, as the divine is the cause or principle of the subject matter of metaphysics. So in one respect, Aquinas agrees with Avicenna (and others, including Roger Bacon): God is not the proper subject of metaphysics. However, Aquinas refuses to subsume God under the category which is the proper subject of metaphysics, namely being *qua* being, because God cannot be just another being under a particular genus (*ens commune*) and yet also be the cause or principle of being itself.[11]

In adopting this position concerning the subject matter of metaphysics and the science which might appropriately undertake a proof of God's existence, Aquinas is navigating a path somewhere between Avicenna and the later Spanish Islamic theologian Averroës (1126–1198). For Averroës, a metaphysical proof of God's existence is not possible.[12] He considers Aristotle's method of establishing the various subject matters and tasks of the different sciences and concludes that, because God is a subject of metaphysics and it is impossible for a science to prove the existence of its own subject, the proof of God's existence cannot belong to metaphysics. First philosophy has beings which are devoid of matter and motion as its subject,[13] so it cannot set about proving the existence of such beings any more than physics sets about proving the existence of moving bodies. According to Averroës, the most that metaphysics can undertake is to demonstrate that God, as its subject of enquiry, possesses certain attributes.

Averroës therefore locates the proof of God's existence within the science of physics rather than metaphysics.[14] However, some careful distinctions need to be observed. In the first instance, if, as Aristotle claims, a science may not prove the existence either of its subject matter or of its principles (in the double sense of the science's basic premises and the causes of its subject matter), how is physics to seek to prove God's existence if the divine is the principle of corporeal and moving being? In answer to this apparent difficulty, Averroës points out that, whereas no science can prove the existence of its own subject matter (because a science cannot say anything until it has a subject), it is possible for a science to prove, rather than demonstrate by syllogism, the existence of its principles.[15] Syllogistic demonstration is not possible because such demonstration works from cause to effect. This is to say that, if one were to attempt to prove the existence of, for example, the heavenly bodies (one of the principles of physics) by syllogism, one would begin with a cause which reasons *to* the heavenly bodies as that cause's effect. So one would need to construct such a syllogism not in physics nor even in astronomy (for the heavenly bodies are the subject matter of astronomy and therefore cannot be proved by this science), but in a prior and still more general science, namely metaphysics. If one were to demonstrate by syllogism the existence of the principle of a particular science, there must be a cause of that principle from which to construct one's reasoning. In the case of God, there is nothing lying behind his existence from which one might construct a syllogism, so his existence cannot be demonstrated in this way, and certainly not by physics. However, as Davidson states, Averroës is willing to allow a notion weaker than syllogistic demonstration, namely 'proof', which, rather than proceeding from cause to effect, proceeds from effect to cause in a fashion more reminiscent of natural theology. The effect from which one might reason back to the cause is motion, the subject matter of physics.

Meanwhile, for Averroës, metaphysics, in being concerned with being *qua* being, is therefore concerned with substance. All being is either a substance or an accident which partakes of substance. Therefore, as Etienne Gilson states, Averroës does not consider 'existence' as a problem separate from substance, and still less does he consider, as Avicenna had done, existence to be an accident of essence.[16] Substance is stipulated as the particular and individual in such a way that being (now coterminous with substance) cannot be predicated univocally. Neither can being be predicated equivocally, because each substance is a being of a particular kind, and these kinds can be gathered to form analogously similar substances. Now metaphysics studies substance in its prime instance, namely the separate substance which is the origin and final cause of all things. In studying this first substance, one tangentially studies all substance or being. Yet Averroës does therefore seem more inclined to include God as the subject of metaphysics. He does indeed claim that first philosophy deals with those substances which include God in their definition.[17]

Aquinas therefore blends elements of Avicenna and Averroës. He agrees with the former concerning the subject matter of metaphysics, being *qua* being, but disagrees concerning the place of God in relation to this science.[18] For Aquinas, God does not fall under the category of *ens commune*. Aquinas agrees with Averroës concerning the role of physics in proving the existence of God, yet he also assigns this function to metaphysics and clearly rejects Averroës' tendency to make the divine substance the subject matter of metaphysics.

Meanwhile, in Avicenna's restriction of the proof of God's existence to metaphysics one finds an early hint of a dualism between first philosophy and physics. For Avicenna, it is clearly wholly inappropriate to suggest that physics might seek to prove the existence of God because this science deals exclusively with motion and matter. In the argument for the intimacy of God in each and every motion outlined in the previous chapter, the contrast with Aquinas (and, one might argue, Aristotle) becomes clear. For Thomas, motion can be related to God for the divine is motion's beginning and end. In being the *telos* of motion, God therefore answers the question concerning why things move. Meanwhile, metaphysics also proves the divine existence, but this time with regard to being. Yet these two proofs are not unrelated, for motion is an ontological category: the motion of a creature, its direction and form are integral to its being. Are there, however, any further indications of the separation of God from creation and metaphysics from physics in the thought of Avicenna? I now turn to describe briefly Avicenna's cosmogony in order to examine his Plotinian emphasis and a further move towards the distancing of God from motion.

Avicenna on cosmogony and the agent intellect

The order of creation for Avicenna is predicated upon a hierarchy of being.[19] This begins with the First Cause which is necessary and in which existence subsumes essence in the perfection of its simplicity. The divine has no quiddity, but is simple being itself.[20] Below this necessary being, there are those beings which are only possible in themselves but which, by their proximity to the First Cause, gain the quality of necessity. These beings, the intelligences and separate substances, are eternal for they have no beginning or end and emanate by necessity from the First Cause. In fact, they differ from the first and necessary existent only in being second in the existential order.[21] Below this second category of being lies those which are merely possible. These beings, namely those of the sublunar world, are subject to generation and corruption.[22]

As Davidson points out, this tripartite division of being allows Avicenna to describe three, and not just two, aspects of intellection, the third providing an account of the emanation of body and soul within the celestial realm.[23] This can be explained as follows. The First Intellect emanates necessarily and in eternity from the First Cause or Pure Being. For Avicenna, it is intellection which is creative, so it is in the self-knowledge of the First Cause that the First Intellect emanates. This emanation of a unity from a unity does not compromise Avicenna's crucial principle that 'from the one insofar as it is one, only one can come into existence'.[24] Moreover, this emanation is not 'intended' because this would entail the First Cause possessing something within itself, namely an intention, which is for the sake of, and less real than, that which is intended in such a way that the simplicity and perfection of the First Cause would be compromised.[25] As for the attributes of the First Intellect, it is possible in essence but gains necessity by virtue of the First Cause. It generates multiplicity, for its intellection of the First Cause forms the second intellect. It is the generation of multiplicity from unity that is the central, and characteristically Plotinian, philosophical problem which Avicenna is seeking to address.[26] So in addition to the emanation of the second intellect, the soul of the first heaven emanates from the First Intellect's self-knowledge as a being necessary by virtue of the divine. Finally, the body of the first heaven emanates from the First Intellect's knowledge of itself as a merely possible being.[27]

This hierarchical pattern of emanation proceeds to the second intellect from which, in turn, emanates the third intellect, the soul of the second heaven and body of the second heaven. Upon reaching the ninth heaven and the tenth intellect one finds the principle which governs the motions of the sublunar, visible realm, whereupon there is too much impurity and potentiality to generate another intellect, soul and body. Avicenna refers to this last intelligence as the agent intellect, the *data formarum*. From this intellect radiate all forms and they are thereby brought to exist in matter and to be known by the human passive intellect.[28] As Gilson states,

> Every time one of these forms happens to find a matter fittingly disposed to receive it, the corresponding sublunary being comes to be; when its matter ceases to be so disposed, the same being loses its form and, consequently, ceases to exist.[29]

Gilson goes on to comment on Avicenna's negative appraisal of matter in this context. Matter is not that which may participate in the realisation of a form, but that which actively prevents such realisation.[30] Moreover, as Conor Cunningham succinctly points out, for Avicenna it is not the case that, as in Aquinas, form is that whereby something is created, but it is now itself created by the agent intellect.[31]

What is the role and importance of this agent intellect, wholly separate from the human soul, in Avicenna's cosmology?[32] In particular, Avicenna is anxious to account for the passage of the human intellect from potency to act in knowing. In order for this to take place, there must be a mover who is already in act, namely a separate intellect which brings the passive human intellect to the actuality of knowledge. The agent intellect, from its inventory of intelligible forms, gives to the separate human 'material' intellect the forms which then constitute knowledge. If the agent intellect were part of the human soul, then the soul would already be in actuality in possessing all knowledge. This not being the case, Avicenna surmises that the agent intellect must be a substance entirely separate from the human soul.

Avicenna's understanding of the action of the separate agent intellect on the human soul hints at later representational theories of knowledge which will first be exemplified in the work of Duns Scotus and which were discussed above with regard to the thought of Roger Bacon.[33] As P. Lee and Gilson point out, according to Avicenna to reduce from potency to act is merely for the actual being to impress what it possesses upon the passive recipient which has been appropriately prepared to receive the requisite form.[34] Thus human knowing is merely a matter of preparing ourselves through our imagination and the consideration of sense images to receive in passive fashion the abstract forms of things from a separate and actual intellect. Therefore, it seems that the encounter of the human intellect with material nature is merely a precursor to actual knowledge, for such an encounter is now a mere preparation to receive forms from the agent intellect.

The forms which come to reside in the soul are only representations of the world around us.[35] Indeed, the incidental nature of the material world for human knowledge is exemplified in Avicenna's allegory of the man born blind who flies in a vacuum, unable to hear or touch anything, not even aware of his own corporeality; he is still able to attain a knowledge of being and his essential nature as a soul.[36]

Avicenna's position is frequently likened to that of Plato, for whom intelligibility exists in the separate realm of being.[37] Yet the two are different in a crucial respect, for material nature, as was argued in Chapter 1 of this essay, can mediate to us, through time's motion, the truth that is properly contained in the Forms. So, contrary to many dualistic readings of Plato, one can surmise that corporeal being causes knowledge through its participation in, and reflection of, the Good. It is not that matter is somehow always unintelligible and 'dark' for Plato, but that its intelligibility and value, in common with the entire realm of becoming, have a transcendent source in the Good. For Avicenna, on the other hand, knowledge must entail an escape from corporeality. The role of material nature in human knowledge is now relatively peripheral and possibly detrimental; our corporeality is only a beginning point for our learning. There is no genuine 'motion' in human learning, for this process is now understood as the stamping of forms by the agent intellect onto the blank slate of the human possible intellect. In order to realise the implications of Avicenna's view, I now turn once again to contrast it with that of Aquinas.

For Aquinas, knowledge is intimately concerned with material nature. He states that,

> in order to understand, we need the sensitive powers not only to acquire knowledge, but also to use the knowledge we have acquired. For we cannot actively consider even those things which we know except by turning to phantasms, even though Avicenna teaches the contrary.
> (Aquinas, *Quaestiones de Anima*, 15.responsio)[38]

For Thomas, the agent intellect is part of the human soul, for it is not in act in the sense of already possessing the forms which constitute knowledge, but in being immaterial in such a way that it is able to abstract the species which migrate from all material natures. In other words, the actual intellect is not a separate substance imprinting knowledge upon us from without, but is rather that which prepares the species from material things to be received by the passive intellect.[39]

For Aquinas, there is a visceral motion, quite circular in form, involved in this acquisition of knowledge.[40] Once the form migrates from the hylomorphic compound, it becomes a more abstract 'species' and enters the mind to arrive initially at the passive intellect before being properly illuminated and expressed by the agent intellect. In turn, this intelligible species in the mind becomes the concept or 'inner word' of the thing known.[41] Acting like a sign, it points us back to what is known to learn and abstract more in such a way that the motion of the knowledge of the material nature recommences. During this motion, however, form is not seized away from matter so as to leave behind a material which is otherwise detrimental to knowledge. For Aquinas, it is not only form which is abstracted from the object known, but also the intelligible species which enter the human mind and are the essence of the thing in question; in so far as matter is of the essence of a thing, it is included within that intelligible species. More particularly, the abstraction of the intelligible species is from individuated and sensible matter rather than common matter. Thus a particular human being has as part of his or her essence a particular body, flesh and bones. From the particularity of this human body, a 'common matter', abstracted from the particular, migrates to reside in the human intellect and constitute knowledge of the person concerned.[42] Moreover, because matter itself proceeds from the divine mind for Aquinas, and God's knowledge is inseparable from his causality, it is not due to the weakness of matter that we need to abstract, but due to the weakness of our minds. Matter is integral to human knowledge and does not lie beyond intelligibility in any dark or sinister fashion.

The Avicennian doctrine of the separate agent intellect is therefore consonant with a more general denigration of sciences such as physics which are concerned primarily with material nature. If knowledge is concerned with the retreat from matter towards a separate agent intellect, then natural philosophy, and with it the science concerned primarily with motion, will be of lesser importance. Moreover, because matter is now only a preparatory starting point for knowledge, it appears that a science which is concerned particularly with material nature will likewise be a mere beginning point, later to be discarded, in human learning. Thus one can see early intimations of the separation of physics from higher sciences and, particularly, from theology.

Avicenna on nature

Is there, however, any way in which Avicenna's physics isolates motion itself? Weisheipl has described how the Persian thinker, in his summary of Aristotle's *Physics* known as the *Sufficientia*, describes nature as an efficient cause of motion in such a way that motion becomes self-explanatory.[43]

In the previous analysis of Aristotle and Aquinas on the motion of inanimate bodies, it was emphasised that nature is regarded as a principle of motion or rest in those things to which it belongs properly rather than incidentally. In the motion of inanimate bodies, nature is not the mover. Rather, the mover is the generator of the given body or, in a secondary sense, anything which may remove an impediment to a body's natural motion. This understanding of nature means that a body's motion always refers beyond itself to an exterior cause, hence the motor-causality principle that 'everything that is moved is moved by another'. Motion is not self-explanatory and the lowly, natural motions of inanimate beings require reference to a higher cause. This 'pointing upwards' implies that physics will not be autonomous, but will seek the origins and ultimate, causal explanation of its subject matter (namely, the motion of bodies) elsewhere. However, there is a tendency in Weisheipl's work to add a different kind of autonomy to moving bodies, namely the notion that, once a body is generated, it has all it requires for its proper motion and the only immediate cause of the motion of inanimate bodies to which one might refer is the remover of an obstacle. I argued that, for Aquinas, at no point is any motion self-explanatory, for one must always refer to causes higher in the cosmological hierarchy, and ultimately to God, in order to understand motion. God is 'constantly conjoined' to every motion.

By contrast, Avicenna deviates from the Aristotelian understanding of the natural motion of inanimate bodies in regarding nature as a force or internal efficient cause of motion. If nature is understood in this way, and the particular nature of each thing is regarded as integral to its being, then all motion becomes, in a sense, self-motion, for heavy bodies, for example, are said to move themselves downwards. Moreover, there does not seem to be any means of differentiating between certain kinds of motion. Following Weisheipl, one might ask how, for example, is it possible to distinguish between a person walking down the stairs and a person falling down the stairs?[44] If the body of a person moves itself even when falling, it seems that both these motions are instances of self-motion. Avicenna, and Scotus later, reply that a person walking down the stairs *knows* that they are walking down the stairs. But this is not a distinction between different kinds of motion. The qualitative element of motion is no longer of importance: there is no significant difference between a stone falling and an eagle swooping. Moreover, if motion is understood as undifferentiated and self-explanatory in this way, that is, as not pointing beyond itself and ultimately to God, it is not surprising that Aquinas's proof of God's existence from motion, which relies on motion having a transcendent origin and *telos*, is rejected by Henry of Ghent and Scotus.[45] For these latter thinkers, motion requires no explanation beyond 'nature' which is now the efficient cause of all motion.

In drawing together these elements of Avicenna's thought, it is possible to see that in at least three different ways his approach issues in a dualism between physics and the higher sciences on the one hand, and the separation of God from motion on the other. First, it is proposed that physics can only know God as that which is beyond matter and motion, so the task of making evident, or proving, God's existence belongs exclusively to metaphysics. The subject matter of metaphysics is being *qua* being and, for Avicenna, God is subsumed under this category of *ens commune*, for the divine is the type of being which is uncaused. The inclusion of God within the subject matter of metaphysics therefore betokens an understanding of being as univocal. As Cunningham states, 'The upshot of this univocity is the loss of the sensible realm. As creation emanates from the creator, being is only given to intelligence, to the extent that the creator does not give being directly to the sensible realm'.[46] This comment points to the second distancing of God from motion, for through an interminable plane of being which stretches univocally between God and creation, Avicenna seeks to insulate the wholly transcendent and absolute One from the vagaries of the moving cosmos by placing a hierarchy of ten intelligences between the divine and the universe as a kind of 'buffer'. It is only the tenth intelligence, the separate agent intellect, the *dator formarum*, who donates being to the universe. The separate agent intellect also imprints forms onto the passive human intellect in such a way that knowledge, rather than being a visceral motion which is utterly entwined in the motion of corporeal being, is estranged from the material world and sought almost exclusively in the transcendent. Finally, with the natural realm now distanced from its divine origins, nature is understood as the efficient cause of all motion in such a way that motion is a self-explanatory category which is restricted more particularly to the realm of the material and the subject of physics.

Avicenna's thought, issuing in a dualism of physics and metaphysics and the separation of God from the science which studies motion, constitutes an early intimation of profound changes which were to take place in natural and first philosophy in the later Middle Ages. At a time when motion was understood as an analogical term which could be related to the different sciences, it seemed apposite to place physics alongside theology and to search for the principles of motion in divine science. However, once motion is distanced from God and the higher sciences which might deal with divine matters, it makes much greater sense to restrict the discovery of the principles of motion to a science in closer proximity to physics, namely mathematics. However, before outlining the mathematical and quantitative study of motion in the fourteenth century, I turn first to examine an important theory of projectile motion known as impetus. Like their forerunner Avicenna, the proponents of the theory of impetus believed in a theoretically indefinite and permanent internal source of projectile motion which begins to point towards Newton's principle of inertia.

The theory of impetus and the quantification of motion

The origin of impetus theory can be traced as far back as John Philoponus of Alexandria (*c.*490–570) who attempted to demonstrate that the medium around the projectile cannot be the cause of the projectile's motion.[47] Philoponus argued that if it is the medium which moves the stone, as Plato and Aristotle suggest, this seems to render the hand throwing the projectile somewhat superfluous. Why can one not move the stone simply by fanning the surrounding air? If one moves a projectile by giving the medium the power to move the body, why touch the stone at all? Philoponus postulated that, far from being a cause of a projectile's motion, the medium is a hindrance. He explained the motion of a projectile in terms of the thrower giving to the projectile an incorporeal motive power (*energeia*) which, because 'borrowed' from the thrower, diminishes by the natural motion of the body (for example, the stone's tendency to fall) and the resistance of the surrounding medium.

As early as the 1320s, the theory of impetus was well known and given serious consideration and credence within the schools. The Franciscan Franciscus de Marchia (*c*.1290–*c*.1344) understood impetus as a *virtus derelicta*, a force given to a body which naturally dissipates and which is capable of moving a projectile against its nature.[48] Contrary to the theory of Philoponus, however, Franciscus maintained that the medium also receives a *virtus* which assists in the motion of the projectile. In being self-dissipating, this *virtus* is not permanent but extrinsic and, in the case of projectile motion, contrary to the nature of a body and therefore violent. Later in the fourteenth century Jean Buridan (*c*.1300–*c*.1358), a student of William of Ockham and later rector of the University of Paris on two occasions, was to formulate a more sophisticated impetus theory very similar to that of Philoponus.[49] He extended the theory to encompass all motion: each variety of motion (rectilinear and circular, natural and violent) is understood as an interaction between impetus and the natural inclination of a body. In support of the theory of impetus, Buridan frequently appealed to *experimenta*, the technique commended in the previous century by Roger Bacon. One example is as follows:

> And you have an experiment (*experimentum*) [to support this position]: If you cause a large and very heavy smith's mill [i.e. a wheel] to rotate and you then cease to move it, it will still move a while longer by this impetus it has acquired. Nay, you cannot immediately bring it to rest, but on account of the resistance from the gravity of the mill, the impetus will be continually diminished until the mill would cease to move.
>
> (Buridan, *Quaestiones super libris quattuor de caelo et mundo*)[50]

Thus impetus is understood to diminish and dissipate because of the competing resistance of the heaviness of the mill.

In articulating his theory of impetus, Buridan recognised the importance of the speed and mass of a body for measuring the amount of impetus which gives rise to local motion. He explained this by comparing the motion of a heavy body, which has a greater quantity of matter, and a body having the same volume and shape but of less density and consequently of less weight. Taking the case of one body made of iron and another of wood, both traversing a space at the same speed, the metal body would traverse a greater distance before its motion begins to fail.[51] This, Buridan postulated, is because its greater weight would enable it to receive more impetus and resist the medium more effectively.[52] Thus the quantities which were later to define momentum within Newtonian mechanics, namely the quantity of matter and speed, are here used by Buridan to identify the amount of impetus. The difference lies in Newtonian momentum being a measure of the quantity of motion, whereas impetus is understood as the cause of motion.

Buridan did, however, take an important step towards the analysis of 'idealised' or 'purified' motion in a way that anticipates Newtonian physics.[53] As Edward Grant points out, like Avicenna, impetus for Buridan possesses a particular permanence in such a way that, if the projectile motion of a body were to remain unhindered, that motion would continue indefinitely. Impetus itself is 'static' for Buridan because once a projectile, for example, has been released by a thrower, no additional impetus could be added. Moreover, it is only by means of the resisting medium or the natural tendency of the moving body towards its proper place that the quantity of impetus is dissipated. This implies that, if all resisting factors can be removed to form an 'idealised' or 'unhindered' motion, a body's motion will continue indefinitely in the same way that, for Newton, the inertial motion of a body through a void would continue indefinitely. Yet Buridan did not draw an inertial theory of motion from his understanding of impetus. In important ways, the theory of impetus was to remain an Aristotelian theory because it maintains the distinction between natural and violent motion. The movement of a projectile is understood as an interaction between the impetus possessed by a body and that body's natural tendency to move towards its proper place within the cosmos. Projectile motion also maintains a necessarily extrinsic source because the impetus of an inanimate body, because it is inanimate, must always find its source in another. Moreover, the notion of an idealised motion in which all resisting factors are removed so as to produce something akin to Newton's inertial motion in a void remains highly counter-intuitive within a finite Aristotelian cosmos in which nature is necessarily saturated with factors which make motion an intricate weaving of numerous competing or co-operating elements. For example, although Buridan maintains that the surrounding medium provides resistance to motion, this resistance is a constant. It is not a variable like friction which can hypothetically be reduced to zero and thus theoretically removed to form a more idealised and simpler consideration of pure motion.

Buridan did, however, deploy his theory of impetus to explain various motions in new ways. In particular, he considered the downward acceleration of a falling heavy body.[54] In most Aristotelian discussions, such motion is considered only with regard to the generator of the falling body or the remover of an impediment to its motion. Yet how could one explain the observation that a body's velocity increases as it falls?[55] Buridan explains and dismisses a number of theories, including that of Averroës who suggested that a body accelerates because of an increasing desire for the goal of its motion and because of the body's heating and consequent rarefying action on the surrounding medium. By contrast, Buridan observes that, in the case of a stone falling downwards, the weight of the stone remains constant, as does the medium through which the stone falls. These factors, in being constant, cannot explain the change in velocity of the stone as it falls. Therefore, the acceleration of the stone must be due not only to that which is acquired from its principal mover, namely its natural tendency to downward motion, but also to the acquisition of an ever greater impetus. 'This impetus,' argues Buridan, 'has the power of moving the heavy body in conjunction with the permanent natural gravity'.[56] It is the body's gravity which initiates the acquisition of ever more impetus in proportion to the weight and density of the body. The greater the weight and density of the body, the greater the amount of impetus that can be acquired by that body as it falls. Thus the velocity of the fall of a body is now understood as a function of the weight of the body and the duration of the fall which together give rise to the gradual acquisition of a certain moving force or impetus.

Buridan drew a further important consequence from his impetus theory:

> it is unnecessary to posit intelligences as the movers of celestial bodies. … For it could be said that when God created the celestial spheres, He began to move each of them as He wished, and they are still moved by the impetus which He gave to them because, there being no resistance, the impetus is neither corrupted nor diminished.
>
> (Buridan, *Quaestiones super libris quattuor de caelo et mundo*, II.12)[57]

Thus the continual intellectual movement of the heavens becomes a superfluous and unnecessary hypothesis for Buridan. To the modern mind, this view seems like so much common sense. However, if one recalls the importance of the doctrine of the angelic intellectual movement of the heavens for Aquinas, it is possible to see that its later abandonment is not merely the innocent ushering in of common sense to bring about an improved physics: there is a significant shift in how one understands motion and its relation to matters of theology. For Thomas, the notion that intelligences move the celestial realm is a consequence of the ontological hierarchy of motion. In other words, according to his cosmology, motions are not univocal any more than being is univocal, because motion is ontological and there are different, but analogically related, varieties. By contrast, for Buridan it seemed that, rather than thinking of motion in terms of hierarchy, it would be simpler to think of motion as a series of mere variations of one particular kind, namely local motion which is efficiently caused by imparting impetus. Thus the celestial rotation of the spheres is a variety of local motion univocal with that seen in the sublunar realm, and God causes the motion of the planets by imparting impetus in the same way that a thrower causes the motion of a projectile. For Aquinas, matters are not as straightforward. The rotation of the celestial spheres has a constancy and completeness which makes it of a different kind to terrestrial local motion, and this requires explanation. Moreover, the motion of the planets, although apparently merely local, is able to cause and sustain lesser terrestrial motion because of its constancy and unity. So, Aquinas might ask, how can apparently lowly local motion be exalted to take on a quality which can cause and sustain all lesser motions within the cosmos and thus participate in the teleological fulfilment of being? Because of the common principle that like produces like, the celestial motion must have a cause which can impart this quality of constancy and unity, a cause which similarly creates and sustains. The creating and sustaining source of celestial motion is ultimately the intellection of God which is the universal and primary cause of all things, but this is imparted to a material and moving cosmos through the participation in the divine thought of separate intelligences. That which creates and sustains and, in this regard, is most akin to the celestial motion, is the motion of thought. As was seen in the discussion of Aquinas's understanding of the immanent divine life, the ultimate creative action is the thought of God which is the emanation of the persons of the Trinity. The celestial motion is an embodiment of creative and sustaining intellective motion: if we could 'see' thought, it would look like celestial rotation.

Therefore, on Aquinas's view, one looks for an explanation of celestial motion that encompasses the spheres' exalted cosmological status and points directly to an immaterial source of the being which is imparted via the heavens' motion. So one can see that Aquinas's understanding of motion is at once also a doctrine of creation: physics and theology are intertwined. By contrast, Buridan's theory counts God as the source of the impetus of the motion of the planets, not because the planets, in having an exalted cosmological status because of their motion, must have an exalted source of that motion, but because God seems to be the only agent sufficiently powerful to move such masses. Motion is no longer the mediated embodiment of something integral to the divine life – namely, the intellection of being – but is now the result only of the supreme divine power and will. Buridan's impetus theory, although more simple, constitutes a more isolated physics which deals with the divine as a mover in the same way that it deals with the thrower of a projectile as a mover.

Buridan's understanding of motion was to have further important influences on later medieval physics. In particular, qualitative change was to be reduced to a variety of quantitative change. In order to see how this came about, it is necessary to recall that for Aquinas a qualitative motion arises through a greater or lesser participation in form. Thus an apple becomes more red by participating more completely in the form of redness. However, an alternative view was proposed by, for example, Duns Scotus, Gregory of Rimini and Nicole Oresme, in which a change in quality is not understood in terms of participation, but rather in terms of the addition of new and formally distinct parts of the existing quality.[58] Qualities could be diminished in similar fashion. Just as one could increase or decrease the weight of a body by the addition or removal of parts of the body, so one could increase or decrease redness, justice or charity by the addition or removal of 'parts' of these qualities. An analogy was drawn with line segments: if one adds a segment to an existing line, the line becomes longer and, in a sense, the old line is preserved within the new, longer line. Likewise, in adding to justice, the former quality of justice is preserved within the new, more intense justice. In this way, qualities came to be understood as quantities and, as

Grant states, 'Scholastic natural philosophers who treated this subject became interested primarily in the mathematical aspects of qualitative change and less interested in the theological and ontological aspects that had been prominent earlier'.[59] Moreover, the study of qualitative and local motion converged under this doctrine of the 'intension and remission of forms'. Thus just as qualitative change could be intensified or diminished by the addition or subtraction of quantities of the requisite form, so too the motion of a body could be intensified or diminished by the addition or subtraction of quantities of impetus. If the quantity of impetus remained the same, so too did the motion.

This quantitative understanding of motion naturally suggested an increasingly mathematical study of all change. Such an approach is to be found in Gerald of Brussels' *Book on Motion*.[60] In this work, motion is not investigated as 'dynamics', that is with reference to its immediate causes, but within 'kinematics'. This latter approach is concerned only with the mathematical description of an existing motion. Those 'Oxford Calculators' of the fourteenth century associated with Merton College, namely Richard Swineshead (*fl.*1340–1355), Thomas Bradwardine (*c.*1300–1349) and others, further developed the distinction between dynamics and kinematics, distinguishing between uniform motion which had a constant velocity, and non-uniform motion known as acceleration.[61] This distinction and emphasis on acceleration, whereupon natural philosophers no longer investigate motion, but *changes* in motion brought about by force or the addition of impetus (namely, acceleration and deceleration or changes in direction), motion being understood as a state equivalent to rest, forms the background to Galileo's and Newton's contribution to the study of motion and their articulations of the principle of inertia.[62]

The Oxford Calculators utilised definitions of uniform speed and uniform accelerated motion which were to be deployed by Galileo two centuries later to explain the acceleration of falling bodies. In particular, they contributed what was to become known as the mean speed theorem. This states that, if a body commences a motion which is uniformly accelerated from rest or from a particular constant velocity, it will traverse a particular distance in a particular time. If the same body were to be in motion during the same interval of time with a constant velocity equal to the mean velocity attained during its uniformly accelerated motion, it would traverse the same distance. Thus uniformly accelerated motion is mathematically equated to a certain uniform motion. It was, however, acceleration, or the change in motion, which was to concern natural philosophers, because motion was understood as a state equivalent to rest. As will be seen in the next chapter, this leads away from a consideration of motion *per se* towards a consideration of the forces which bring about changes in motion.

This brings to a conclusion the brief study of an alternative medieval understanding of motion in physics and theology. Already in the work of Avicenna, one finds a dualism between physics and metaphysics. It is the separate agent intellect which donates being and insulates the transcendent One from a moving cosmos. Furthermore, Avicenna sees motion as self-explanatory because nature becomes an efficient cause of motion within sublunar bodies. This already points towards a rejection of an Aristotelian physics in which any motion always points beyond itself and finds its source in a higher being. This simplification and isolation of motion was taken much further in the fourteenth-century recovery of the theory of impetus which leads to the quantitative study of changes in motion. This increasingly mathematical physics, in which motion is understood as a simple, univocal and physical category devoid of theological significance or meaning, is much more consonant with our common, contemporary notions. By far the most influential and original exponent of this vision is Isaac Newton, the father of modern physics, in whose *Principia* and theological manuscripts one finds a view of motion and God's relation to the cosmos so different from that outlined in the previous chapters of this essay. To his work I now turn, investigating in particular Newton's valiant attempts to keep theology and physics in some kind of proximity. However, it will be seen that Newton's thought constitutes the ultimate separation of God from motion.

6 Newton

God without motion

The publication of Sir Isaac Newton's *Principia Mathematica* in the summer of 1687 was perhaps a critical event within a protracted labour resulting in the birth of modern science. Newton's work on motion, apparently the crucial focus of the *Principia*, was to bring the greatest advances for the scientific community in its investigations of nature. As well as providing a great advance within a tradition of enquiry established by earlier modern natural philosophers such as Johannes Kepler (1571–1630) and Galileo Galilei (1564–1642), Newton's treatise also inaugurated a decisive break with the Platonic and Aristotelian understandings of motion which have been the principal focus of this essay thus far. Gone was the distinction between natural and violent motion and any persistent reference to causes other than the efficient. Newton further clarified the principle of inertia and incorporated it within his three laws of motion. These laws were, by necessity, unchanging and motionless principles of reality. The ability to encapsulate nature within a mathematical formula and bring consensus through the infinitely and identically repeated scientific experiment was to bring nature under control: scientists could now predict its every motion.

Although Newton is chiefly remembered for his contribution to natural philosophy and particularly the understanding of motion, these formed what are arguably only a small part of his intellectual endeavours. Newton's written works, many of them in unpublished manuscripts,[1] display lifelong interests in Biblical interpretation, Church history, prophecy, the nature of divine action, Christology, the history of ancient civilisations and alchemy.[2] Numerous comments of Newton suggest that he perceived these variant interests, particularly his natural philosophy and religious studies, as an integrated whole with a specifically theological purpose. For example, in a much quoted letter of December 1692 to an ambitious clergyman, Richard Bentley, a keen reader of Newton's work and the deliverer of the inaugural series of Boyle Lectures which were established to defend religion from atheism,[3] Newton began by stating that,

> When I wrote my treatise about our Systeme [the *Principia*] I had an eye upon such Principles as might work wth considering men for the beleife

of a Deity & nothing can rejoyce me more than to find it usefull for that purpose.[4]

In the same letter, Newton states that he was forced to ascribe the design of the solar system to a voluntary agent and, moreover, 'ye motions wch ye Planets now have could not spring from any natural cause alone but were imprest by an intelligent agent'.[5] Meanwhile, in a sentence added towards the end of the *General Scholium* as the second edition of the *Principia* was being printed, Newton commented that, 'This concludes the discussion of God, and to treat of God from phenomena is certainly a part of natural philosophy'.[6] It seems, therefore, that Newton perceived his principal treatise in natural philosophy also as an exercise in natural theology and apologetics, particularly as an antidote to the views of Descartes which, so Newton and a number of his contemporaries feared, could only issue in at best an arid deism, at worst outright atheism.[7] In attempting to understand the view of nature and God which forms the intellectual background to the *Principia* and its conception of motion, one must begin with theological convictions which were established much earlier in Newton's intellectual career, beliefs in which, unlike those of his natural philosophy, he barely wavered throughout his life. Most central to this outlook were an Arian conception of the nature of Christ and a thoroughgoing theological voluntarism.

The theological context of Newtonian motion

Newton expounded his Arian views of Christ at least fifteen years before the publication of the *Principia*. He expressed these doctrines in a manuscript dated to the period 1672–1675 which includes a series of statements on religion.[8] In his second and third statements, Newton comments that,

> The word God put absolutly without particular restriction to ye Son or Holy ghost doth always signify the Father from one end of the scriptures to the other. ... When ever it is said in the scriptures that there is but one God, it is meant of ye Father.
> (Newton, Yahuda MS, var. 1, 14, f. 25)[9]

An equally clear position is articulated in a later manuscript:

> When therefore the father is called God & the son is called Lord (as is done in the Creed) it signifies that the father is the highest Lord & the son is Lord next under him and that ye son sits at the right hand of God. And when the Son is also called God it signifies that the name of God is in him and that he is Lord over all things under the father. And yet they are not two Gods, because a king and his viceroy are not two kings, nor is the name of God to be understood of both together.
> (Newton, Yahuda MS, var. 1, 15, f. 98 recto)

There are two principal reasons why Newton held such an Arian view of God. The first relates to studies in Biblical interpretation and religious history which he initially undertook in earnest between the late 1660s and the mid-1680s and to which he was to return in the early part of the eighteenth century.[10] Through his studies, Newton became convinced that the earliest Christian Church held an uncorrupted non-Trinitarian faith which understood Christ as an exalted and yet created mediator between God and the universe. During the fated fourth century, however, certain corrupting self-interested forces within the Church, most particularly Athanasius, introduced what Newton believed to be the idolatrous doctrine of the Trinity, the belief in the consubstantiality of the Father, Son and Holy Spirit. In letters intended for his contemporary and Arian sympathiser John Locke, Newton wrote of the corruption of scripture by the Trinitarians with particular reference to 1 John 5 and 1 Timothy 3.16, arguing that, because supporters of Trinitarianism prior to or during the fourth-century disputes made no reference to these otherwise obviously useful texts, they must be fourth-century or later corruptions.[11] Moreover, Newton was sure that the central message of *The Revelation to John* was the prophecy of this corruption of the true and ancient faith by the deceitful Roman Church.[12] In a wider context, through his studies of ancient history, Newton was sure that there was once a *prisca fides et sapientia*, an ancient true religion found particularly in the figure of Noah (this religion also included a pure natural philosophy of which his own was a rediscovery) which had over time fallen into idolatry. He traced idolatrous traditions amongst the Chaldeans, Egyptians and Assyrians. The Lord God had intermittently sent prophets to draw people back from this idolatry, Christ being just such a prophet. According to Newton, the idolatries of the fourth century onwards constituted yet another relapse from the pure ancient Noachian faith and thereby another fall away from a true natural philosophy.[13] In forceful terms which reflect well the rage he felt at what he saw as the corrupt Trinitarian idolatry, Newton states that,

> The year 381 is therefore w^th^out all controversy that in wch this strange religion of ye west wch has reigned ever since first ye overspread world, & so ye earth wth them that dwell therein began to worship ye Beast & his Image, y^t^ is ye church of ye western Empire & the afforesaid Constantinopolitan Counsel its representative.
>
> (Newton, Yahuda MS, var. 1, 14, f. 50)

Christianity in its pre-fourth-century, pre-lapsarian form was therefore a mere reinstantiation of an ancient faith to which Newton wished to return and of which he was himself a tiny remnant. The essence of this faith was expressed in Newton's version of the Creed stripped of its Nicene accretions.[14] A corollary of this sense of history is that Newton did not view himself as an innovator in natural philosophy, but, in both his theological and scientific guise, as a nostalgic prophet returning the world to a once pristine knowledge and faith.[15]

The second reason for Newton's Arianism, and one which was at the same time a consequence of this Christology, is more explicit and, although this view was undoubtedly formulated much earlier, it appears in the *General Scholium* of the second and third editions of the *Principia* itself. This was the belief in the utter supremacy, power and freedom of the will of the Lord God of Dominion.[16] Newton writes,

> He [God] rules all things, not as the world soul but as the lord of all. And because of his dominion he is called Lord God Pantakrator [here Newton adds a note: 'That is, universal ruler']. ... We know him by his properties and attributes and by the wisest and best construction of things in their final causes, and we admire him because of his perfections; but we venerate and worship him because of his dominion.
>
> (Newton, *Principia Mathematica*, pp. 940 and 942)

Newton shows his view of the supremacy of the divine will in a manuscript dated to a similar period:

> And as yᵉ wisest of men delight not so much to be commended for their height of birth, strength of body, beauty, strong memory, large fantasy, or other such gifts of nature as for their good and great actions the issues of their will: so yᵉ wisest of beings requires of us to be celebrated not so much for his essence as for his actions, the creating and preserving and governing all things according to his good will and pleasure.
>
> (Newton, Yahuda MS, var. 1, 21, f. 2)

It was a supremely free and sovereign will which, for Newton, was the supreme attribute of God. Because this will was supremely free, this entailed its inscrutability and arbitrary character. It was because of God's omnipotent wilful dominion alone that he was worthy of worship. This voluntarism featured a dualistic distinction between God's *potentia ordinata* and *potentia absoluta*. It was by the former that God ordained and preserved the regular workings of the laws of nature. However, in the latter was enshrined the absolute power of God's will to suspend or change these laws at any moment. This was a kind of arbitrary 'addition' to God's *potentia ordinata*. This is most clearly expressed in an unpublished text dated by J. E. McGuire to the early 1690s, just a short time after the publication of the first edition of the *Principia*:

> That God is an entity in the highest degree perfect, all agree. But the highest idea of perfection of an entity is that it should be one substance ... by his will effecting all things possible ... and constantly cooperating with all things according to accurate laws, as being the foundation and cause of the whole of nature, *except where it is good to act otherwise*.
>
> (Newton, David Gregory MS, 245, f. 14a)[17]

For Newton, therefore, the laws of nature and the 'final ends' of which he speaks in the *General Scholium* are not immanent in nature, part of creation's ontology, but merely imposed from without by a God whose rule is supreme. A theological interpretation of the laws of nature within this voluntaristic context would see them not as something integral to the universe, but as measures of an otherwise inscrutable divine will. The early modern scientist is therefore charged with forever confirming the constancy of the laws and activity of nature by repeated experimental practice in order to judge the currents in the divine will as it replenishes a decaying creation. This goes some way to explain why Newton rejected the necessitarian rationalism of Leibniz for whom experimental observation of nature was of less importance: for Newton, Leibniz's position is a rejection of the divine will's sovereign power and freedom to change the otherwise constant laws of nature at any moment.[18]

In a recent study, Stephen Snobelen has pointed out further details of Newton's theological thought, arguing that he was influenced by the Socinians, a Polish unitarian movement also popular with Newton's · contemporaries John Locke and Samuel Clarke.[19] Newton himself owned a number of Socinian works and had access to many others in the library of Trinity College. Yet Socinianism was not merely a unitarian movement; there were many emphases besides, including a commitment to believers' baptism, mortalism, the separation of church and state and the support of religious tolerance.[20] A particular aspect of their thought is expressed in Johann Crell's *De Deo ejus attributes* where he argues that the name 'God' does not refer to the divine essence but rather to 'a name of power and empire'.[21] As Snobelen argues, Newton shares this desire to remove talk of God's essence from theology.[22] Instead, God is known only by his relations with his creatures, namely by his power and dominion over them. Furthermore, the term God, in being relative, can be bestowed upon lesser beings (such as angels, kings or Christ) without rendering them 'very God of very God'. It is simply the case that God himself, the 'One', has an absolute dominion, whereas lesser gods share a univocal but diminished power.[23] In declining to emphasise talk of God's perfection, infinity or eternity (for Newton, these cannot be relative terms) and referring more particularly to the power that God wields over other things, Newton is reducing all theology to natural theology and, ultimately, to his universal mechanics which will be discussed below. What revelation amounts to is not a mystical glimpse into a divine life in which we might participate, but an account of God's wilful power, this being known only through the divine governance, replenishment and manipulation of creation. We are mere objects of God's supreme subjectivity. Moreover, this leads to an inevitable anthropomorphising of the divine: if God is known only by his relations to us (namely, his power), the categories by which God is named and known can only be those of human time and history, for this is the realm in which we are dominated and ruled. Furthermore, as we will see, those relations by which God is known take on an arbitrary character because relationality, for Newton, is not contained within the divine life itself. Whereas in a Trinitarian understanding of God the relations between the persons are the eternal basis of God's relation to creation (and indeed of *all* relationality, as for Aquinas), for Newton there is no such eternal relation within the One: divine relationality, by which God is known, is only temporal and exists within a univocity between God and creatures.

Along with other commentators on Newton's work, Snobelen has also pointed out important connections between Newton's theological and scientific methodology.[24] The first relates to the principle of simplicity which was prevalent in seventeenth- and eighteenth-century science.[25] For Newton, not only is nature a simple system whose mysteries can be unlocked by the human mind, but also the book of scripture is straightforward and, because written for the vulgar, its full meaning is attainable by the application of straightforward hermeneutical techniques. Natural philosophy and scriptural interpretation are brought to accommodation through a dualism between the absolute and relative meaning of terms. As will be seen below, this distinction is central to Newton's understanding of space, time, place and motion. He is clear that one should not confuse the absolute variety of these notions with the more commonly understood relative variety. Similarly, one should not think that scripture contains a description of the absolute understanding of things that one finds in natural philosophy:

> Relative quantities, therefore, are not the actual quantities whose names they bear but are those sensible measures of them ... that are commonly used instead of the quantities being measured. But if the meanings of words are to be defined by usage, then it is these sensible measures which should properly be understood by the terms 'time', 'space', 'place', and 'motion', and the manner of expression will be out of the ordinary and purely mathematical if the quantities being measured are understood here. Accordingly those who interpret these words as referring to the quantities being measured do violence to the Scriptures.
>
> (Newton, *Principia Mathematica*, p. 414)

It seems that Newton is suggesting that scripture and natural philosophy are related just as the relative and the absolute notions of time, space, place and motion are related. The former is the 'vulgar' notion of the latter.

Although the literal meaning of the books of nature and scripture was Newton's aim and, so he believed, both interpretative enterprises could proceed 'outwards' from essentially straightforward principles to explain the apparently more complex, nevertheless he draws a distinction between simple truths which are accessible to all and those which can only be understood by the more experienced and educated.[26] Newton adopts the distinction from Hebrews 5 between 'milk for babes' and 'meat for elders'.[27] There are certain straightforward truths with which one should begin before progressing to more profound and difficult matters. This distinction informed Newton's attitude to both his natural philosophy and theology. On the one hand, it is clear that Newton believed his natural philosophy to be 'meat', a return to an understanding of nature which had once been part of the *prisca fides et sapientia*. He seems to have thought that his contemporaries were not ready to receive his work in natural philosophy, for he deliberately made it abstruse so that only those with proper training would be able to understand. For example, the Cambridge mathematician Gilbert Clerke complained that the *Principia* was impenetrable. To Newton he wrote, 'you masters doe not consider ye infirmities of your readers, except you intended to write only to professors or intended to have your books lie, moulding in libraries or other men to gett the credit of your inventions'.[28] Numerous others, including John Locke and Richard Bentley, struggled with the treatise, as did numerous Heads of Colleges in Cambridge to whom the work was distributed.[29] Appeal was made to specialist mathematicians, for example Huygens and John Craig, to assist in decoding the *Principia*. The obscurity surrounding the *Principia* (not only was it written in Latin and the language of mathematics, but also copies were very rare and expensive) added to the mystique and godlike status of Newton himself, something which he himself fostered.[30] His treatise was regarded as if it were itself divine revelation. Yet Newton was also reluctant to reveal his theological beliefs, not only because of heresy and the fear of losing his status and Fellowship at Trinity College, but also because he believed the common mind ill-prepared to receive the 'meat' of his doctrines. Thus his theological thought remained unpublished, and veiled in many layers of meaning whenever it did appear in published form, not least the *General Scholium* to the second and later additions of the *Principia*. Taken together, this approach to both natural philosophy and theology reveals the continued restriction of both realms of enquiry to the specialist or *virtuosi*. It was left to the apologists and entrepreneurs (the latter ran theatrical experimental demonstrations of the 'wonders' contained within the *Principia*) to 'democratise' Newton's work, while he himself confined truth to a private realm distinct from the general public, and expressed himself in an increasingly specialised language. One can see that this leads to a division between a 'common and vulgar' or public understanding of such concepts as motion, while the real understanding of nature was confined to the 'priest–scientists'.[31]

This drama of obscurantism, then, was the theological context and initial metaphysical basis from which Newton's understanding of motion arose and came to be expressed in the *Principia*. Turning to the opening *Scholia* of this text in which Newton lays the philosophical basis of his universal mechanics, it will be seen that Newton held a view of a marginalised God in such a way that motion becomes not the means of participation in the divine life, but a purely incidental aspect of creation which precipitates a universe of conflictual *stasis*.

Motion in the *Principia*

Newton identifies the purpose of his *Principia* as follows: 'to determine true motions from their causes, effects, and apparent differences, and, conversely, of how to determine from motions whether true or apparent, their causes and effects. For this was the purpose for which I composed the following treatise'.[32] The first two books of the *Principia* are designed to form a written systematic presentation of dynamics, the laws of which are supposed to yield the 'system of the world' outlined in Book Three. Newton distinguishes between the real and the apparent, the absolute and the relative, the mathematical and the common. His method is not one of mere observation, for such observation implies a relativity based on the standpoint of the observer. For Newton, reality is deeper than experience or any agreement among scientists could establish, hence his total commitment to the incorporation of exact geometry within mechanics and experimental practice. The visible realm for Newton provides a mere indication of that which lies beneath as the substructure of the phenomenologically given.

Having established his aim and the purpose of natural philosophy as the investigation of the forces of nature 'from the phenomena of motions ... and then from these forces to demonstrate the other phenomena', Newton begins with eight definitions: the quantity of matter, the quantity of motion, inertia, impressed force, centripetal force and finally the three measurements of centripetal force. These definitions develop from the simple quantity of matter to that which measures not only this quantity but the quantity of every movement and change in the universe. It is important to note that, at this stage, Newton has only defined inertia and force. Matter, motion and centripetal force have not been defined, only their measure has been identified. Newton comments that 'time, space, place and motion are very familiar to everyone'.[33] However, he does wish to distinguish absolute space, as well as absolute, true and mathematical time, place and motion, in order to remove 'certain preconceptions' which arise from the common understanding of these categories only in relation to sense perception.

Beginning with the definition of absolute, true and mathematical time, it can be seen that Newton separates time from any physical motion:

> Absolute, true, and mathematical time, in and of itself and of its own nature, without relation to anything external, flows uniformly and by another name is called duration. Relative, apparent, and common time is any sensible and external measure (precise or imprecise) of duration by means of motion — for example, an hour, a day, a month, a year — is commonly used instead of true time.
>
> (Newton, *Principia Mathematica*, p. 408)

Although motion is the measure of time, there may not be a sufficiently perfect motion by which time can be measured. All motion could be accelerated, but the flow of absolute time is unchanging. The duration of the existence of things remains the same regardless of motion. Duration, or absolute time, would remain in a universe of no motion. Therefore, Newton distinguishes duration from that which is merely its measure, namely motion. Yet even if a perfectly uniform periodic motion were possible which would thereby form an exact measure of time, it would not itself be absolute time, for absolute time is not reducible to any of its observable measures, however accurate. Similarly, absolute space, in contrast to relative space, is 'without relation to anything external' and 'remains homogeneous and immovable'.[34] Although absolute and relative spaces are identical in species and magnitude because, in contrast to motion, space is homogeneous, they do not remain numerically identical at all times. Newton provides the following example: a space of air within the atmosphere of the moving earth does not move relative to the earth and yet it does move relative to absolute space due to its participation in the earth's motion.[35] However, at any one moment in which motion is absent from the earth the species and magnitude of the space of air on the earth are identical to its absolute space. Absolute space is mathematical and the true location of a body in relation to absolute space is its absolute location or absolute place. Therefore, place or location can similarly be distinguished into absolute and relative.

It is at this point that Newton introduces the distinction between absolute and relative motion: 'Absolute motion is the change of position of a body from one absolute place to another; relative motion is change of position of one relative place to another'.[36] He uses the example of a sailor on a ship to describe the way in which the motion of the sailor is the vector sum of three motions: the sailor's motion relative to the ship, the ship's motion relative to the earth, and the motion of the earth relative to absolute space. Thus, if the earth is assumed to be at absolute rest (that is, not moving through absolute space), then the sailor who is at relative rest in relation to the ship will nevertheless move truly and absolutely through absolute space with the velocity with which the ship is moving across the earth, namely by participating in the motion of the ship. Later in the *Scholium* Newton argues more specifically that

> whole and absolute motions can be determined only by means of unmoving places and relative motions to moving places. Moreover, the only places that are unmoving are those that all keep given positions in relation to one another from infinity to infinity and therefore always remain immovable and constitute the space that I call immovable.
>
> (Newton, *Principia Mathematica*, p. 412)

Newton states that, without the need to resort to sense perception, absolute and relative rest and motion are distinguished from each other because they have logically different properties, causes and effects.[37] It is important to note that throughout the *Scholium*, Newton is not searching for means to identify instances of absolute time, space and motion in order to justify empirically their distinction from the relative varieties. Absolute time, space and motion are logical axioms within the dynamic system of the *Principia* and Newton is only concerned to show how these absolute categories are distinguished from the relative and also how they operate as interpretative tools within his schema. Therefore, he is anxious to show that absolute motion and rest are not simply special or particular instances of relative motion and rest. After pointing out that certain properties of absolute motion and rest can be adopted as predicates within his overall scheme of demonstrating their logical differentiation from relative motion and rest, Newton proceeds to the second means of making this differentiation, namely by their causes. He states that, 'True motion is neither generated nor changed [that is, caused] except by forces impressed upon the moving body itself, but relative motion can be generated and changed without the impression of forces upon this body'.[38] Newton is here describing the kind of relative motion which one experiences when an adjacent train leaves the platform. No force is impressed upon the train in which one sits, yet one appears to be moving and indeed one is moving relative to the train which is leaving the station. True motion, however, has no such cause and is, by definition, only the result of a force impressed upon the body in question. The difficulty, however, lies in identifying exactly which body is the subject of an apparently non-perceptible force.

For these reasons, Newton must complete a distinction between absolute and relative motion and rest with particular reference to their effects. It is here that he introduces the famous experiment of the rotating bucket and begins by stating that,

> The effects distinguishing absolute motion from relative motion are the forces of receding from the axis of circular motion. For in a purely relative circular motion these forces are null, while in true and absolute circular motion they are larger or smaller in proportion to the quantity of motion.
> (Newton, *Principia Mathematica*, p. 412)

For example, if a bucket of water is attached to a coiled cord (see Figure 1), the bucket will rotate when released. At first the bucket moves relative to the water which remains at rest. Gradually the motion of the bucket is imparted to the water and eventually, when the water is rotating at the same rate as the bucket, the bucket and the water are at rest relative to one another. However, as the water rotates it will take on a concave shape as it recedes from the axis of rotation.

Here Newton remarks that,

> The rise of the water reveals its endeavour to recede from the axis of motion, and from such an endeavour one can find out and measure the true and absolute circular motion of the water, which here is the direct opposite of its relative motion.
> (Newton, *Principia Mathematica*, p. 413)

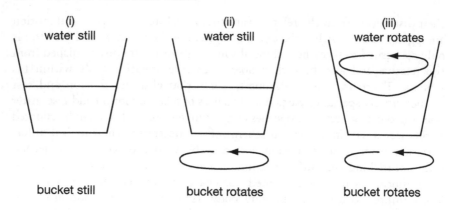

Figure 1 Newton's experiment of the rotating bucket.

It is the effect of the recession of the water from the axis of rotation shown in the concave shape of the water that, for Newton, demonstrates the occurrence of true motion: the extent of the recession of the water from the axis is positively correlated with the quantity of absolute circular motion. It must be noted once again, however, that Newton is not here proving the existence of absolute motion, for his whole inertial system presupposes this absolute frame of reference. Rather, he is merely using a thought experiment to illustrate that in this instance absolute motion is distinguished from relative motion by an endeavour to recede from its axis of circular rotation.[39] Thus when the water remained flat, motion belonged properly to the bucket and only relatively to the water. Once the bucket imparts its motion to the water, the water is then able to reveal that it is the subject of a motive force, and therefore possesses absolute motion, by the effect of its recession from the rotational axis.

The next stage in Newton's *Scholium* is to remark that, 'It is certainly very difficult to find out the true motions of individual bodies and actually to differentiate them from apparent motions, because the parts of immovable space in which the bodies truly move make no impression on the senses'.[40] It seems that absolute motion and rest occur relative to immovable absolute space, but there is no way of detecting absolute space. However, all is not lost as Newton introduces the case of two revolving globes which are attached to the ends of a cord and rotated at their common centre of gravity. As the globes attempt to recede from their axis of rotation a tension will be placed on the cord. This tension can be measured and is the indication of true and absolute motion. Once again, Newton has defined absolute motion by identifying bodies which are the subjects of motive force. The cord which ties the two rotating globes, like the endeavour of recession of the water in the case of the revolving bucket, reveals the centripetal motive force which results in the recession from the axis of rotation. However, in the case of the rotating globes it is also possible to tell in which absolute direction they are rotating. Initially, one places an equal force on alternate faces of each globe to increase or decrease their motion, this being indicated by an increase or decrease in the tension of the cord which holds them together. This would indicate which faces of the globes must be the subjects of a motive force for a maximum increase in motion, 'that is, which were the posterior faces, or the ones that are in the real circular motion'.[41] This establishes both the quantity and direction of the motion, clockwise or anticlockwise, without relation to any other body. It seems that Newton has potentially found a way of identifying whether or not a single body in an empty universe is rotating or not.

These experiments have, however, been the subject of considerable criticism because it seems that one can account for the phenomena in a number of ways: there are different 'readings' of the experiment. It remains open to question whether absolute rotation – that which is relative to absolute space – is directly causally related to the observed forces. It may be the case that the forces exerted on the water are not caused by motion *relative* to the bucket, yet this does not mean that they are independent of *all* relative motion. Ernst Mach levelled such a criticism, suggesting that one still requires a relative context for motion. He argued that in the case of the bucket and globe experiments Newton does not view motion in relation to adjacent or contiguous bodies, but the fixed stars. Even if one could calculate how bodies moved in the absence of the fixed stars, there is no way of guaranteeing that Newton's laws of motion hold in such a scenario. Mach is pointing out that motion is contextual, and in being so will always contain a crucial 'relative' element.[42] In any complex system, namely one which is not an 'idealised' motion of a body or bodies in a vacuum, there will always be competing possible explanations of such phenomena. Mach is employing the principles of a thoroughgoing empiricist, arguing that Newton's scenarios simply do not perforce pertain; there is no way of making the empirical observations because, by necessity, we can only ever experience relative motions, namely motions which are relative to us. The introduction of an observer crucially alters, influences and makes more complex the phenomena in question.

However, as was stated above, and as S. Toulmin and R. Rynasiewicz have argued at length,[43] Newton is not attempting to prove by observation the existence of absolute space and motion, nor that the centrifugal forces on the water and the globes are caused by rotation relative to absolute space. The centrifugal forces which are witnessed in the recession of the water and the extension of the cord holding the globes are what already define absolute rotation. Newton has defined this theoretical element of his inertial system by demonstrating how it is to be observed and measured. He is not taking an observation and explaining its cause. Rather, he is taking a theory or series of axioms concerning absolute space and motion and demonstrating how certain phenomena might define those theoretical quantities. Therefore, as long as Newton can demonstrate that the universe obeys his axioms and laws, that is, his theory has an identifiable empirical content, one can postulate absolute motion and space. Yet on Mach's criticism, it seems doubtful that there could be a strictly empirical content to a physics which defined absolute space, time and motion, because observation is necessarily relative.

Having discussed these opening propositions within Newton's *Scholium*, I now turn to examine in more detail the concepts which appear to underlie his understanding of motion. It can be seen from the above discussion that 'force' and 'space' are crucial elements in the delineation of true motion and rest, either as their cause, effect or determinate context. Before considering Newton's understanding of absolute space and its theological import as the famed *sensorium dei*, one may ask: what is this 'force' which is so crucial to Newton?

Motion and force

In Definition IV of the opening of the *Principia*, Newton states that, 'Impressed force is the action exerted on a body to change its state either of resting or of moving uniformly straight forward'.[44] Michael Buckley remarks that 'Seventeenth and eighteenth century mechanics tended to think of force as a characteristic of a body in motion. But the Newtonian mechanics offer "quite a different definition of force, an external action on a body producing a change in motion"'.[45] Newton has therefore identified absolute motion with reference to force which produces a change in motion or rest, this leading to his famed first law of motion: 'Every body perseveres in its state of being at rest or of moving uniformly straight forward, *except insofar as it is compelled to change its state by forces impressed*'.[46] This first law is known as the principle of inertia and was described by Herbert Butterfield as 'the great factor which in the seventeenth century helped to drive the spirits out of the world and opened the way to a universe that ran like a piece of clockwork'.[47] There are three aspects to this principle of inertia which have been thought to distinguish it from the medieval account of motion. First, motion is simply a state, and rest is describable merely as motion reduced numerically to zero. That is, motion and rest are quantitatively different instances of the same state. Secondly, if a body is in a state, it does not at the same time undergo a change. Motion and rest are states which express basic facts about the nature of a body requiring no further explanation. This distinguishes Newton's account of motion from that of Aquinas and Aristotle given earlier: for the latter two, motion constituted a constant change of state in such a way that motion could never be a primitive fact requiring no further reflection. Finally, the explanation of any change in a given body is given by reference to a force acting upon that body to change its state of motion or rest. It is assumed that a body will continue in its state of rest or motion continually in a straight line unless it is the subject of a force which changes that state. Therefore, a corollary of this principle of inertia is that Newton's *Principia* is not so much concerned with motion *qua* motion as with the forces which change the state of motion.[48]

However, J. E. McGuire has suggested that the principle of inertia has its roots in a certain interpretation of the Aristotelian position which he traces in the work of medieval and early modern writers including Jacopo Zabarella, Nicole Oresme and Duns Scotus.[49] The proposed connection is dependent upon ascribing to Aristotle the view that nature is an efficient cause of the motion of bodies. This has been discussed at length elsewhere in this essay, but in order to understand the connection with Newton's thought it is necessary to refer to Definition III of the introduction to the *Principia*. Here Newton makes a statement that seems to anticipate and clarify the principle of inertia as expressed in the first law of motion: 'Inherent force of matter (*vis insita materiae*) is the power of resisting by which every body, so far as it is able, perseveres in its state either of resting or of moving uniformly straight forward'.[50] Newton goes on to explain that this 'force of matter' is both a force of resistance to a change in motion (a *vis inertiae* or 'force of inactivity') but also an impetus:

> resistance insofar as the body, in order to maintain its state, strives against the impressed force, and impetus insofar as the same body, yielding only with difficulty to the force of a resisting obstacle, endeavours to change the state of that obstacle.
>
> (Newton, *Principia Mathematica*, p. 404)

On the one hand, bodies have a passive resistance to change (*vis inertiae*), while on the other hand they have an impetus to resist the change imposed by another body by changing the state of that body (*vis insita*).[51] Within the interpretative tradition outlined by McGuire, these principles of resistance and impetus are linked to Aristotelian motion in a fashion which ignores any differentiation between natural and violent motion. With reference to the beginning of Book 2 of the *Physics*,[52] it is suggested that the Aristotelian definition of natural things as those which possess 'an innate principle for change' may be interpreted as in some way equivalent to Newton's innate impetus or force outlined in Definition III. It is proposed that Aristotle refers to nature not only as a 'source of change and nonchange, but also as a *cause* of change that is in the thing it changes *per se* and not accidentally'.[53] Thus it is maintained that 'the exercise of a natural power can be described as the operation of an internal efficient cause'.[54]

However, the association with Newtonian *vis insita* and *vis inertiae* cannot be maintained for two further reasons. First, it is not the case that for Aristotle the motion of a heavy body upwards could be due to an impetus or principle within the body '*per se* and not accidentally', for this motion is contrary to the body's nature and therefore violent in such a way that its cause must be from without. Secondly, the identification of Newton's *vis insita* and *vis inertiae* with Aristotle's principle of change fails because the former two describe motion as the outcome of competitive conflict rather than the outcome of a participative co-operation between a mover (which is more properly described as the cause of a motion) and a body's inner passive principle of change. In Aristotelian natural motion, far from being a force of resistance, the inner principle of change in a body is a 'passive force' of co-operation and enhancement of the motion imparted by a mover. Motion for Newton is therefore the mere product of a more ontologically basic resistance and competition between an external force and the internal forces of *vis insita* and *vis inertiae*. Strangely, in Aristotelian terms, all Newtonian change (that is, an alteration in the state of motion) is violent. Meanwhile, bodies themselves are merely indifferent to different motions. Because a body is indifferent to its motion, this motion can play no part in telling us anything significant about the nature of the body in question and, moreover, motion plays no part in a body's fulfilment in a *telos*. Thus motion becomes an ontologically insignificant category which does not refer to final causes. Motion is therefore not 'from something to something' or the means of actualising a potentiality as it had been for Aquinas and Aristotle, but a mere state to which a body is impartial. This cannot be the case within Aristotelian motion because here bodies are not indifferent to motion but possess within themselves a 'natural principle for change' only with regard to their natural motion which is defined not as the outcome of conflictual forces but as a co-operation between mover and moved, namely that which is the means of the attainment of their *telos* and is thus integral to their ontology.

McGuire goes further in outlining the links between a certain medieval interpretation of Aristotle and the work of Newton. Is it the case that, with reference to natural motion, Aristotle could ascribe a body's motion not to an external mover but solely to the body in question such that inanimate objects may be described as 'self-movers'? McGuire refers particularly to the work of Jacopo Zabarella (1533–1589), an ardent interpreter of Aristotle who was Professor of Philosophy at the University of Padua. According to Zabarella in his *Liber de natura* (which forms part of the treatise *De Rebus Naturalibus*), Aristotle's principles do allow that elements can be described as self-movers for 'Aristotle does not deny absolutely that heavy and light things are moved of themselves, but he denies that they are moved of themselves in the way in which animals are moved of themselves'.[55] As was argued earlier in this essay, if any body is to be regarded as self-moving in an Aristotelian sense then some distinction must be made between a mover and something that is moved while both are contained in the one being.[56] This distinction between mover and moved is understood by Zabarella through a body's composition of form and matter. Thus the form of heaviness would be the mover in the fall of a heavy object. However, for Aristotle an inanimate body only possesses the passive, rather than the active, principle of its own motion. The form of heaviness in a body is not therefore the active source of the motion of the body, but the passive principle of motion, namely the ability to be the recipient of downward motion. The form of heaviness in an inanimate body cannot be the mover which moves the body. Rather, the generator of the body is the mover (or in a weak and secondary sense, that which removes any impediment to such a motion), because the generator supplies the body with its nature and thereby all that is required for its natural motion.[57] Moreover, Aristotle preserves a distinction between animate and inanimate by arguing that properly animate bodies determine their own motion and thus display the possibility of a variety of motions. Inanimate bodies, however, have their motion fully determined.

These attempts to draw connections between Newton's *vis inertiae* and *vis insita* and a particular interpretation of the Aristotelian tradition within late medieval natural philosophy reveal two aspects of the transition from the properly classical to the early scientific and modern understanding of motion. First, through the collapse of the distinction between its natural and violent varieties, motion and rest for Newton become mere states to which bodies are indifferent. Secondly, because motion and rest are merely primitive but numerically different states, Newton's *Principia* focuses primarily on force as that which changes a state of motion or rest. As has been seen, change is understood not as the result of co-operation and participation of motions but as the outcome of violent forces. Moreover, the principles of *vis insita* and *vis inertiae* blur the distinction between self-moved beings and those which are inanimate in such a way that all things can be said to be in some sense 'self-moving'.[58] Newton therefore describes not a hierarchy of motions which can stretch up to participate qualitatively in a transcendent source of motion, but rather a flattened, extended, quantified and monadic universe of discrete objects whose motion does not require any explanation by reference to other beings. This is well expressed by McGuire:

> Newton ... conceives natural motion along a right line as a physically possible condition of bodies that could and would obtain in the absence of opposing forces. In other words, the *vis insita* of bodies is an *absolute* feature of what they are, one that does not depend on their relationships to one another.
>
> (McGuire, 'Natural Motion and Its Causes', p. 327)[59]

Having outlined the importance of force in Newtonian motion, it is now possible to assess the context in which absolute motion can be said to take place, namely absolute space. As an initial step in the consideration of the importance of absolute space in Newtonian motion, it is necessary to consider whether, given the *vis insita* and *vis inertiae* of a body and the fact that Newton has described a universe of discrete objects which do not require relationality in the explication of their ontology, God has any purpose within a universe of such motions.

Absolute space, Christ and motion

Newton's voluntarist Lord God of Dominion as described above was utterly remote and transcendent. This concept of the divine fitted neatly with the notion of a universe filled with discrete objects whose particular motion required no reference to a relation with any other being. Interaction between discrete objects constituted change brought about by conflictual forces. Through the natural resistance to change possessed by bodies, the universe exhibited some degree of stability and changelessness, this being a reflection of the divine nature itself.[60] However, this left a theological gap for Newton which was somewhat unpalatable: how was he to describe a mode of divine action within such a world so as not to make God incidental to cosmology? He gave two apparently different answers to this question. The first, in typical Arian fashion, saw the divine as utterly remote and acting through Christ as an intermediary. God and Christ were not one in substance, but one in unity of will and dominion. Newton states that,

> If any think it possible that God may produce some intellectual creature so perfect that he could, by divine accord, in turn produce creatures of a lower order, this so far from detracting from the divine power enhances it; for that power which can bring forth creatures is exceedingly, not to say infinitely, greater.
>
> (Newton, *Unpublished Scientific Manuscripts*, pp. 108 and 142)[61]

On this view, Christ is understood as the 'viceroy' of God, putting into action the dictates of the divine will. The second means of divine action, however, is direct within absolute space. McGuire has argued that this latter form of divine action shows that Newton's Arianism was limited in its effect upon his cosmology.[62] However, I will suggest that the latter notion of divine action is also the result of Newton's Arianism and that this conception of God reinforces the understanding of motion discussed above. It will then be possible to see how Newton's voluntarism and Arianism fit with the view of motion outlined above.

Absolute space is the context and basis for motion in Newton's universe. He outlined his notion of space in *De Gravitatione et Aequipondio Fluidorum*, a treatise which was to form the basis of many arguments in the first edition of the *Principia*.[63] Newton explains that space is neither substance nor accident, but rather 'an eminent effect of God, or a disposition of all being'.[64] Space is ultimately characterised as extension. We are able to abstract 'the dispositions and properties of a body so that there remains only the uniform and unlimited stretching out of space in length, breadth and depth'.[65] This space extends infinitely in all directions and it is a 'disposition of being *qua* being', for 'no being exists or can exist which is not related to space in some way'.[66] It is also 'eternal in duration and immutable in nature, and this because it is the emanent effect of an eternal and immutable being'. In a fashion which appears to consider space as 'begotten' of God, Newton explains that, 'If ever space had not existed, God at that time would have been nowhere; and hence either he created space later (in which he was not himself), or else, which is less repugnant to reason, he created his own ubiquity'.[67] Finally, it is explained that the infinite extension of space includes also an infinite and integral pattern of imperceptible geometrically defined solids which differ in size and shape. Thus space has the capacity to receive within itself any size or shape of bodies which, unlike space, are compositions of substance and accident. In contrast to space, the bodies exist as effects of the sovereign will of God. Space, therefore, is the recipient of the effects of the divine will. It is by divine volition that regions of absolute space are inscribed with materiality, thus forming bodies. Newton therefore states that, 'If there are bodies, then we can define bodies as determined quantities of extension which omnipresent God endows with certain conditions'.[68] Those conditions are motion, the inability for two bodies to fill one absolute space and the ability to excite various perceptions of the senses. Newton concludes that extension and the divine will are sufficient for the existence of a material universe. This is further explicated in Query 31 of the 1717 and later editions of the *Opticks* in which Newton notoriously described space as God's 'boundless uniform Sensorium', a description which provoked the mocking criticism of Leibniz.[69] Newton states that,

> [the intricate aspects of the parts of animals] can be the effect of nothing else than the Wisdom and Skill of a powerful ever-living Agent, who being in all Places, is more able by his Will to move the Bodies within his boundless uniform Sensorium [that is, absolute space], and thereby form and reform the Parts of the Universe, than we are by our Will to move the Parts of our own Bodies. ... And since Space is divisible in infinitum, and Matter is not necessarily in all places, it may be also allow'd that God is able to create Particles of Matter of several Sizes and Figures, and in several Proportions to Space, and perhaps of different Densities and Forces, and thereby to vary the Laws of Nature, and make Worlds of several sorts in several Parts of the Universe. At least, I see nothing of Contradiction in all this.
>
> (Newton, *Opticks*, pp. 403–404)[70]

Thus it can be seen that in the absence of a fully divine Christ, absolute space becomes the basis of creation, forming the 'disposition of being *qua* being', for such space is 'eternal in duration and immutable in nature, and this because it is the emanent effect of an eternal and immutable being'. While space might not be literally God's sensory medium, it is difficult to avoid the conclusion that Newton has described a spatial and three-dimensional Godhead. Whereas, for Aquinas, God creates and sustains the world through Christ's emanation from the Father, so for Newton, God creates the world *in* a co-eternal and uncreated absolute space through the exercise of his will. Moreover, as T. F. Torrance comments when writing of Newton, 'If God himself is the infinite container of all things He can no more become incarnate than a box can become one of the several objects it contains'.[71]

In concluding this discussion of Newton, let us consider the implications of what we have found. It seems that absolute space coupled with the action of the divine will is the ontological precondition of all being. It is by means of co-eternal and infinite space that God is able to operate and instantiate a material cosmos. This may seem to reflect the view of divine creative action in Aquinas, for Newton understands God's action in creating the world as one of formal causation – the divine actualises in matter a form which exists eternally in the divine mind. However, whereas for Aquinas the motion of a body is itself a participation and effect of the knowledge of the body's form in the perfect 'motionless motion' of God, namely in the emanation of the Son from the Father, for Newton creation occurs through the inscrutable and arbitrary 'motions' of the divine will. This is expressed in a recent article by McGuire in which he states that for Newton,

> God does not recreate similar conditions in successive regions of space; he maintains the same formal reality in different parts of space through a succession of times. In this way the continuity of motion is the real effect of God's motion.
>
> (McGuire, 'The Fate of the Date', p. 282)[72]

Yet what divine motions can these be within Newton's Arian voluntarism? They can only be the motions of an arbitrary and inscrutable divine will. Whereas, for Aquinas, the 'motionless motion' of the divine emanation was able to provide the ontological basis and goal of all motion, for Newton, who has already discounted the possibility of relationality within the Godhead, motion can only be the effect of the imposition of divine volition. The lack of Trinitarian relationality in Newton's conception of God means that the universe cannot be thought of as a hierarchy and system of related motions which are images of the divine life, but rather as the action of one being, God, within absolute space to instantiate a material body, whereupon the created being retains a primitive state of motion which is discrete and self-explanatory. There is a stark contrast with the cosmology outlined in the opening chapters of this essay. In the work of Plato, Aristotle, Grosseteste and Aquinas, one finds motion analogically ascending the created hierarchy to link created being with eternal being. Newton, whose work constitutes a significant moment in the protracted sundering of natural philosophy from metaphysics and questions of theology, describes a more discrete cosmology and, consequently, a more autonomous physics. It is therefore not surprising that as Newtonian science became the basis of enquiry into nature and motion in the decades following Newton's death, the role of God came to be seen as incidental to any explanation of the cosmos in natural philosophy.

Having described Newton's understanding of motion, force and divine action, in the final section of this chapter I will describe briefly the ambivalence which existed concerning mechanistic construals of nature and motion even within Newton's own work and amongst his contemporaries. It seems that the beginning of the demise of the strict Newtonian physics was present even in the moment of its birth and triumph at the end of the seventeenth century. A brief description of some of the central themes of physics in the nineteenth and early twentieth centuries will suggest that science was not bound to proceed along a Newtonian path. I will conclude with a short analysis of the mysterious phenomena presented by the most recent developments in physics, namely quantum theory, and make some speculative suggestions concerning the possibility of a return to a more traditional ontology as the basis for understanding this science of moving bodies.

The fate of mechanistic motion

Throughout his career, Newton maintained a certain ambivalence concerning the status of mechanical accounts of the natural phenomena that he described in the *Principia*. As Alan Gabbey has pointed out, Newton's early alchemical work was not conceived in mechanistic terms.[73] It was not until the early 1670s that Newton began to combine mechanistic explanation with alchemy.[74] In a paper read to the Royal Society in 1675 entitled 'An Hypothesis explaining the Properties of Light discoursed of my severall Papers', Newton sought to explain light by means of an 'aethereall Medium' similar to air, but much rarer and more elastic.[75] As Gabbey points out, this hypothesis had to explain 'a wide range of phenomena, such as surface tension, the cohesion of solids, animal motion, the phenomena of static electricity and magnetism', and, of course, gravity.[76] How might the aether, which Newton was to describe in the *Principia* as 'a certain very subtle spirit pervading gross bodies and lying hidden in them', explain gravity?[77] In a paper composed at the same time as the *Principia*, Newton states that, beyond the observable motions of the cosmos, there are also innumerable motions among hidden particles. He writes that, 'If any one shall have the good fortune to discover all these, I might almost say that he will have laid bare the whole nature of bodies so far as the mechanical causes of things are concerned'.[78] Newton goes on to state that, as there are forces of attraction and repulsion between larger bodies, for example magnetism, electricity and gravity, so there is no reason to suppose that such forces arise from similar forces existing between lesser bodies. This is to say that gravity is not 'action at a distance', but is a force mechanically transmitted between bodies by lesser bodies, ultimately through an insensible aethereal medium.

However, Newton was very aware that any mechanical explanation of forces such as gravity presented a particular difficulty: there appears to be an infinite regress. If the aether is material (as Newton supposed it to be), what is the cause of the interaction between its particles? Is this to be explained by *another* aether, yet more subtle and rare? Are the aether's actions material or immaterial? Gabbey points to Newton's ambivalence concerning this issue.[79] On the one hand, in Newton's treatise 'Of Natures obvious laws & processes in vegetation', there is a mechanical interaction underlying all alchemical activity which is consonant with his understanding of motion caused by impact. On the other hand, in 'An Hypothesis explaining the Properties of

Light', Newton writes of non-material 'secret principles of (un)sociableness' that account for the repulsion and attraction between bodies. Even within this latter treatise, there is some doubt in Newton's mind:

> God who gave Animals self motion beyond our understanding is without doubt able to implant other principles of motion in bodies wch we may understand as little. Some may readily grant this may be a Spiritual one; yet a mechanical one might be showne, did not I think it better to passe it by.
>
> (Turnbull *et al.*, *The Correspondence of Isaac Newton*, vol. 1, p. 370)

The influences on Newton were numerous as he developed his views concerning gravity, the aether and mechanical causation. Prominent amongst these were the moderate Puritan Cambridge Platonists, including Henry More (1614–1687) and Ralph Cudworth (1617–1688).[80] Although More, along with other Cambridge Platonists, attempted to interpret and, with subtle nuances, assimilate the new Cartesian philosophy within a broadly Platonic philosophy (in particular, the notion that there was something existent other than brute matter), as J. E. McGuire and P. M. Rattansi claim, by the end of the 1650s More was finding difficulties in the mechanistic natural philosophy.[81] In common with Newton, he believed Cartesianism to lead to atheism. Cudworth, in his *True Intellectual System of the Universe*,[82] proposed a much more delicate and less dualistic approach which was also very different to the prevalent mechanistic philosophy. He emphasised a hylarchic principle known as 'plastic nature', a formative agency which acts as an intermediary between God and nature. This concept is very similar to the World Soul in Plato's *Timaeus*. Cudworth's 'plastic principle' is a spiritual agent, although it operates unconsciously in performing the directives of God with immediacy. For Cudworth, plastic nature is the means whereby God acts within creation, informing the cosmos with his wisdom and goodness and, like the World Soul, rendering the cosmos intelligible.

In postulating plastic nature, Cudworth was treading a delicate line between, on the one hand, the hylozoism of thinkers such as Spinoza, and, on the other hand, a purely mechanistic natural philosophy and occasionalism. For Cudworth, God was immanent in the plastic principle of nature, yet nature maintained a certain autonomy in which the concept of natural motion, allied with a system of final as well as efficient causes, was preserved. Therefore, unlike Descartes, for whom matter is mere *extensio*, Cudworth's matter is no longer simply inert: through plastic nature there is an inner source of motion to all things as well as genuinely ensouled and animate beings. God is not, as for Newton, the cause of all non-mechanical occurrences in nature, because such events seem often to be arbitrary or absurd. For a voluntarist such as Newton, this understanding of divine action merely served to underline the freedom of the divine will. For Cudworth, God's action within the cosmos was less capricious and, at the same time, mediated through the constancy of the principle at the very heart of creation, namely plastic nature. Indeed, Cudworth stresses, against the voluntarists, in *A Treatise Concerning Eternal and Immutable Morality*, that God's action in the world is itself subject to the principle of non-contradiction.[83]

Already in the work of a contemporary, Cudworth, who bore some influence on Newton's thought, one finds an alternative view to a mechanistic universe. Clear contrasts may be drawn. On the one hand, Newton, as we have seen, tends towards a pantheism in which space is the *sensorium dei*. On the other hand, Cudworth maintains God's immanence through the mediation of plastic nature because of which there is a non-conscious dynamic aspect of the cosmos whereby all things, by nature, seek their own fulfilment. Therefore, for Cudworth, all motion is traceable to an active principle which is not found uniquely in the moving body itself: motion implies a hierarchy of causes which, via plastic nature, reach ultimately to God: 'body and matter has no self-moving power, and therefore it is moved and determined in its motion by a higher principle, a soul or mind'.[84] By contrast, for Newton motion may be idealised in such a way that the movement of a body is discrete and self-explanatory.

So the mechanical Newtonian cosmology (which Newton himself espoused only periodically in his career), which was based on an understanding of motion only as an impact between two bodies and viewed any motion as potentially discrete and idealisable, was brought into question almost from the moment of its inception. However, it was not until the early decades of the nineteenth century that more significant and decisive blows were dealt. In 1864, James Clerk Maxwell presented to the Royal Society his treatise *A Dynamical Theory of the Electromagnetic Field*.[85] This work, to which I will return shortly, drew widely on the research of others, including Michael Faraday (1791–1867) and André-Marie Ampère (1775–1836). Einstein was later to claim that, 'The greatest change in the axiomatic basis of physics, and correspondingly in our conception of the structure of reality, since the foundation of theoretical physics through Newton, came through the researches of Faraday and Maxwell on electromagnetic phenomena'.[86]

In the mid-1800s, when Maxwell proposed his unification of the theories concerning electrical and magnetic forces, physics was presented with three central questions: what are magnetism and electricity; what is the relation between them; and what is the nature of the aether? The aether, however mysterious and 'occult', was crucial for Newton in understanding the action of bodies upon one another when separated by space. Although, as was seen above, he developed his understanding of the nature of the aether and the centrality of mechanical causation, he never believed space to be a passive void through which bodies moved. It was certainly clear for Newton at a certain stage in his career that force, including gravity, was transmitted between bodies through the mechanical interaction of whatever constituted the aether.[87] As David Park points out, this general notion of an aether began to present difficulties in the early nineteenth century.[88] The French physicist and mathematician Augustin Jean Fresnel (1788–1827) discovered that light understood as a transverse rather than longitudinal wave better explained the phenomena of polarised light (the process of 'filtering' light of certain wavelengths).[89] Fresnel's discovery had stark implications because a transverse wave can only travel along the surface of, and not in, a 'liquid' such as the aether was assumed to be. If a transverse wave were to travel through the body of the aether, the aether would have to be highly elastic and solid like jelly. Yet how can solid bodies, for example the planets, move through such an aether without the aether having some noticeable effect on their motion? This was a question that had already occurred to Newton. Although an answer was not immediately forthcoming, a number of scientists, including Claude-Louis Navier (1785–1836) and Augustin-Louis Cauchy (1789–1857) developed mathematical equations for motion in a solid and elastic aether.

Attempts to explain the nature of the aether faded in the nineteenth century. A rapidly developing mathematics was deployed to explain natural phenomena, such as gravitation and electricity, within a basically Newtonian system. However, the question of the aether remained particularly important with regard to magnetism and electricity because these forces most obviously give rise to what appears to be action at a distance. What was the nature of the aether which transmitted these forces? In 1600 William Gilbert had published his treatise *De magnete* in which he argued that magnetism emerges principally from the form of an object, while electrical impulses emerge from matter. Magnetism is allied to form because it is long-lasting and acts through intermediaries such as glass or paper. By contrast, electrical impulses are weaker, fade quickly and can only be produced by, for example, rubbing amber or combing one's hair rapidly on a dry day. It was clear that magnetic and electrically charged bodies emit some kind of attractive force in the immediately surrounding area. Gilbert believed this to be an aether akin to a current of air that floats around the body.

It was not until the late nineteenth century, and the researches of Michael Faraday, that some genuine progress was made in understanding the nature of magnetic and electrical forces. According to Faraday, every atom of matter has an atmosphere of force surrounding it. This was later to be known as a 'field'. Remarkably, one could see the lines of this force surrounding a magnet when one sprinkled iron filings on paper covering the magnet. Rather than considering them merely in the abstract, Faraday postulated that these lines of force, the field, possess a physical reality: 'Without departing from or unsettling anything then said, the inquiry is now entered upon of the possible and probably physical existence of such lines'.[90] Wide-ranging experiments showed Faraday that these lines of force were curved as well as straight and that fields of force exerted an influence upon one another and thereby transmitted effects across space. Such action was, so Faraday believed, a function of the surrounding aether.[91] However, as Park states, by 1852 Faraday no longer considered these lines of force to be some kind of material substance.[92] Where mathematicians had seen centres of force acting instantaneously at a distance, Faraday's field theory postulated a mechanism of continuous transmission of force at a finite speed. This constituted a significant break with the Newtonian view of force, for force is no longer understood as a push or a pull, namely an action, but more like something which saturates the cosmos.[93] Bodies are constantly interacting with each other and, as it were, participating in each other's motions.

Faraday's findings remained largely confined to experimental reproduction and lacked the clarity and certainty afforded by mathematical expression and proof. William Thompson (1824–1907), later Lord Kelvin, was the first to give Faraday's theories and findings concerning the electrical and magnetic field theory some degree of mathematical support. He did so in a remarkable way: he suggested that there is an analogy, mediated by mathematics, between electrostatic lines of force and the lines of heat flow in thermal conduction. In the case of a bar magnet, there are opposite poles, north and south, between which lines of magnetic force pertain. Thompson realised that, if one replaces the magnet with a solid bar that is hot at one end and cold at the other, the flow of heat and cold (if one can put it that way) can be described mathematically in terms that are identical to the field of force of a magnet. The pattern of heat flow is exactly that produced by iron filings when they are placed near a magnet. Given such a result, it would be tempting, on a literalist reading of nature, to conclude that magnetism is a kind of heat. But Thompson drew no such conclusion. Rather, he maintained that there is an analogy. Faraday, commenting on Thompson's views, stated

> Professor W. Thompson, in referring to a like view of lines of force applied to static electricity, and to Fourier's law of motion for heat, says that lines of force give the same mathematical results of Coulomb's theory ...; and afterwards refers to the 'strict foundation for an analogy on which the conducting power of a magnetic medium for lines of force may be spoken of'.
>
> (Faraday, *Experimental Researches in Electricity*, p. 831)

The power of analogy in scientific reasoning was to be developed by James Clerk Maxwell. He postulated an analogy between the magnetic field and gas moving through a resisting substance (for example, tightly bound steel wool).[94] In an important paper written when Maxwell was aged just 25, he used analogies between different sciences to explain how the ideas and methods of Faraday might be made mathematical. Maxwell stated that, 'In order to obtain physical ideas without adopting a physical theory, we must make ourselves familiar with physical analogies. ... A partial similarity between the laws of one science and those of another ... makes each of them illustrate the other'.[95] Most crucially, Maxwell was to bring the laws of electricity and magnetism, which seemed to be analogously related but hitherto treated quite separately, into a single theory of electromagnetism. In his treatise *A Dynamical Theory of the Electromagnetic Field*, he combines the mathematical equations which 'connect electric field with electric charges, magnetic fields with electric currents, and also those that connect electric and magnetic fields with each other'.[96] There is no mention of experiment (Faraday had conducted much of this work) and, although Maxwell was adamant concerning the existence and importance of the aether, there is no speculation concerning its nature. Rather, there is an emphasis on the relations between electricity, magnetism and light. Maxwell suggests that these three phenomena are all varieties of electromagnetic wave which move through space as a wave moves across the sea. In comparing the similarity in the value he obtained for the velocity of light when using electricity and magnetism compared with that of Foucault who considered only visible light, Maxwell states that, 'The agreement of the results seems to shew that light and magnetism are affections of the same substance, and that light is an electromagnetic disturbance propagated through the field according to electromagnetic laws'.[97]

Maxwell's general approach in scientific investigation can be found in a revealing short essay written in 1856 entitled 'Are there Real Analogies in Nature?'[98] Here Maxwell argues that analogous relations are indeed real in nature and that they cannot be reduced to a kind of literal equivalence, what he calls 'transformed identities'.[99] So, for example, one cannot, in Hobbesian fashion, reduce all phenomena to matter in motion and seek to understand nature by mastering the fundamental laws of physical motion.[100] This would not be a view of the cosmos based upon analogous relations, but one based upon a reductive literal equivalence. Maxwell goes on to state his view of analogy:

> Whenever they [observers of nature] see a relation between two things they know well, and think they see there must be a similar relation between two things less known, they reason from the one to the other. This supposes that although pairs of things may differ wildly from each other, the *relation* in the one pair may be the same as that in the other.
>
> (Maxwell, 'Are there Real Analogies in Nature?', p. 242)

Although this clearly depends upon an analogical model, it should be noted that the relation in question is one of equivalent proportion, rather than a reappropriation of Aquinas's modification of Aristotle's *pros hen* predication, or analogy of attribution. Such an understanding of analogy lends itself easily to a reduction to a univocal, mathematically identifiable proportion between pairs of phenomena.

However, there can be no doubt that analogy is deployed within the process of scientific reasoning and not merely in the description of an already established set of phenomena. In the discussion above, it was seen how

Thompson saw an analogy between the lines of force around a magnet and the transference of heat energy, yet this is not to say (after the fashion of a 'transformed identity') that magnetism is a form of heat. A sense of analogous, but not literally identical, causes in nature is revealed when Maxwell, considering motion in relation to human activity, writes that,

> Some of our motions arise from physical necessity, some from irritability or organic excitement, some are performed by our machinery without our knowledge, and some evidently are due to us and our volitions. Of these, again, some are merely a repetition of a customary act, some are due to the attractions of pleasure ... and a few show some indications of being the results of distinct acts of the will. Here again we have a continuation of the analogy of Cause.
>
> (Maxwell, 'Are there Real Analogies in Nature?', p. 240)

In the midst of the writings of one of the originators of the 'revolution' which took place in physics in the early twentieth century, one finds a mode of scientific reasoning which is more akin to pre-Newtonian natural philosophy than a reductive mechanistic physics. However, it was to be the advent of another science, emerging parallel to electromagnetism, which was to call into fundamental question the classical, mechanistic Newtonian world. The term 'thermodynamics' – claiming to be a new science unifying physics and chemistry – was first used in 1854 by William Thompson. In 1867, Thompson and Peter Guthrie Tate (1831–1901) published their *Treatise on Natural Philosophy* in which they proposed a science of the dynamics of energy rather than the Newtonian dynamics of force (thermodynamics was then known as 'energetics'). One particular way in which the science of thermodynamics sought to unify physics and chemistry while at the same time establishing a break with Newtonian mechanistic science lay in circumventing the distinction between reversible and irreversible phenomena. At the beginning of the nineteenth century, it was thought that physical phenomena, for example the motion of bodies, light, heat and electromagnetism, were reversible and, in principle, perfectly predictable using mathematical analysis of atomic and molecular units. By contrast, chemistry was characterised by laws of constant and identifiable proportions between elements, along with specific atomic and molecular weights. The chemical processes which were studied were irreversible: once a gas had been produced and dissipated, the reaction was complete and the results could be analysed. As M. J. Nye notes, for chemistry, issues of velocity or time were of no particular concern because the chemical reaction was not going to proceed any further. A point of completion and rest could be established.[101]

Nye goes on to note that, by the mid-nineteenth century, the notions of reversibility and irreversibility were no longer defining the boundary between physics and chemistry.[102] Mathematical studies of the behaviour of gases, which introduced notions of probability that were to lead to the

indeterminacy of quantum theory, marked a clear break from the mechanistic, determinate universe of Newtonian physics. It was realised by
Thompson, Rudolf Clausius (1822–1888) and others that energy cannot be
transferred back into an earlier state without any loss. Meanwhile, in chemistry the importance of the rate of chemical change introduced time as a
factor in understanding chemical changes, along with temperature, mass and
pressure. Reactions were no longer regarded as pertaining exclusively to
static equilibrium, but now constituted an on-going redistribution of chemical reactants which could, in some instances, be reversed.

One aspect of the science of thermodynamics is of particular interest. By
1914 three laws of thermodynamics were established: the internal energy of
an isolated system is constant; a spontaneous change is accompanied by an
increase in the total entropy of the system and its surroundings; and
Nernst's heat theorem which entails that the entropy of a pure perfect crystal
at a temperature of zero kelvin is itself zero. The second law states that
energy tends to flow from being concentrated in one place to becoming
diffused or dispersed or spread out. It hardly seems fanciful to point to a
similarity with Grosseteste's cosmogony as elucidated in the third chapter of
this essay. On this view, the cosmos emanates and dissipates from a single
point of light to form a sphere in which all things are more or less rarefied
forms of light. Each level of the cosmological hierarchy is analogically
related to the others, and finally to the source of the cosmos itself. Similarly,
the science of thermodynamics describes the 'natural' dissipation or motion
of energy away from concentration. Very different events – for example, the
explosion of a tyre or the heating of a pan of water – possess particular causes
but also a common and more general (and therefore more potent) cause in
the natural dispersal of energy. In this case, motion is not indifferent to its
direction and is not straightforwardly reversible. Rather, the dispersal of
energy described in the second law of thermodynamics indicates what had
already been hinted at in the cosmologies of Plato, Aristotle, Grosseteste and
Aquinas: that is, the motions of the universe tend in particular directions
and are characterised by diffusion. For Grosseteste, all things are a more or
less rarefied form of spontaneously dissipating light. For Thompson and
Maxwell, all things are a more or less rarefied form of spontaneously dissipating energy. Yet the second law of thermodynamics is not suggesting the
demise of the universe towards a state of stasis when all energy is dissipated.
Likewise, Aristotle's physics of bodies moving to their natural place does not
suggest that the universe will eventually come to stasis when all bodies have
reached their natural place. Rather, both theories recognise the complex relationality of the cosmos, namely the co-operative interaction between its
many systems in such a way that the exchange and motion of energy is
perpetual throughout nature.

In the early twentieth century, the sciences of electromagnetism and thermodynamics were to give rise to a new type of mysterious physics known as
quantum theory. The term 'quantum' relates to the hypothesis that energy is

parcelled into discrete 'quanta'.[103] This view was proposed by Max Planck (1858–1947) in order to account for the different ways in which bodies emit electromagnetic radiation. Following the researches of such scientists as Niels Bohr (1885–1962), Ernest Rutherford (1871–1937) and Albert Einstein (1879–1955), the new physics was to develop into a cosmology which presented very unexpected phenomena, amongst them a strange view of motion which is at odds with the mechanistic philosophy of a previous century. In concluding this discussion of physics and motion after Newton, I turn now to consider just one of the strange paradoxes of quantum theory and the recent suggestion that a Thomist ontology fits better with such phenomena than a classical Cartesian reductionism which has been the underlying assumption throughout the developments in physics described above.

One aspect of quantum strangeness arises from what is known as the superposition principle. Following Wolfgang Smith, one might imagine a physical system which consists of just one particle, and two states of that system in which the particle is located first in region A and then in region B.[104] Quantum physics stipulates that in a linear combination of these two states, the particle is mysteriously located in both regions. This is not to say that the particle is in one or other region, but we cannot tell which. Physicists can reproduce experimentally the linear combination of states in such a way that interference effects are observed which would not be present were the particle situated in either region A or region B. As Smith states, 'In some unimaginable way the particle seems thus to be actually in A and B at once'.[105] It is possible, along the lines of probabilistic calculations relating to heat in thermodynamics, to calculate probabilities for the particle being in different positions. Quantum theory therefore possesses a double mystery: first, the bi-location of particles that is unimaginable to a mind confined to a classical Newtonian view of the cosmos, and, secondly, a purely probabilistic event in nature. However, in a situation where a particle bi-locates in this way, its position may be fixed by experimental observation. The act of measuring sends the system, in which the particle is at position A and B, into a new state in which the location of the particle can be identified. Smith emphasises that, although this result is strange, the situation is fully coherent mathematically.

Such strange quantum phenomena appear to take place only at a subatomic, particulate level. Apparently, they do not impinge at the level of macroscopic nature. However, the famous thought experiment known as 'Schrödinger's Cat' shows how quantum effects bear upon occurrences at a macro level. Imagine a cat confined to a box with a canister of poisonous gas. The release of the gas will be triggered via some appropriate mechanism by the decaying of a radioactive nucleus. According to quantum theory, however, the radioactive nucleus, which is unobserved, is in a superposition state; that is to say, as it decays one cannot be sure whether its location will be such as to trigger the release of the poisonous gas. Just as in the case of the bi-located particle above, it is in the act of measurement that the state of the system (whether the cat is dead or alive) is determined. As Smith points out, the radioactive nucleus is in a superposition state and transfers this to a macro level by placing the cat in a similarly indeterminate state. The act of measurement then determines the state of the cat. However, what is more curious still is that, whatever one observes in the box, a cat which is alive or dead, it is not possible to provide a compelling reason why that particular state, rather than its alternative, pertains. Nature appears to be governed, even at a macro level, by probability.

Given that this thought experiment shows that superposition states can transfer to macro-nature, this suggests one final puzzle regarding quantum theory to which one might point. There seems to be no discernible reason why we do not experience the superposition of macroscopic bodies. If photons can bi-locate, it seems reasonable to suppose that bodies composed of photons should also bi-locate. As Roger Penrose states, 'Why, then, do we not experience macroscopic bodies, say cricket balls, or even people, having two completely different locations at once? This is a profound question, and present-day quantum theory does not really provide us with an answer'.[106]

There thus appears to be a double ontology within quantum mechanics. On the one hand, there is the microscopic realm, where superposition states remain and probability rules over certainty and stability. On the other hand, there is the macroscopic world where the motion of bodies is apparently stable and predictable (although, as we have seen, it is not clear that indeterminacy should not also belong to the macroscopic world). These two worlds, one of quantum indeterminacy, the other of predictable certainty, seem to be unrelated and our familiarity with the latter entails that the former appears strange and even paradoxical.

Smith argues that the mysteries of quantum physics arise not because of a defective physics or mathematics, but because the ontology which underlies the interpretation of that physics is still Cartesian and fit only for a mechanistic cosmology in which objects in nature can be reduced to their more basic physical components. Such a view cannot tolerate the paradoxes of quantum theory. Following Whitehead,[107] Smith refers to this ontology as 'bifurcation'. This approach is akin to Descartes' distinction between *res extensae* (extended bodies which do not have sensible qualities) and *res cogitantes* (thinking beings wherein reside sensible qualities).[108] Smith argues that this is essentially a distinction between a 'physical' object and a 'corporeal' object. Every corporeal object is associated with a physical object; the latter is the molecular 'real' object which gives rise to certain qualities in the mind, namely the corporeal object. However, for the bifurcationist (John Locke included), this distinction between the corporeal and physical object does not properly pertain because the corporeal object does not really exist independently of the mind. Hence there is a tendency to reduce all qualities of the corporeal object to the purely physical attributes of extended, physical bodies.

Quantum physics, argues Smith, insists on a proper ontological distinction between the corporeal and the physical because the mysterious science perceives a difference between the motions and operations of the 'physical' microscopic realm and the 'corporeal' macroscopic realm. The process of measurement reinforces the necessity of this distinction because it displays and constitutes a transformation from the physical to the corporeal. Smith comments that,

> No wonder, therefore, that quantum theory should be conversant with two very different 'laws of motion', for it has now become apparent that Schrödinger evolution operates within the physical domain, whereas projection has to do with the transit out of the physical into the corporeal.[109]

It is indeed the case that if cricket balls and cats were made of aggregates of particles, then they would be able to bi-locate. But if one were to stipulate a distinction between the physical and corporeal in such a way that the latter is not plainly reducible to the former, then cricket balls and cats are not made simply of aggregates of particles. There is something else which raises the physical to the corporeal; there is some aspect of the corporeal realm which renders bi-location impossible.

How should one distinguish the physical and the corporeal in such a way that a dualistic ontology is avoided and the quantum realm brought to a more comfortable relation with the corporeal? To answer this question, Smith returns to Aquinas and, by association, Aristotle.[110] He proposes that one understand the distinction between the physical domain (which features quantum bi-location) and the corporeal domain (which features a more stable certainty) in terms of the distinction between potency and act. The realm of the microscopic particle from which comes quantum indeterminacy is in potency to becoming a corporeal being of stability and intelligibility. One may extend and clarify Smith's analysis in terms of further distinctions between potency and act which were first introduced in the second chapter of this essay. It was seen that a being in first potentiality, for example someone who does not speak French, is in a kind of radical potency to its act, namely actually speaking French at the present moment. One may refer to this as first potentiality. Someone is in second potentiality when they are learning to speak French, and in first actuality when they can speak French but are not exercising this skill at the present moment. Someone is in second actuality with regard to speaking French when they are actually speaking French. Applying this Aristotelian analysis of motion to the distinction between the physical realm and corporeal realm, one can see that individual particles are in first potentiality to constituting a corporeal entity. An aggregate of physical particles is in second potentiality to becoming a corporeal entity. This aggregate of particles may be raised to first actuality, but the aggregate is constituted in second actuality as a fully determinate corporeal entity when it is perceived, for example, within experimental observation.

What raises an aggregate of particles from second potentiality to first actuality? Smith stipulates that it is substantial form. It is this form which bestows upon an aggregate of particles its 'whatness' and transfers it to an intelligible, corporeal entity.[111] For the mechanistic natural philosophy of the seventeenth century, typified in the work of Descartes and advocated with varying conviction by Newton, substantial form was unnecessary. All that was required was the interaction of material particles which, when aggregated, would form the corporeal entities of our common experience. However, quantum theory betokens a return to an ontological hierarchy which was a characteristic of the cosmology described in the opening chapters of this essay. At a base level, physics has discovered a mysterious world of great potency where motion is indeterminate and unpredictable, akin to a kind of 'wandering' described as the 'khora' or realm of necessity in Plato's *Timaeus*. Such a realm is persuaded towards greater actuality, and hence stability and intelligibility, by an intensification of its participation in substantial form. This is to say that form becomes the mediating principle between what Smith refers to as the physical and corporeal realm, between the quantum and macroscopic worlds. This ontology is neither dualistic, on the one hand, nor bifurcationist (that is, reductionist), on the other. Rather, it places substantial form back into nature, re-institutes a kind of ontological hierarchy and suggests the participation of one level of the hierarchy in the next.

Finally, one can suggest that because quantum theory, as Smith states, appears to deal with different 'laws' of motion at different ontological levels, so one might suggest a more analogical relation between those levels, and hence a more analogical view of their different motions. The motions of the quantum realm seem indeterminate and 'wandering'. Once participating in substantial form, those motions become ordered towards a particular *telos*. Yet one may speculate that there are yet higher realms, such as intelligible thought, where motion becomes yet more akin to the real in the sense that it is increasingly circular and contained within its own bounds. Just as quantum theory requires us to stipulate a delicate boundary between the physical and corporeal (and neither obliterate nor ossify that boundary), so too there may yet be the possibility of further ontological distinctions, and with them analogical distinctions, between different kinds of motion. It is possible that the hierarchy and distinctions reintroduced by quantum theory may yield distinctions between different kinds of motion which were so characteristic of earlier cosmologies.

The return of substantial form into nature in order to account for the transformation of the physical to the corporeal returns us in many ways to the ontology and physics of the first four chapters of this essay. Of course, Plato, Aristotle, Grosseteste and Aquinas understood form in very different ways. Within a Thomist perspective, in which form exists in the mind of God and is known in the emanation of the persons of the Trinity, one may yet be led to concur with Etienne Gilson: 'If a physics of bodies exists, it is because there is first a mystical theology of the divine life'.[112]

Conclusion

The orthodox Christian tradition has understood God as beyond change. Expressed in Aristotelian terms, one might say that God is fully actual and devoid of all potency. Yet Aristotle defines nature as a principle of motion and rest. This appears to establish a radical disjunction between God and creation: how might God have anything to do with nature which is defined by motion? This essay has sought to answer this question by examining the nature of motion.

For Plato, motion is the very means of the participation of the realm of becoming in the realm of being, for in motion's succession the cosmos participates in the fullness of the eternity of the Forms and, ultimately, the Good. The *Timaeus* is a mythic cosmology about a universe which is itself 'mythic'. Therefore, this natural philosophy is only ever provisional: it is itself subject to the motion of which it speaks. Meanwhile, Aristotle examines nature and motion from definite metaphysical principles. Motion lies between potency and act which divides being, the subject of metaphysics. This category is therefore mysterious, by definition having no fixed ontological status. Motion is thus identified as the realisation of a potentiality, *qua* potentiality. As for Plato, motion for Aristotle is a crucial precondition for all cosmology, for the motion of cosmology and human knowing is itself part of the very motion of the cosmos. There is a hierarchy of motion, extending from the local, rectilinear motion of bodies to the circular and complete motion of the heavens which in turn results from the creative psychic motion of separate substances. Ultimately, because motion requires a source which is beyond potency and in full actuality, physics points towards the divine origin of all motion.

One of the earliest attempts to synthesise the Platonic and Aristotelian traditions of natural philosophy arose in the early thirteenth century through the work of Robert Grosseteste. In Grosseteste's cosmogony of light, one finds an underlying physics in which all motion is analogically related to the 'motionless motion' of the first uncreated point of light. Moreover, for Grosseteste all human knowledge is ontological: being is a more or less rarefied form of light, and likewise all knowledge is a more or less scintillating illumination in the divine light. In this context, I argued that Grosseteste's advocacy of the *experimentum* has a subtly different motivation to modern science. He is not attempting to overcome a dark chasm of induction in any proto-Humean fashion, but is aware that human knowledge is discursive and therefore requires the motion of sense experience. As for Plato and Aristotle, cosmology is entwined in the motions of nature. Also, I argued that the difference between the various sciences could be understood by the motion involved in their exposition. For example, the science of physics requires a discursive enquiry which proceeds by stages and involves the study of the most dense form of light, namely physical bodies moving through space. The science of theology, by contrast, in studying the most perfect, actual and spiritual divine being, is least discursive and exists in the 'motionless motion' of illumination by divine revelation. While Roger Bacon repeats and extends Grosseteste's endorsement of experiment and mathematics in the study of nature, he nevertheless takes important steps towards modern notions of knowledge as representation (thus forming a dualism between beings in nature and our representative knowledge of them) and the deployment of experiment for the purposes of natural theology and technology.

In the thought of Thomas Aquinas, one finds a most exacting synthesis of the Platonic and Aristotelian traditions manifest in an understanding of motion which extends analogically from physics to ethics and *sacra doctrina*. All motion has as its ultimate origin the dynamic Trinitarian life, and *esse ipsum* reaches to the depths of cosmic motion as the first, most general and therefore most potent cause of all motion. No motion is simple for Aquinas: all motion ultimately points to a more actual source and thus is part of a hierarchical pattern of shared and co-operative cosmic motion. I examined the way in which, for Aquinas, God is said to move creation: from the emanation of creation originating from the eternal emanation of the persons of the Trinity; by the teaching of the Old and New Law; by the impartation of grace through the sacraments; and as the ultimate *telos* of all things. Most particularly, one finds the unification of all cosmic motion in the sacrament of the Eucharist. Crucially, I argued that there is no fundamental division within the sciences. Although one might separate the sciences by means of such concepts as motion (for example, motion might be thought not to belong in any way to theology because God is beyond motion and motion belongs only to physical bodies and space), for Aquinas terms such as motion are applied in all sciences and are related by analogy which is ultimately based upon the analogical participation of *esse commune* in *esse ipsum*.

An early intimation of the separation of the sciences, and the concomitant isolation of physics, can be found in Avicenna's restriction of the proof of God's existence to metaphysics. For the Persian thinker, motion cannot attain to the divine. Whereas for Aquinas the lower sciences participate in the higher and all truth is found ultimately in God's own knowledge which he shares with the blessed, for Avicenna physics and metaphysics become sundered because of a restriction of motion to the former with no analogical application in higher sciences. I also argued that Avicenna denigrates the lower sciences which deal with the physical realm because, in understanding knowledge as representation (the 'stamping' of ideas in the mind rather than a visceral motion originating essentially from the migration of the species in the mind to be illuminated by the agent intellect), sense experience becomes a mere 'occasion' for knowledge which now arrives more immediately from a separate agent intellect or *dator formarum*.

This more isolated physics, which studies a simple, local and self-explanatory motion, is seen clearly in the work of Jean Buridan and his recovery and development of the theory of impetus to explain projectile motion. Because motion is now understood in terms of an interior self-contained impetus, the origins of the motions of bodies are no longer sought exteriorly in the more actual. In fact, because all motion is now understood in terms of the quantity of impetus, this implies that God moves not as the final cause of all being, but as an efficient imparter of motion to the cosmos. Thus one finds an intimation of Newton's emphasis on purely quantitative, local motion which refers exclusively to efficient causality.

In this final chapter I have examined what might be regarded as the clearest exposition of a view of physics and theology which begins centuries earlier but which inaugurates a view of motion which was to remain predominant until the late nineteenth century. By contrast to the more traditional Platonic and Aristotelian view, Newton understood motion to be a self-explanatory state confined to bodies in space. As with Buridan, God is an efficient cause of motion. Yet because Newton excludes relationality within the Godhead, he requires, for reasons at once both scientific and theological, an absolute space eternally begotten of God in and through which the divine might act (with efficient causality) on a moving cosmos to replenish its decaying motion. This absolute space takes on the role of a more orthodox Christ in Newton's doctrine of creation, such as it is. It does appear that God has become three dimensional for Newton. Yet through the *vis insita* and *vis inertiae* of bodies, motion becomes self-explanatory. Thus physics no longer refers beyond itself. The sciences become isolated, their division marked by the restriction of concepts of motion to natural philosophy. This was the view of motion bequeathed to the eighteenth and nineteenth centuries.

Throughout the first four chapters of this essay, I sought to identify the way in which motion, when understood analogically, can relate the different sciences one to another. On such an understanding, motion is not a simple category restricted to a single science; there exists a more delicate boundary between, for example, the motion of bodies through space, the motion of thought, and the motionless motion of the divine life. The basis of this understanding of motion is an analogical and participatory metaphysics which sees all motion in relation to its origin and *telos* in the divine life. This analogical understanding of motion is not prevalent throughout the medieval tradition; I cited a number of instances where the boundaries between physics and metaphysics, body and soul, and knowledge and reality are more strictly delineated in such a way that certain concepts, including motion, are confined to particular sciences. This division reaches a clear expression at a time when the link between theology and natural philosophy is articulated in a particularly self-conscious way, namely during the advent of Newtonian physics in the seventeenth and eighteenth centuries. The relationship between God and creation is no longer one of ontological participation whereby motion is integral to creaturely fulfilment, but is expressed in functional terms: God merely replenishes a decaying mechanism and legislates laws of nature which are, as it were, 'enacted' by Christ, his 'viceroy'.

These final brief remarks concerning post-Newtonian investigations of nature suggest that science did not inevitably follow a mechanistic path towards a flattened cosmology. Rather, the reintroduction of analogy into scientific thinking is suggestive of the approach described earlier in this essay in the work of Aristotle, Grosseteste and Aquinas. This vision of the cosmos, which is even akin to a metaphysics of participation rather than dualism or an ontology of discrete layers of being, suggests a more delicate continuum between different kinds of analogically related motion, for example the motion of bodies, the motion of thought and the motion of the human will. Why might this be the case? Because in the electromagnetic understanding of the cosmos in which fields constantly interact, and the thermodynamic cosmos in which the energy of natural entities is constantly emanating to others, beings seem to 'participate' in each other's motions in such a way that they are intertwined. The boundaries between different entities in the cosmos are more delicately drawn. Indeed, a being can no longer be properly understood outside of its electromagnetic or thermodynamic system. The notion of a discrete body moving through a void is no longer indicative of a physics of motion. Moreover, the collapse of the mechanistic view of the universe with the implication that phenomena are reversible and therefore indifferent to their motion, and its replacement with thermodynamics and the 'emanation' of energy, is highly conducive to the cosmology outlined in the first part of this essay. The being of things once again has a natural and established motion. There is a tendency to dissipate and 'share' being within the cosmos (a genuine *esse commune*) rather than contract into individual atomic units which are part of a mechanism.

Over the past century, the physics of motion has been presented with some further perplexing and counter-intuitive phenomena. As we have seen, Wolfgang Smith has suggested that the mysterious aspect of quantum dynamics is due not to physics, but to a defective underlying ontology. The return of a notion of substantial form, along with an ontological hierarchy which at once outwits both dualism and reductionism, suggests that categories such as motion may no longer be considered simple and self-explanatory. The boundaries between the various ontological realms of the cosmos remain delicate yet interwoven through participation. Likewise, the motions of these realms may remain distinct but interpenetrating. The possibility of an account of motion which is at once both scientific and theological is open once again.

Notes

1 Plato's *Timaeus* and the soul's motion of knowing

1 'Through its Latin translations, it [*Timaeus*] had an influence on Mediaeval philosophy and, after the rediscovery of its original text at the time of the Renaissance, it played a crucial role in the thought of scientists such as Kepler and Galileo', L. Brisson and F. W. Meyerstein, *Inventing the Universe: Plato's* Timaeus, *the Big Bang and the Problem of Scientific Knowledge*, Albany, NY: State University Press of New York, 1995, p. 4.

2 See, for example, T. M. Robinson, 'The Relative Dating of the *Timaeus* and *Phaedrus*', in L. Rossetti (ed.), *Understanding the* Phaedrus: *Proceedings of the II Symposium Platonicum*, Sankt Augustin: Academia Verlag, 1992, vol. 1, pp. 23–30.

3 Plato, *Timaeus*, 35b–36c.

4 See, for example, *Timaeus*, 38c ff. for Plato's description of the motion of the heavenly bodies.

5 These *physiologoi* (a term apparently coined by Aristotle) included, for example, Thales, Heraclitus of Ephesus and the atomist Leucippus of Miletus. See G. Vlastos, *Plato's Universe*, Oxford: Clarendon Press, 1975, ch. 2, and his 'Equality and Justice in Early Greek Cosmologies', in D. J. Furley and R. E. Allen (eds), *Studies in Presocratic Philosophy*, vol. 1, London: Routledge, 1970, pp. 56–91. As Vlastos points out, the *physiologoi* would, of course, defend their system against Plato's charge of impiety by maintaining that they endowed universal substance with the ability to design and order the cosmos and even regarded this as a god. As will be seen particularly in considering the work of Newton, the tendency of the *physiologoi* to view the cosmos in mechanistic terms and to make the explanation of the universe a purely immanent enterprise resurfaces with a vengeance in early modern physics.

6 On Plato's harsh sanction against the impiety of the *physiologoi*, see *Laws*, Book X, 888c ff.

7 The relation of any cosmological account to the Good is vital for Plato. He assumes that, in forming the cosmos, the Demiurge, the craftsmen, follows the Form of the Living Creature and thus makes the best possible likeness. He denounces the *physiologoi* for not invoking the Good in their explanations of, for example, the shape of the earth. The Ionians believed that the earth was a flat disc floating in water without considering that a sphere would be a more perfect shape. Plato does not separate facts and values in the manner of the *physiologoi* or much modern science. See Plato, *Phaedo*, 97c f. and the discussion of this passage below.

8 See Plato, *Sophist*, 252e ff.

9 Plato, *Timaeus*, 29a–b. All translations are from F. M. Cornford, *Plato's Cosmology: the* Timaeus *of Plato translated with a running commentary*, London: Routledge & Kegan Paul, 1937. See also *Republic*, Book VI, 509b ff. Although Plato does not mention the Good as a distinct Form in the *Timaeus*, one of his most consistent appeals is to the 'goodness' of the realm of becoming. See *Timaeus*, 30a ff. See also *Phaedo*, 97c–99d, in which Socrates criticises purely immanent explanations of nature, such as those of Anaxagoras, which are not teleological and make no reference to the Good.

10 Plato, *Timaeus*, 29d ff.
11 In *Republic*, Book X, 597a ff., Plato refers to a craftsman who makes a bed with reference to the 'real bed', a Form which he does not himself create and which transcends all visible created beds. The Form of the bed is, nevertheless, existent in the nature of things. A bad craftsman is one who takes a created object as his model. Such a craftsman merely creates an image of an image, the product being at a third move from reality.
12 Plato, *Timaeus*, 29c.
13 Some commentators, notably Cornford, do not take the Demiurge or the 'chaos' and disorderly motion which preceded the formation of the universe in a literal fashion. See Cornford, *Plato's Cosmology*, pp. 34, 37, 176, 203, *et passim*. As well as historical arguments (the testimony of the Academy), Cornford argues that Plato could not have meant the pre-existent disorderly motion to be taken literally because motion cannot antecede the creation of time or soul. Others, notably Gregory Vlastos (see 'Disorderly Motion in the *Timaeus*' and 'Creation in the *Timaeus*: Is it a Fiction?', in R. E. Allen (ed.), *Studies in Plato's Metaphysics*, London: Routledge & Kegan Paul, 1965, pp. 379–399 and 401–419), have argued more convincingly to the contrary, claiming that there are no strong textual grounds for Cornford's view and that Plato would have considered even a mechanistic immanentist 'motion' prior to *genesis* to be disordered. What the work of the Demiurge and the advent of the World Soul provide is a teleology and measure of motion and therefore an order to the cosmos, namely a deepening participation in the intelligible Forms. 'What makes the state pre-cosmic is precisely the absence of purpose. The state is not some sort of primordial chaos, but an envisioning of how being, becoming and space affect one another necessarily. Their interaction is, of course, at its best when it is directed to an end', A. F. Ashbaugh, *Plato's Theory of Explanation: A Study of the Cosmological Account in the 'Timaeus'*, Albany, NY: State University of New York Press, 1988, p. 113 and n. 47.
14 Plato, *Timaeus*, 29a–b, 30c–31a.
15 Plato, *Timaeus*, 31a–b.
16 Plato, *Timaeus*, 33b–34a.
17 Plato, *Timaeus*, 31b–33b.
18 On the proportions of the elements, see Cornford, *Plato's Cosmology*, pp. 44–52.
19 Plato, *Timaeus*, 33b.
20 On the shape and formation of the elements, see *Timaeus*, 53c ff. It is important to remember that in referring to, for example, fire, Plato means not only that which burns but also things that produce light or heat. Similarly, 'water' refers to liquids, 'earth' to solids and 'air' to gases.
21 Plato, *Timaeus*, 58a–c.
22 Plato, *Timaeus*, 35a.
23 Plato, *Timaeus*, 36e.
24 Plato, *Sophist*, 248e.
25 Plato, *Timaeus*, 57d–58c. See H. J. Krämer, *Plato and the Foundations of Metaphysics*, Albany, NY: State University of New York Press, 1990. Krämer argues that 'Being is ... essentially *unity within multiplicity*' (p. 79). He outlines Plato's basic ontological conception as the generation of being from the principles of unity and multiplicity. He comments that, 'Everything that is exists in the measure in which it is something limited, determined, distinct, identical, permanent, and insofar as it shares in the basic unity, which is the principle of every determination. Nothing is *something*, so far as it is not *one* something. But it can be, precisely, something and one and share in unity only because, at the same time, it shares in the contrary principle of multiplicity, and because of this, it is *another* with respect to unity itself' (ibid.). Krämer goes on to trace briefly the importance of the principles of unity and multiplicity in the *Timaeus* (pp. 110 ff.). He argues that the two kinds of triangle (right-angled isosceles and equilateral) from which the corporeal bodies of earth, air, fire and water are formed comprise a multiplicity and yet also a unity in their simplicity, symmetry, harmony and equilibrium. The rotational

movement of the cosmos as a whole, which, in its perfection, might be regarded as indistinguishable from rest, thereby forms a unity from the multiplicity of rest and motion. Thus, in the same vein as Krämer, it might be said that the harmony between Sameness and Difference in the World Soul provides a unity and multiplicity which makes possible the transmission of motion to the multiplicity of the cosmos and at the same time keeps the cosmos analogically connected to the Good in its unity. Proper rational motion requires difference, that is multiplicity, and yet that motion also requires a unity in its teleology. Thus the World Soul, as an intermediate between the realms of being and becoming, must, in its motion, be both unity and multiplicity, Sameness and Difference.

26 Proclus remarks that, in the traditional 'Tables of Opposites', Right stood in the column of superior things, Left in the column of inferior (Proclus, *Procli Diodochi in Platonis Timaeum commentaria* ii, 281 (19), noted in Cornford, *Plato's Cosmology*, p. 74).

27 Plato, *Timaeus*, 36c–d.

28 See also Plato, *Phaedrus*, 245c ff.

29 Plato, *Timaeus*, 36e.

30 Plato, *Laws*, 896a.

31 Plato, *Laws*, 893a ff.

32 I am indebted to Brisson and Meyerstein (see *Inventing the Universe*, pp. 28 ff.) for the clarity of their discussion of this intricate aspect of Plato's cosmology.

33 Plato, *Timaeus*, 35b–c.

34 For a more detailed explanation of the harmonic and arithmetical means into which the World Soul is divided, see Brisson and Meyerstein, *Inventing the Universe*, pp. 34 ff.; Cornford, *Plato's Cosmology*, pp. 66 ff.

35 This interpretation appears to be the most straightforward and follows L. Brisson, *Le même et l'autre dans la structure ontologique du Timée de Platon*, Paris, 1974, pp. 40 ff.

36 Cornford claims that this division of the World Soul has little to do with music (*Plato's Cosmology*, p. 69). However, this music is a metaphysical category that is not restricted to the purely modern aesthetic understanding of music. For a discussion of the cosmological character of music with particular reference to Augustine's *De Musica*, see C. Pickstock, 'Music: Soul, City and Cosmos', in J. Milbank, C. Pickstock and G. Ward (eds), *Radical Orthodoxy: a new theology*, London: Routledge, 1999, pp. 243–277.

37 Plato, *Laws*, 897.

38 One might think that the body, being inert, should not affect the motion of the soul. However, Plato certainly appears to talk of the 'disordered motions of the body'. All bodies, whether alive (that is, 'en-souled') or not, possess some kind of motion by virtue of their place in the universe. Bodies which are not alive are at the lowest point in the hierarchy of motions. They are, however, subject to the motion of the World Soul to some degree for, as Plato says, the motion of the World Soul leaves nothing in the universe untouched. Bodies which do not possess their own soul are recipients of a disordered motion which is at many removes from its origin in the circle of the Different.

39 Plato, *Timaeus*, 37c–38c.

40 Ibid.

41 Plato, *Timaeus*, 29a.

42 Plato, *Timaeus*, 47e ff.

43 Plato, *Phaedo*, 97a ff.

44 All translations are by Harold North Fowler in the Loeb Classical Library edition volume 1, Cambridge, MA: Harvard University Press, 1914.

45 Plato, *Phaedo*, 99b.

46 See Aristotle, *Physics*, Book II, 198b, 17.

47 Plato, *Timaeus*, 46d.

48 Plato, *Timaeus*, 48a.

49 On the role of persuasion in the *Timaeus*, see also G. R. Morrow, 'Necessity and Persuasion in Plato's "Timaeus"', in R. E. Allen (ed.), *Studies in Plato's Metaphysics*, London: Routledge & Kegan Paul, 1965, pp. 421–437.

50 Plato, *Phaedrus*, 230e–234c.
51 Plato, *Phaedrus*, 234e–235a.
52 Plato, *Phaedrus*, 235b.
53 Plato, *Phaedrus*, 237a–238c. On Phaedrus's desire for the possession of written, exhaustive speeches and the alignment of writing with capital and trade, see C. Pickstock, *After Writing: On the Liturgical Consummation of Philosophy*, Oxford: Blackwell, 1998, ch. 1, particularly pp. 6–10.
54 See Plato, *Phaedrus*, 236e, 242a and 243e.
55 Plato, *Phaedrus*, 242d. All translations are by Harold North Fowler in the Loeb Classical Library edition volume 1, Cambridge, MA: Harvard University Press, 1914.
56 See Plato, *Phaedrus*, 260b–e.
57 Plato, *Phaedrus*, 244a–257b.
58 Plato, *Phaedrus*, 271d.
59 Plato, *Phaedrus*, 271a–b.
60 Plato, *Phaedrus*, 271b.
61 Plato, *Phaedrus*, 245c–246a.
62 Plato, *Phaedrus*, 246e.
63 Plato, *Phaedrus*, 247c.
64 Plato, *Phaedrus*, 248a.
65 See Plato, *Phaedrus*, 244a ff.
66 Plato, *Phaedrus*, 257a.
67 Plato, *Phaedrus*, 265e.
68 Plato, *Phaedrus*, 266b.
69 Plato, *Phaedrus*, 268c–e.
70 Plato, *Phaedrus*, 260c f. and 271a.
71 I borrow this term from Wayne Hankey, *God in Himself: Aquinas's Doctrine of God as Expounded in the* Summa Theologiae, Oxford: Oxford University Press, 1987, p. 56. For its deployment with regard to Aquinas, see below p. 115.
72 Plato, *Timaeus*, 34b–38c. At 34b Plato clearly outlines why soul is prior to body. He states that this is not strictly a temporal priority, but that the soul is the body's 'mistress and governor'.
73 Plato, *Timaeus*, 47e–49a.
74 See Plato, *Phaedrus*, 235d where Phaedrus determines to erect a statue at Delphi both of himself and Socrates should the latter's speech prove successful.
75 Plato, *Critias*, 109b–c. All translations are by R. G. Bury in the Loeb Classical Library edition volume 9, Cambridge, MA: Harvard University Press, 1929.
76 See Plato, *Critias*, 118–119.
77 Plato, *Critias*, 120e–121a.
78 Plato, *Critias*, 119c–120d.
79 Plato, *Critias*, 121a–b.
80 On this aspect of the *Timaeus*, see also Ashbaugh, *Plato's Theory of Explanation*, pp. 65–71 and G. R. Carone, 'The Ethical Function of Astronomy in Plato's *Timaeus*', in T. Calvo and L. Brisson (eds), *Interpreting the Timaeus–Critias: Proceedings of the IV Symposium Platonicum; selected papers* (International Plato Studies vol. 7), Sankt Augustin: Academia Verlag, 1997, pp. 341–348.
81 Plato, *Timaeus*, 29a.
82 Plato, *Timaeus*, 51e.
83 Plato, *Timaeus*, 47a. See also 44b, 47c, 87b.
84 Plato, *Timaeus*, 47b.
85 Plato, *Timaeus*, 47c.
86 In the light of a much debated passage in Book VII of the *Republic*, around 528e–530c, one may question what Plato means by 'astronomy'. In this passage it appears that Plato regards astronomy as an abstract study of mathematical proportion and idealised motion which turns from any observation of the motions of the heavenly bodies. On the basis of

this passage, it has been argued, for example by Alexander Mourelatos ('Plato's "Real Astronomy": *Republic* VII. 527d-531d', in J. P. Anton (ed.), *Science and the Sciences in Plato*, New York: Caravan Books, 1980), that Plato regards astronomy as akin to geometry and rejects any observation of the material world as being without worth. However, Eríkur Smári Sigurdarson has argued that one should not regard this passage from the *Republic* as a statement of Plato's 'philosophy of Science' (see 'Plato's Ideal of Science', in Erik Nis Ostenfeld (ed.), *Essays on Plato's* Republic, Aarhus: Aarhus University Press, 1998, pp. 85–90). Sigurdarson points out that Plato, in his discussion of astronomy in this passage from the *Republic*, is particularly concerned with the improvement of the soul of the guardians, not with the improvement of our knowledge of the stars or any attempt to lay aside observation for a while in order to concentrate on developing better theoretical explanation (as is argued by D. R. Dicks, *Early Greek Astronomy to Aristotle*, London, 1970, pp. 106–108). This is amply indicated in the passage from the *Timaeus* currently under discussion: Plato notes that an observation of the motions of the heavenly bodies can provide a crucial stepping stone to the contemplation of the more exact but abstract 'astronomy' of the Forms. In other words, observation is excellent and wholly satisfactory for the formation of 'opinion' and the establishment of the problems of astronomy. However, is it the case that Plato is ultimately advocating a turn from the material to a purely interior and abstract reasoning? One will answer in the affirmative only if one already assumes that Plato operates with a fundamentally dualistic epistemology (and therefore ontology) in such a way that contemplation of the Forms and contemplation of the realm of becoming can be ultimately sundered one from the other, with the latter rendered valueless. However, it is vital to recall that knowledge for Plato is only possible of Forms and not of the visible realm of becoming and change. In order to understand the imperfect motions of the material heavens, one must understand them in relation to the perfect and eternal realm because the former participate in, and have their being from, the latter. As can be seen from the *Timaeus*, contemplation of the material realm is of great value in the training of the soul, yet the final aim is always a contemplation of the Forms. Moreover, a contemplation of the Forms enables a yet more apposite contemplation of the physical cosmos, because the cosmos is only properly known in its relation to the Forms from which it emanates and has its being. On this see also Gregory Vlastos, 'The Role of Observation in Plato's Conception of Astronomy', in J. P. Anton (ed.), *Science and the Sciences in Plato*, pp. 1–31.
87 Plato, *Timaeus*, 90c.
88 Plato, *Timaeus*, 90d. See also 43a–e and 47b–c.
89 Plato, *Timaeus*, 47a–b.
90 Plato, *Timaeus*, 90d.
91 See also Plato, *Laws*, Book VII, 809c–d and 817c ff. on the need for all to study astronomy.
92 On the ordering motion of the Receptacle of Becoming (the term which might cautiously be interpreted as 'space') see Plato, *Timaeus* 52d ff.

2 Aristotle: ecstasy and intensifying motion

1 See, for example, W. Jaeger, *Aristotle: Fundamentals of the History of His Development*, trans. R. Robinson, Oxford: Clarendon Press, 1934; W. Wians (ed.), *Aristotle's Philosophical Development: Problems and Prospects*, London: Rowman and Littlefield, 1996; E. Frank, 'The Fundamental Opposition of Plato and Aristotle', *American Journal of Philology* 61, 1940, 34–53 and 166–185; F. Solmsen, 'Platonic Influences on Aristotle's Physical System', in I. Düring and G. E. L. Owen (eds), *Aristotle and Plato in the mid-Fourth Century: papers of the Symposium Aristotelicum held at Oxford in August, 1957*, Göteborg: Elanders Boktryckeri Aktiebolag, 1960, pp. 213–235. For the view that Plato and Aristotle are more closely aligned, see H. G. Gadamer, *The Idea of the Good in Platonic–Aristotelian Philosophy*, trans. P. Christopher Smith, New Haven, CT: Yale University Press, 1986.

2 See below, pp. 32 ff., for a discussion of chance and spontaneity.
3 Aristotle, *Physics*, II.1.192b. All translations are by P. H. Wicksteed and F. M. Cornford in the Loeb Classical Library edition (2 vols), Cambridge, MA: Harvard University Press, 1929 and 1934. Aristotle discusses other meanings of *physis* in the work of his philosophical predecessors in *Metaphysics*, V.4.1014b–1015a.
4 Aristotle, *Physics*, II.1.193b.
5 Ibid.
6 Strictly speaking, Aristotle does not regard absolute generation and corruption as forms of motion. Motion requires a subject, and simple generation from nothing has no initial subject. Because motion takes place between opposites, and corruption is the opposite of generation from which motion cannot take place, so too corruption cannot be a form of motion. See *Physics*, V.1.225a.
7 On the interpretation of Aristotle's definition of motion, see L. A. Kosman, 'Aristotle's Definition of Motion', *Phronesis* 14, 1969, 40–62.
8 Aristotle, *Physics*, III.1.201a. See also *Physics*, V.1.
9 See Aristotle, *Physics*, II.8 and III.4. On pre-Socratic Atomist philosophy, see C. C. W. Taylor (ed.), *The Atomists Leucippus and Democritus: fragments: a text and translation*, Toronto: University of Toronto Press, 1999; M. R. Wright (ed.), *Empedocles: the extant fragments*, New Haven, CT: Yale University Press, 1981; D. Sider (ed.), *The Fragments of Anaxagoras*, Meisenheim am Glan: Hain, 1981.
10 J. Lear, *Aristotle: the desire to understand*, Cambridge: Cambridge University Press, 1988, pp. 36–42.
11 Some have attempted to interpret Aristotle as arguing for the compatibility of mechanistic and teleological explanation. See, for example, W. Wieland, 'The Problem of Teleology', in J. Barnes, M. Schofield and R. Sorabji (eds), *Articles on Aristotle*, London: Duckworth, 1975–1979; M. Nussbaum, 'Aristotle on Teleological Explanation', in *idem*, *De Motu Animalum*, Princeton, NJ: Princeton University Press, 1985, pp. 59–61.
12 See Aristotle, *Physics*, II.4–8.
13 Aristotle, *Physics*, II.7.197b.
14 Aristotle, *Physics*, II.6.197b.
15 Ibid.
16 Aristotle, *Physics*, II.5.197a. See also *Physics*, II.4.199b.
17 Aristotle, *Physics*, II.8.198b.
18 Aristotle, *Physics*, II.5.197a.
19 See Aristotle, *Physics*, I.9.192a.
20 Aristotle, *Parts of Animals*, II.1.646a. All translations are by A. L. Peck in the Loeb Classical Library edition, Cambridge, MA: Harvard University Press, 1937.
21 For a discussion of such an interpretation of Aristotelian teleology, see S. Broadie, 'Nature and Craft in Aristotelian Teleology', in D. Devereux and P. Pellegrin (eds), *Biologie, Logique et Métaphysique chez Aristote* (Actes du Séminaire C.N.R.S.-N.S.F.), Paris: Éditions du Centre National de la Recherche Scientifique, 1990, pp. 389–403.
22 Aristotle, *Physics*, II.8.197a.
23 The four causes are best understood as aspects under which cause may be understood. Aristotle is not claiming that four distinct causes must be enumerated for every effect. On the coincidence of the four aspects of cause in nature, see Aristotle, *Physics*, II.7.198a.
24 Aristotle, *Physics*, II.3.194b.
25 Aristotle, *Physics*, VIII.1.252a.
26 Aristotle, *Metaphysics*, I.1.980a.
27 Aristotle, *Physics*, II.1.
28 See Aristotle, *Physics*, IV.8.215a. See also *Physics*, V.6, where Aristotle claims that the classifications 'natural' and 'violent' apply not only to all kinds of motion (qualitative and quantitative as well as local), but also to rest. The holding of a ball in the air is a 'violent' rest.
29 Aristotle, *Physics*, VII.1.241b. The medieval appropriation of this principle and its role in the thought of Aquinas will be discussed in Chapter 4.

30 J. Weisheipl, *Nature and Motion in the Middle Ages*, ed. W. E. Carroll, Washington, DC: Catholic University Press of America, 1985, pp. 25–48 and 75–120. See also A. Moreno, 'The Law of Inertia and the Principle "Quidquid movetur ab alio movetur"', *The Thomist* 38, 1974, 306–331.

31 W. D. Ross, *Aristotle's* Physics, Oxford: Oxford University Press, 1936, p. 722, cited in Weisheipl, *Nature and Motion in the Middle Ages*, p. 80. Such an interpretation is suggested by numerous translations which ignore the middle passive sense of 'κινεῖσθαι'. For example, 'If a thing is in motion it is, of necessity, being kept in motion by something', *Physics*, VII.1.241b, trans. P. H. Wicksteed and F. M. Cornford, Cambridge, MA: Harvard University Press, 1934.

32 Aristotle, *Physics*, VIII.5.256b.

33 Aristotle, *Physics*, VIII.4.254b.

34 See, for example, S. Waterlow, *Nature, Change and Agency in Aristotle's* Physics, Oxford: Oxford University Press, 1982, pp. 204–257. It seems that on Waterlow's interpretation, all nature becomes 'self-moved'. See also W. Charlton, *Aristotle's* Physics: *Books I and II*, Oxford: Clarendon Press, 1970, p. 92. For Aristotle's clear statement that simple bodies are not self-moving, see *Physics*, VIII.4.255a. For a critical assessment of the suggestion that a differentiation between mover and moved is possible in relation to matter (as the moved) and form (the mover), see Weisheipl, *Nature and Motion in the Middle Ages*, pp. 81–82.

35 Aristotle, *Physics*, VIII.4.255a; Aristotle, *De Anima*, II.5.417a. For references to *De Anima*, see Aristotle, *On the Soul/Parva Naturalia/On Breath* (Loeb Classical Library), trans. W. S. Hett, Cambridge, MA: Harvard University Press, 1957.

36 For a discussion of the distinction between principle and cause, see Weisheipl, *Nature and Motion in the Middle Ages*, pp. 103–108.

37 See, for example, D. Graham, *Aristotle's* Physics, Book 8: *translated with a commentary*, Oxford: Oxford University Press, 1999, pp. 74–89.

38 The definitive account of gravity comes in Newton's *Principia Mathematica*: 'Gravity exists in all bodies universally and is proportional to the quantity of matter in each'. Isaac Newton, *The* Principia: *Mathematical Principles of Natural Philosophy* ('Philosophiae Naturalis Prinicipia Mathematica', third edition, 1726), trans. I. Bernard Cohen and Anne Whitman, London: University of California Press, 1999, Book 3, Proposition 7, Theorem 7, p. 810.

39 H. Lang, *The Order of Nature in Aristotle's* Physics: *Place and the Elements*, Cambridge: Cambridge University Press, 1998, pp. 253 ff.

40 For the suggestion that place is a final cause of, for example, the downward motion of heavy bodies, see R. Sorabji, *Matter, Space and Motion*, London: Duckworth, 1988, p. 222; Pierre Duhem, *Le Système du Monde: Histoires des doctrines cosmologiques de Platon à Copernic*, Paris: Hermann, 1988 edition, Tome 1, ch. 11. For the argument that place is not exclusively a final cause, see Lang, *The Order of Nature in Aristotle's* Physics, p. 252.

41 See Aristotle, *De Motu Animalum*, 698a7–10 and 700a7–12; Aristotle, *Physics*, VIII.4.254b14–33. For references to *De Motu Animalum*, see M. Nussbaum, *Aristotle's* De Motu Animalum: *text and translation, commentary and interpretive essays*, Princeton, NJ: Princeton University Press, 1985.

42 Aristotle, *Physics*, VII.1.241b ff.

43 Aristotle, *De Anima*, I.4.408b ff.

44 D. J. Furley, 'Self-Movers', in Mary Louise Gill and J. G. Lennox (eds), *Self-Motion: from Aristotle to Newton*, Princeton, NJ: Princeton University Press, 1994, p. 13.

45 However, the order in which Aristotle describes the occurrences in the soul leads to the conclusion that the soul of an animal is only moved by the object of desire in a secondary sense. There is a distinction (which will be discussed in more detail below) between what Aristotle terms *energeia* ('operation' or 'activity') and motion proper. As will be seen, *energeia*, unlike motion, does not have an end or goal outside itself. *Energeia* would include, for example, thinking or seeing. By contrast, phenomena such as learning or weaving are more properly described as motions because they are not ends in themselves;

these particular motions aim at knowledge or the production of cloth. Contemplation or seeing are instances not of motion but of *energeia*, whereas desire, because it has a goal outside itself in the satisfaction of that desire, is a motion proper. The object perceived by the animal initially provokes in the animal's soul not the motion of desire, but the *energeia* of seeing or contemplation. In other words, the object in question provokes the soul to its characteristic operation and this operation then leads to the recognition of the object as desirable, and this desire in turn moves the animal from the potency of attaining the object to the actuality of the attainment of the object. Thus the external object only 'moves' the soul by provoking the soul's operation of seeing or contemplating which in turn gives rise to a proper motion in desire. This is why Aristotle was keen to maintain the belief that the soul is a mover but is not *moved* in the strict sense. Furley ('Self-Movers', p. 14) ends his article by stating that 'Although he [Aristotle] could plausibly retain the proposition that animals are self-movers, I am not sure that it would be worth struggling to retain the concept of the animal soul as unmoved mover. The point is that external objects are not in themselves sufficient causes for the voluntary movements of animals. But they do have some effect on the soul, and it would be obstinate of Aristotle to deny the effect can be called a movement'. Yet within the *energeia–kinesis* framework it is possible to see why Aristotle claims that the soul is unmoved. The 'motion' of the soul is the *energeia* of 'seeing' or 'understanding' an object as significant so as *then* to initiate *kinesis*. However, interestingly, and with particular reference to Plato's soul and Aristotle's first unmoved mover, it is at this point that Aquinas seems to detect little difference between the two as he writes, 'According to Plato ... that which moves itself is not a body. Plato understood by motion any given operation, so that to understand and to judge are a kind of motion. Aristotle likewise touches upon this manner of speaking in the *De Anima*. Plato accordingly said that the first mover moves himself because he knows himself and wills or loves himself. In a way, this is not opposed to the reasons of Aristotle. There is no difference between reaching a first being that moves himself, as understood by Plato, and reaching a first being that is absolutely unmoved, as understood by Aristotle', St Thomas Aquinas, *Summa Contra Gentiles*, trans. A. C. Pegis, London: University of Notre Dame Press, 1975, I.13.10. I will concur with Furley below in arguing that Aristotle maintains too great a wedge between *energeia* and *kinesis* as, unlike Plato, he denies relationality (a fundamental characteristic of motion) at the highest level.

46 Just as there is a hierarchy of motion, so too for Aristotle there is a hierarchy of rest (see, for example, Aristotle, *Physics*, V.6.230b–231a).

47 For detailed discussions of this distinction see, for example, J. L. Ackrill, 'Aristotle's Distinction Between *Energeia* and *Kinesis*', in R. Bamborough (ed.), *New Essays on Plato and Aristotle*, London: Routledge & Kegan Paul, 1965, pp. 121–141; S. Menn, 'The Origin of Aristotle's Concept of Ενεργεια: Ενεργεια and Δυναμις', *Ancient Philosophy* 14, 1994, 73–114; D. Graham, 'The Development of Aristotle's Concept of Actuality: Comments on a Reconstruction by Stephen Menn', *Ancient Philosophy* 15, 1995, 551–564; G. A. Blair, 'Unfortunately, It Is a Bit More Complex: Reflections on Ενεργεια', *Ancient Philosophy* 15, 1995, 565–580.

48 Aristotle, *Metaphysics*, IX.6.1048b23–25.

49 By 'thinning' Aristotle is referring to the slimming of the body or the thinning of a piece of wood.

50 J. Protevi describes motion as 'me-ontological interweaving' in *Time and Exteriority: Aristotle, Heidegger, Derrida*, London and Toronto: Associated University Presses, 1994, pp. 49–51.

51 Cf. Aristotle, *De Anima*, II.5.417a21 ff. In the following analysis, I am indebted to L. A. Kosman, 'Aristotle's Definition of Motion', *Phronesis* 14, 1969, 40–62.

52 I borrow these terms relating to degrees of potentiality from Kosman, 'Aristotle's Definition of Motion'.

53 It is for this reason that Kosman objects to Ross's understanding of Aristotle's definition of motion. Ross claims that the motion of building is the realisation of the potentiality of

bricks and stones of being fashioned into a house (see W. D. Ross, *Aristotle's* Physics, Oxford: Oxford University Press, 1936, p. 537, cited in Kosman, 'Aristotle's Definition of Motion', pp. 43–44). However, this would render Aristotle's definition vacuous and equivalent to the claim that motion is the realisation of the potential to be in motion (Kosman, p. 44). Thus if one defines the motion of building as the realisation of the potentiality of bricks to become a house, one has *not* identified the process of building, but rather the process of that process coming into being. However, Aristotle is adamant that there is no such 'motion' by which motion begins (cf. *Physics* V and VI). However, Kosman claims that it is not the *potentiality* which is realised. Rather, *the bricks and stones* are realised *qua* potentially a house. In other words, the motion of building is not the realisation of the potentiality of bricks and stones to *become* a house, but rather the realisation of bricks and stones to be potentially a house. Kosman's understanding of Aristotle avoids the sense that Aristotle has unhelpfully defined motion as the realising of a potential to be in motion. With regard to the example of the motion of building, it is when the buildable is being built that it fully manifests itself as actually buildable. The motion of building is *not* the actualisation of a potentiality *to become* a house (this identifies the process whereby building begins), but the actualisation of bricks and stones as potentially a house.
54 Aristotle, *De Anima*, II.5.417a.
55 Cf. Aristotle, *Physics*, IV.12.221b3. See P. Aubenque, *Le Problème de L'Être chez Aristote*, Quadrige: Presses Universitaires de France, 1962, pp. 433 ff.
56 Aristotle, *Metaphysics*, IX.6.1048b ff.
57 Kosman, 'Aristotle's Definition of Motion', p. 57; Protevi, *Time and Exteriority*, p. 38.
58 See Plato, *Republic*, 502d ff.
59 See Aristotle, *Physics*, VIII.1.
60 See Plotinus, *Enneads*, VI.9.
61 See especially below, pp. 110 ff. of this work.
62 Aristotle, *Physics*, VIII.7.260b.
63 Aristotle, *De Caelo*, I.2. For references to *De Caelo*, please see Aristotle, *On the Heavens* (Loeb Classical Library), trans. W. K. C. Guthrie, Cambridge, MA: Harvard University Press, 1939.
64 Aristotle, *Physics*, IV.11.219b.
65 See Aristotle, *Physics*, VIII.9.
66 Aristotle, *Physics*, IV.2.209a.
67 Aristotle, *Physics*, IV.4.210b.
68 Aristotle, *Physics*, IV.4.211b69. For an interpretation of Aristotle which claims an equation of place with the two-dimensional surface of a body, see, for example, R. Sorabji, *Matter, Space and Motion: theories in antiquity and their sequel*, London: Duckworth, 1988, pp. 127–128, 187. Such an interpretation is implied in numerous translations (for example, *Physics*, IV.4.211b9 of the Loeb translation). For a contrasting account of place (including an analysis of the translation of διαστημα, see H. Lang, *The Order of Nature*, ch. 3, especially pp. 105 ff. In the present discussion I am indebted to Lang's clear account of Aristotle's understanding of place as a motionless limit.
69 Aristotle, *Physics*, IV.4.212a20; cf. IV.1.209a5–7: 'Now a *topos*, as such, has the three dimensions of length, breadth, and depth which determine the limits of all bodies; but it cannot itself be a body, for if a "body" were in a "place" and the place itself were a body, two bodies would coincide'.
70 Place may be likened to both form and matter. On the one hand, both form and place are limits, yet form is the limit of the thing contained, whereas place is the containing limit of the whole body. Both matter and place 'hold' motion, yet matter cannot be separated from or surround the body (cf. *Physics*, IV.4.212a ff.).
71 Aristotle, *Physics*, III.5.206a.
72 Aristotle, *De Caelo*, I.8.276a23 ff.; Aristotle, *Physics*, IV.1.208b15 ff. The nature of place as a cause of motion is contentious. Sorabji argues that place is a final cause of a body's motion for the body aims to reach its proper resting place by its motion up or

down. See Sorabji, *Matter, Space and Motion*, p. 222. Lang argues that place is a cause, but not of the final, formal, material or efficient variety. See Lang, *The Order of Nature*, pp. 72 and 252.

73 All translations are by H. Rackham in the Loeb Classical Library edition, Cambridge, MA: Harvard University Press, 1934. Hereafter, this work is cited as 'NE'.

74 Aristotle, *NE*, I.7.

75 Aristotle, *NE*, X.4.8.

76 See, for example, Aristotle, *Metaphysics*, IX.8.1049b24; Aristotle, *Physics*, VIII.

77 See Aristotle, *De Caelo*, II.13.293b11.

78 Aristotle, *Metaphysics*, XII.10.1075a12 ff.

3 Light, motion and *scientia experimentalis*

1 For a general account of the reception of Aristotle into the Latin west, see N. Kretzmann, A. Kenny, J. Pinborg (eds), *The Cambridge History of Later Medieval Philosophy: from the rediscovery of Aristotle to the disintegration of Scholasticism, 1100–1600*, Cambridge: Cambridge University Press, 1988, part II. For a summary of the contemporary medieval Latin translations of Aristotle, see A. de Libera, *La Philosophie Médiévale*, Paris: Presses Universitaires de France, 1993, pp. 358 ff.

2 See, for example, E. Grant, *The Foundations of Modern Science in the Middle Ages*, Cambridge: Cambridge University Press, 1996: 'Aristotle's natural books formed the basis of natural philosophy in the universities, and the way in which medieval scholars understood the structure and operation of the cosmos must be sought in those books' (p. 54). In his analysis of Grosseteste, Steven Marrone makes the rather implausible claim that Grosseteste wholeheartedly rejected the Neoplatonism of his upbringing in favour of the newly available Aristotelian philosophy. More particularly, Marrone claims that Grosseteste 'eliminated the divine element from his definition of simple truth' so that 'the path lay open for him to explain human knowledge and the procedure by which men came to recognize the truth in a way that emphasized objects and powers in the created world', S. Marrone, *William of Auvergne and Robert Grosseteste: New Ideas of Truth in the Early Thirteenth Century*, Princeton, NJ: Princeton University Press, 1983, p. 161. See especially pp. 157 ff. Aside from the arguments below which suggest otherwise, it seems prima facie most unlikely that any medieval theologian would forsake the Augustinian tradition in which they were educated in favour of a 'new' natural philosophy. Accommodation would be the most radical option.

3 This victory is located principally in the use of mathematics to understand nature and motion. As was seen in Chapter 1, the expression of the order of the cosmos as a series of harmonic and geometric ratios expressible in number was integral to the *Timaeus*. See Plato, *Timaeus*, 35b–36b.

4 A. Koyré, *Galileo Studies*, trans. J. Mepham, Hassocks, Sussex: The Harvester Press, 1978, p. 202. See also A. Koyré, 'Galileo and Plato', in *idem, Metaphysics and Measurement: essays in the scientific revolution*, London: Chapman & Hall, 1968, pp. 16–43; A. Funkenstein, *Theology and the Scientific Imagination*, Princeton, NJ: Princeton University Press, 1986, particularly sections 2 and 3; S. Shapin, *The Scientific Revolution*, Chicago: Chicago University Press, 1996, ch. 1, particularly pp. 57 ff.

5 P. Duhem, *Le Système du Monde: Histoires des doctrines cosmologiques de Platon à Copernic* (10 vols), Paris: Librairie Scientifique Hermann, 1913–1959. Duhem focuses particularly on the continuity between Aristotelian science and early modern physics. For a précis of Duhem's view, see his essay 'History of Physics', in *idem, Essays in the History and Philosophy of Science*, ed. and trans. R. Ariew and P. Barker, Indianapolis: Hackett, 1996, pp. 163–221. Of those who contest the 'continuity thesis', a prominent example is A. Maier, *On the Threshold of Exact Science: selected writings of Anneliese Maier on late medieval natural philosophy*, ed. and trans. S. D. Sargent, Philadelphia: University

of Pennsylvania Press, 1982. For a recent discussion of many pertinent issues, see M. J. Osler (ed.), *Rethinking the Scientific Revolution*, Cambridge: Cambridge University Press, 2000.

6 The influence of Plato's *Timaeus* and later Neoplatonism is particularly clear in the work of Kepler. See R. Martens, *Kepler's Philosophy and the New Astronomy*, Princeton, NJ: Princeton University Press, 2000, especially pp. 32–38 and 40–50.

7 The suggestion that Bacon, in particular, was the founder of experimental science is found in the comments of the nineteenth-century mathematician and scientist William Whewell who wrote that 'Roger Bacon's works are not only so far beyond his age in the knowledge which they contain, but so different in the temper of the times, in his assertion of the supremacy of experiment, and in his contemplation of the future progress of knowledge, that it is difficult to conceive how such a character could then exist', W. Whewell, *History of the Inductive Science* (2 vols), New York, 1858, third edition, vol. 1, p. 245. It became commonplace in the later nineteenth and early twentieth centuries to develop this assessment of Bacon which was itself derived from the praise given by his namesake Francis Bacon, a writer who was otherwise disparaging in his assessment of medieval natural philosophy. For particularly enthusiastic appraisals of Roger Bacon as an experimental scientist, see, for example, the introduction to Robert Bridges' translation of Bacon's *Opus Majus*, in R. Bridges (trans.), *The Opus Majus of Roger Bacon* (3 vols), London, 1900; E. Charles, *Roger Bacon: Sa vie, ses ouvrages, ses doctrines*, Paris: L. Hachette, 1861, pp. 102 ff.; Charles Singer, 'The Dark Ages and the Dawn', in F. S. Marvin (ed.), *Science and Civilization*, London, 1923, pp. 139–143; Joseph Kupfer, 'The Father of Empiricism: Roger not Francis', *Vivarium* 22, 1974, 52–62; J. Hackett, 'Roger Bacon on *Scientia Experimentalis*', in *idem* (ed.), *Roger Bacon and the Sciences: commemorative essays*, Leiden: Brill, 1997, pp. 277–315. For the view that Bacon is not a precursor of modern experimental science, see, for example, D. Lindberg, *Roger Bacon's Philosophy of Nature: A Critical Edition, with English Translation, Introduction, and Notes, of* De Multiplicatione Specierum *and* De speculis comburentibus, Oxford: Oxford University Press, 1983, pp. liii ff., and M. Heidegger, *The Question Concerning Technology and Other Essays*, trans. W. Lovitt, New York: Harper & Row, 1977, p. 122: 'If, now, Roger Bacon demands the *experimentum* – and he does demand it – he does not mean the experiment of science as research; rather he wants the *argumentum ex re* instead of the *argumentum ex verbo*, the careful observing of things themselves, i.e. Aristotelian *empeiria*, instead of the discussion of doctines'. As will be discussed below, Bacon was deeply indebted to the work of Grosseteste who has himself been understood as a forerunner to modern experimental science. This view receives its most sustained defence in A. C. Crombie, *Robert Grosseteste and the Origins of Experimental Science 1100–1700*, Oxford: Oxford University Press, 1953. Hereafter this work is cited as 'Crombie, *Origins*'. Crombie's thesis has proved controversial and is discussed in detail below. For further comments, see, for example, A. Koyré, 'The Origins of Modern Science: A New Interpretation', *Diogenes* 16, 1956, 1–22.

8 See, for example, Pseudo-Dionysius, *De Divinis Nominibus*, IV.697c ff.; Plotinus, *Enneads*, I.6.3; III.8.5 and 11; IV.3.11; V.5.7; VI.7.41, *et passim*.

9 See, for example, St Augustine, *De Trinitate*, II.2, IV.27, VII.3 to 5, VIII.2 and 3, XII.15; St Basil, *Hexaëmeron*, II.7 ff., VI, *et passim*. For a detailed description of Augustine's uses of light imagery, see F.-J. Thonnard, 'La notion de lumière en philosophie augustinienne', *Recherches Augustiniennes*, 1962, 124–175, and R. A. Markus, 'Augustine: Reason and Illumination', in A. H. Armstrong (ed.), *The Cambridge History of Later Greek and Early Medieval Philosophy*, Cambridge: Cambridge University Press, 1970, pp. 362–373.

10 On the importance of light in Franciscan spiritual mysticism and its relationship to later natural philosophy, see R. French and A. Cunningham, *Before Science: The Invention of the Friars' Natural Philosophy*, Aldershot: Scolar Press, 1996, chs 9 and 10. Although Grosseteste was closely associated with the Franciscans, becoming their first Lector in Oxford around 1230 in the years before taking the See of Lincoln, he never joined the Order. Bacon took the habit around 1257. See R. Southern, *Robert Grosseteste: The Growth of an English Mind in Medieval Europe*, Oxford: Oxford University Press, 1986, ch. 4, and

A. Little, 'Introduction: On Roger Bacon's Life and Works', in *idem*, *Roger Bacon: essays contributed by various writers on the occasion of the commemoration of the seventh century of his birth*, Oxford: Oxford University Press, 1914, pp. 1–32.

11 For a general overview of light and its relation to metaphysics, see D. C. Lindberg, 'The Genesis of Kepler's Theory of Light: Light Metaphysics from Plotinus to Kepler', *Osiris* 2nd series, 1986, 5–42.

12 R. Grosseteste, *De Veritate*, p. 137: 'Therefore, created truth also shows that which is, but not in its own illumination, but in the light of the Highest Truth, as colour shows the body, but only in the light put upon it. ... In the same fashion is the power of the light of the Highest Truth, which so illumines the created truth that, illumined itself, it reveals the true object'. Grosseteste's twenty-seven philosophical works, including *De Veritate* and *De Luce*, are available in L. Baur (ed.), *Die philosophischen Werke des Robert Grosseteste, Bischofs von Lincoln* (BGPM Bd. IX) (Münster i. W., 1912), available at: http://www.grosseteste.com/baurframe.htm (accessed 28 July 2004). I follow the pagination of Baur's Latin text.

13 Grosseteste, *De Lineis, Angulis et Figuris*, pp. 59–60.

14 For example, Genesis 1; Isaiah 60.19; John 1.1–1.18, 8.12 and 9.5; Acts 22.6 f.; Ephesians 5.14; 1 Timothy 6.16; 1 John 1.5; Revelation 21.23; Revelation 22.5.

15 See Pseudo-Dionysius, *De Divinis Nominibus*, 693b ff.

16 See, in particular, Crombie, *Origins*. For a more tempered view, see J. McEvoy, *The Philosophy of Robert Grosseteste*, Oxford: Oxford University Press, 1982, particularly pp. 206–222.

17 Grosseteste, *De Luce*, p. 51. Unless otherwise indicated, all translations are by C. C. Reidl, *Robert Grosseteste on Light*, Milwaukee: Marquette University Press, 1942. Marking the importance of Grosseteste's *De Luce*, McEvoy comments that this work is 'one of the few scientific cosmologies, and perhaps the only scientific cosmogony, written between the *Timaeus* and modern times' (McEvoy, *The Philosophy of Robert Grosseteste*, p. 151). Many names were used in the Middle Ages for what might be called 'light': *fulgor, ardor, radius, claritas*. In *De Luce*, Grosseteste distinguishes between *lux* and *lumen*. The former is the primordial simple 'light' or 'form' from which the universe emerges; the latter is visible light, a subtle bodily spirit whose motion is characterised by instantaneous self-propagation.

18 This is not, according to Grosseteste, a local motion, for if it were we would perceive illumination to occur in stages. See *Hexaëmeron*, Part II, ch. 10, section 1.

19 See Grosseteste, *De Luce*, p. 52: 'Thus light, which is the first form created in first matter, multiplied itself by its very nature an infinite number of times on all sides and spread itself out uniformly in each direction. In this way it proceeded in the beginning of time to extend matter which it could not leave behind, by drawing it out along with itself into a mass the size of the material universe'.

20 It is not clear to which of Aristotle's texts Grosseteste is referring. McEvoy (*The Philosophy of Robert Grosseteste*, p. 152, n. 10) suggests *De Caelo*, I.5.271b15 ff.

21 Grosseteste, *De Luce*, p. 53.

22 Ibid.

23 Ibid. As an aside, Grosseteste remarks that this principle was well known to both Atomists and Platonists. The former understood all things to be composed of atomic units, while the latter, as was seen in Chapter 1, believed all things to be composed of surfaces, lines and points.

24 See Plato, *Republic*, 524a ff. Grosseteste's line of thought may have interesting origins in Pythagorean science. Of the Pythagoreans, Jacob Klein comments that, 'We may conjecture that they [Pythagoreans] saw the genesis of the world as a progressive *partitioning* of the first "whole" *one*, about whose origins they themselves, it seems, were not able to say anything conclusive. ... This first "one", as well as the subsequent "ones" which were the result of partition, i.e., the "numbers" themselves, they therefore regarded as having *bodily extension*', in *Greek Mathematical Thought and the Origin of Algebra*, Cambridge, MA: MIT Press, 1968, p. 67.

25 Grosseteste, *De Luce*, p. 54.

26 Grosseteste, *De Luce*, p. 55: 'its passing takes place through the multiplication of itself and the infinite generation of light (*lumen*)'. This motion only occurs to the centre of the cosmos because, for Grosseteste, there is no 'outside'.

27 Although McEvoy ascribes only a 'general' influence to the *Timaeus*, this section of *De Luce* is highly reminiscent of Plato's cosmology. On the sources employed by Grosseteste in writing *De Luce*, see McEvoy, *The Philosophy of Robert Grosseteste*, pp. 158–167.

28 Grosseteste, *De Luce*, p. 57: 'Since the inferior bodies participate in the form of the superior bodies (*participant formam superiorum corporum*), the inferior, by participating of the same form of the superior body, is capable of motion by the same incorporeal motive power, by which motive power the superior body is moved'.

29 Ibid.

30 Grosseteste, *De Luce*, pp. 57–58.

31 Grosseteste, *De Luce*, p. 58.

32 Ibid.

33 Grosseteste, *De Luce*, p. 59.

34 Grosseteste, *Hexaëmeron*, (On the Six Days of Creation), trans. C. F. J. Martin, Oxford: Oxford University Press, 1996. All page references are to this edition of Grosseteste's *Hexaëmeron*.

35 Grosseteste, *De Veritate*, p. 134. Unless otherwise indicated, all translations are from R. McKeon (ed. and trans.), *Selections from Medieval Philosophers*, vol. 1, New York: Charles Scribner's Sons, 1957.

36 Ibid. Aquinas was later to adopt a very similar approach to truth. Grosseteste here prioritises 'interior' speech over verbal speech because he understands truth to be predicated on a hierarchy of emanation in which a more immanent emanation implies a more replete and complete communication of being. See Chapter 4, pp. 111 ff., for a discussion of this matter in relation to Aquinas.

37 Grosseteste, *De Veritate*, p. 135.

38 Ibid.: 'But in so far as a thing is as it should be, to that extent it is true. Therefore, the truth of things is for them to be as they should be and is their rightness and conformity to the Word by which they are said eternally'.

39 Ibid.

40 Grosseteste, *De Veritate*, p. 137.

41 Grosseteste, *De Veritate*, pp. 137–138.

42 Grosseteste, *De Veritate*, p. 138. This issue receives greater attention in Grosseteste's *Commentarius in Posteriorum Analyticorum Libros*. For a discussion of this matter in the Commentary see McEvoy, *The Philosophy of Robert Grosseteste*, pp. 323 and 332–334.

43 Grosseteste, *De Veritate*, p. 138.

44 This we learn from Grosseteste when he likens the dependence of created being on God's eternal Word to the dependence of water on its container for its support and form. See Grosseteste, *De Veritate*, pp. 141–142.

45 McEvoy, *The Philosophy of Robert Grosseteste*, p. 326. For McEvoy's full discussion of Grosseteste and the charge of ontologism, see pp. 324 ff. While concurring with much of McEvoy's discussion, it will be evident from what follows that I avoid his description of Grosseteste's theory as 'dualistic' (p. 328). It seems that the whole thrust of Grosseteste's *De Veritate* is towards the delineation of an account of truth which recognises multiplicity and difference without juxtaposing this with the simplicity of the divine *lux* in any proto-modern, dualistic fashion.

46 See, for example, Grosseteste, *Hexaëmeron*, I.6.1 (pp. 53 ff.).

47 Grosseteste, ed. P. Rossi, *Commentarius in Posteriorum Analyticorum Libros*, Firenze: Leo S. Olschiki Editore, 1981, Book I, ch. 7, line 96, pp. 139 ff. All page references, when provided, refer to this edition of Grosseteste's commentary. See also P. Duhem, *Le Système du Monde: histoires de doctrines cosmologiques de Platon à Copernic*, vol. V, Paris: Hermann, 1958, pp. 345–351; McEvoy, *The Philosophy of Robert Grosseteste*, pp. 327–329; S. Marrone, *William of Auvergne and Robert Grosseteste: New Ideas of Truth in the Early Thirteenth Century*, Princeton, NJ: Princeton University Press, 1983, pp. 167–171.

48 Grosseteste, *Commentarius in Posteriorum Analyticorum Libros*, I.7.141 ff.
49 I borrow the term 'motionless motion' from Wayne Hankey. He employs this phrase to describe Aquinas's view of the divine life. See W. Hankey, *God in Himself: Aquinas' Doctrine of God as Expounded in the* Summa Theologiae, Oxford: Oxford University Press, 1987, p. 56. I will employ the same phrase in the next chapter when discussing Aquinas on the Trinity.
50 Grosseteste, *Commentarius in Posteriorum Analyticorum Libros*, I.7.102–103.
51 Some qualification is required here: as was seen in Chapter 2, Aristotle does not regard generation as a motion *per se* because it involves no constant subject.
52 Grosseteste, *Hexaëmeron*, II.7.1 (pp. 94–95).
53 Grosseteste, *De Luce*, p. 57: 'For this reason the incorporeal power of intelligence (*intelligentiae*) or soul, which moves the first and highest sphere with a diurnal motion, moves all the lower heavenly spheres with this same diurnal motion'.
54 Grosseteste makes some very different claims concerning the diurnal rotation of the heavens depending on which 'heaven' he is considering. In his *Hexaëmeron*, he directly opposes the view which states that the first heaven (the resting place of the blessed which is above both the water and the 'second heaven' or 'firmament' as these are described in Genesis) has a motion because the successive attainment of as many positions as possible most directly resembles the actuality of the creator. Mockingly, he states that, 'If [an intelligence] moves the heaven according to the account given, then on the same account human beings, too, ought to be in perpetual motion, wandering over land and sea, so that they might take up every position possible to them and thus become more like the creator who at one and the same time has everything that he can have' (Grosseteste, *Hexaëmeron*, I.17.1, p. 75). Grosseteste therefore concludes that it is most appropriate to say that the heaven is at rest: 'So the state of rest, and an action performed by one at rest, is more like the state of divinity, as has been said' (ibid.). Moreover, he seems to express a functional view of motion which understands the movement of the heavens merely to be necessary for generation which has its ultimate goal in the generation of the body of Christ, the Church. However, three points require consideration. First, it is not clear what kind of 'motion' Grosseteste thinks might be attributed to the first heaven. In the passage quoted above, he does not mention *diurnal* or *circular* motion. Secondly, one must preserve a careful distinction between the first heaven which is the resting place of the blessed and the second heaven or firmament where the celestial spheres reside. Thirdly, and with reference to the second heaven, Grosseteste does not appear to offer such a functional view of motion. Later in the *Hexaëmeron* we learn that the second heaven, or firmament, moves 'by a simple uniform movement, from east to west', thereby '[causing] to turn with it, in its daily movement, the whole that is beneath it, with the sun, moon and stars, down as far as the sphere that lies beneath it' (III.16.3, pp. 118–119). Why does this heaven have a circular motion? Because this motion is 'uniform, ordered, undisturbed, and is the swiftest of all movements. It is in movement *though it is at rest*, since though it partly alters its position, it does not alter its position as a whole, nor does it alter its place' (III.16.4, p. 119, my emphasis). Grosseteste goes on to claim that this singular, simple motion is one of the three modes of unity of this heaven and thereby the entire cosmos. It therefore seems that a particular quality of motion is indeed akin to actuality. Whether this heaven or firmament is moved in turn by an intelligence is unclear. McEvoy believes that Grosseteste opposed this view (McEvoy, *The Philosophy of Robery Grosseteste*, p. 191). However, we have seen in *De Luce*, in addition to other indications in the *Commentary on the Celestial Hierarchy* cited but dismissed by McEvoy, some evidence that Grosseteste held to the notion of an intelligence as the source of heavenly motion.
55 See Grosseteste, *De Luce*, pp. 57–58.
56 Grosseteste, *Hexaëmeron*, VI.1.1–3 (pp. 187–188). Grosseteste describes how different parts of the hierarchy are mixed: 'The air is made thick and gross with the exhalations of the other two [earth and water], and made gross by watery moisture and contains water, and is often included with the heavy liquid of water under the name "water"' (VI.1.2, p. 187).

57 Quoting Augustine, *De Genesi ad litteram*, III.5–6. More particularly, Grosseteste held to the extramission theory of vision propounded by Plato, Aristotle, Ptolemy and Euclid. On this view, the act of vision includes rays pouring forth from the eye. See Grosseteste, *Commentarius in Posteriorum Analyticorum Libros*, II.4.464 ff., p. 386; *De Iride*, pp. 72–73, cited in Crombie, *Origins*, p. 118. As Crombie states, the thirteenth century saw considerable debate about whether light is a real movement in space or merely the result of perception. Roger Bacon appears to opt for the former explanation, adding that light passes in an imperceptible time, and therefore with a motion we cannot perceive. See *The Opus Majus of Roger Bacon*, ed. and trans. Robert Belle Burke, Philadelphia: University of Pennsylvania Press, 1928, part 5, distinction 9, ch. 3, pp. 488 ff. The division of Bacon's *Opus Majus* is not consistent. References will therefore not be abbreviated. Page numbers refer to the edition by Robert Belle Burke.

58 On this doctrine, which received more comprehensive enunciation in the work of Roger Bacon, see D. Lindberg, *Roger Bacon's Philosophy of Nature*. Lindberg remarks that this doctrine has its origins in the work of Arabic philosophy, in particular al-Kindī (p. lii).

59 Grosseteste, *Hexaëmeron*, II.10.1 (pp. 97–98).

60 See McEvoy, *The Philosophy of Robert Grosseteste*, p. 94, n. 73.

61 Grosseteste, *Hexaëmeron*, VII.14.1 (p. 214).

62 My emphasis. Cited in Crombie, *Origins*, p. 73.

63 See Plato, *Phaedrus*, 250e ff.

64 Grosseteste, *Commentarius in Posteriorum Analyticorum Libros*, I.17.340–365.

65 Quoted in Crombie, *Origins*, p. 53. See Aristotle, *Posterior Analytics*, I.13.

66 Crombie, *Origins*, p. 53, n. 4.

67 Aristotle, *Posterior Analytics*, I.2.

68 Aristotle, *Posterior Analytics*, I.18.81b1–10. All translations are available in Aristotle, *Posterior Analytics/Topica* (Loeb Classical Library), trans. H. Tredennick and E. S. Forster, Cambridge, MA: Harvard University Press, 1960. See also I.31 where Aristotle explains that 'scientific knowledge cannot be acquired by sense-perception'.

69 Grosseteste, *Commentarius in Posteriorum Analyticorum Libros*, II.4, pp. 256 ff., pp. 376 ff.

70 Crombie, *Origins*, pp. 64 ff. Grosseteste himself adopts this method in investigating the 'definition' or nature common to horned animals. See Grosseteste, *Commentarius in Posteriorum Analyticorum Libros*, II.4, pp. 381 ff.

71 Crombie, *Origins*, p. 65.

72 Crombie, *Origins*, p. 66.

73 Crombie, *Origins*, p. 71 (and n. 3).

74 The first of these assumptions refers to 'intuitive induction' (one has to intuit a link between cause and effect). The second refers to 'enumerative induction' (one cannot provide a fully exhaustive account of the link of cause and effect between observed particulars and so must assume the uniformity of nature). This distinction is discussed in more detail below.

75 Quoted in Crombie, *Origins*, p. 81.

76 Aristotle, *Generation of Animals*, III.10.760.b31. All translations are by A. L. Peck in Aristotle, *Generation of Animals* (Loeb Classical Library), trans. A. L. Peck, Cambridge, MA: University of Harvard Press, 1963. In the *Posterior Analytics*, Aristotle writes that 'if by observing repeated instances we had succeeded in grasping the universal, we should have our proof; because it is from the repetition of particular experiences that we obtain our view of the universal' (I.31.88a ff.).

77 Crombie, *Origins*, pp. 84 ff.

78 Grosseteste, *De Generatione Stellarum* and *De Cometis*, discussed in detail in Crombie, *Origins*, pp. 87–90.

79 See Grosseteste, *Roberti Grosseteste Episcopi Lincolniensis Commentarius in VIII libros Physicorum Aristotelis, IV*, cited in Crombie, *Origins*, pp. 99–100.

80 See Crombie, *Origins*, pp. 99–102.

81 Grosseteste, *Commentarius in Posteriorum Analyticorum Libros*, I.14.247–271.

82 Grosseteste, *Commentarius in Posteriorum Analyticorum Libros*, I.14.254.
83 E. Serene, 'Robert Grosseteste on Induction and Demonstrative Science', *Synthese* 40, 1979, 97–115.
84 Serene, 'Robert Grosseteste on Induction and Demonstrative Science', p. 100. See Aristotle, *Posterior Analytics*, II.19.
85 Serene, 'Robert Grosseteste on Induction and Demonstrative Science', p. 101, quoting J. Barnes, *Aristotle's Posterior Analytics*, Oxford: Oxford University Press, 1975, pp. 256–257.
86 See, for example, Crombie, *Origins*, p. 71: 'To leap this gap in the logical process of induction he [Grosseteste] envisaged an act of intuition or scientific imagination, corresponding to Aristotle's νοῦς, by which the mind reflecting on the classification of facts produced by induction suddenly grasped a universal or principle or theory explaining the connexion between them'.
87 See Crombie, *Origins*, p. 57.
88 Serene argues that neither Aristotle nor Grosseteste make this distinction. My concurrence with her criticism of Crombie's distinction will be evident shortly. See Serene, 'Robert Grosseteste on Induction and Demonstrative Science', pp. 105–106.
89 See Grosseteste, *De Generatione Stellarum* (ed. Baur, p. 32) cited in Crombie, *Origins*, p. 85.
90 Quoted in Crombie, *Origins*, p. 81.
91 See Crombie, *Origins*, pp. 85 ff.
92 Serene, 'Robert Grosseteste on Induction and Demonstrative Science', p. 102. See Crombie, *Origins*, p. 134.
93 Serene, 'Robert Grosseteste on Induction and Demonstrative Science', pp. 110–112.
94 The following point is made by Serene, 'Robert Grosseteste on Induction and Demonstrative Science', p. 111.
95 See Grosseteste's beautiful illustration of this in *De Veritate*, p. 142.
96 Grosseteste, *Commentarius in Posteriorum Analyticorum Libros*, I.18.205–207 (p. 269), quoted by R. Southern, *Robert Grosseteste: The Growth of an English Mind in Medieval Europe*, Oxford: Oxford University Press, 1986, p. 165.
97 R. Bacon, *Opus Majus*, trans. Robert Belle Burke, Philadelphia: University of Pennsylvania Press, 1928, part 2, ch. 3 (p. 39). See also Augustine, *De Doctrina Christiana*, 2.144 ff. For dedicated treatments of Bacon's understanding of the relationship between theology and philosophy, see J. Hackett, 'Philosophy and Theology in Roger Bacon's *Opus maius*', in R. James Long (ed.), *Philosophy and the God of Abraham: Essays in memory of James A. Weisheipl*, Toronto: Pontifical Institute of Medieval Studies, 1991, pp. 55–69; *idem*, 'Averroes and Roger Bacon on the Harmony of Religion and Philosophy', in J. Hackett, M. S. Hyman, R. James Long and C. H. Manekin (eds), *A Straight Path: Studies in Medieval Philosophy and Culture*, Washington, DC: Catholic University Press of America, 1988, pp. 98–112; and D. Lindberg, 'Science as Handmaiden: Roger Bacon and the Patristic Tradition', *Isis* 78, 1987, 518–536.
98 Bacon, *Opus Majus*, part 2, ch. 15 (p. 65). See also part 2, ch. 4 (p. 43): 'Christians may apply as their own in divine matters whatever is useful in the liberal sciences'. The author is here appealing to the authority of Bede and his commentary on the book of Kings.
99 Bacon, *Opus Majus*, part 2, ch. 1 (p. 36): 'I say, therefore, that one science is the mistress of the others, namely, theology, *to which the remaining sciences are vitally necessary*, and without which it cannot reach its end' (my emphasis). See part 2, ch. 14 (p. 65).
100 Bacon, *Opus Majus*, part 2, ch. 5 (p. 43).
101 See especially Bacon, *Opus Majus*, part 2, chs 5–7.
102 Bacon, *Opus Majus*, part 2, ch. 14 (pp. 64–65). Bacon claims that Plato must have read the books of Genesis and Exodus, the former because it is so similar to the *Timaeus*, the latter because it includes the name of God which Plato could not have otherwise known. On Aristotle's supremacy as a philosopher, see part 2, ch. 13 (pp. 62–63).
103 For the origins of the debate concerning the passive and active intellects, see Aristotle, *De Anima*, III.4–5.

104 Artistotle, *De Anima*, III.5.430a. All translations are by W. S. Hett in the Loeb Classical Library edition (*On the Soul*), Cambridge, MA: Harvard University Press, 1957.

105 Bacon, *Opus Majus*, part 2, ch. 5 (p. 48).

106 On its own, philosophy 'leads to the blindness of hell, and therefore it must be by itself darkness and mist', Bacon, *Opus Majus*, part 2, ch. 19 (p. 74).

107 Bacon, *Opus Majus*, part 4 (pp. 116–418); R. Steele, *Opera hactenus inedita Fr. Rogeri Baconis*, Oxford, 1905–1941, Fasc. XVI (1940): 'Communia Mathematica Fratris Rogeri'. For a more detailed assessment of Bacon on mathematics, see D. Lindberg, 'On the Applicability of Mathematics to Nature: Roger Bacon and his Predecessors', *British Journal of the History of Science* 15, 1982, 3–25; N. W. Fisher and S. Unguru, 'Experimental Science and Mathematics in Roger Bacon's Thought', *Traditio* 27, 1971, 353–378.

108 Bacon, *Opus Majus*, part 4, distinction 1, chs 2–3 (pp. 116 ff.). An example of Bacon's commitment to mathematics is seen in the following characteristic statement: 'These reasons are general ones, but in particular this point can be shown by a survey of all the parts of philosophy disclosing how all things are known by the application of mathematics. This amounts to showing that other sciences are not to be known by means of dialectical and sophistical argument as commonly introduced, but by means of mathematical demonstrations entering into the truths and activities of other sciences and regulating them, without which they cannot be understood, nor made clear, not taught, nor learned' (part 4, first distinction, ch. 3, pp. 126–127).

109 Bacon, *Opus Majus*, part 4, 1st distinction, ch. 2, (p. 120).

110 Ibid.

111 Ibid.

112 Bacon, *Opus Majus*, part 4, 1st distinction, ch. 3 (p. 125).

113 Bacon, *Opus Majus*, part 4, 1st distinction, ch. 3 (p. 126).

114 Bacon, *Opus Majus*, part 4, 2nd distinction, ch. 1 (pp. 128–129).

115 S. Shapin and S. Schaffer, *Leviathan and the Air-pump: Hobbes, Boyle and the Experimental Life*, Princeton, NJ: Princeton University Press, 1985, pp. 25 ff. Shapin and Schaffer discuss the 'material', 'literary' and 'social' technologies deployed by early modern scientists (their attention is particularly focused on Robert Boyle) in the production of scientific matters of fact.

116 See Bacon, *Opus Majus*, part 4, 2nd distinction, ch. 2 (pp. 131ff.); *idem*, De Multiplicatione Specierum (available in D. Lindberg, *Roger Bacon's Philosophy of Nature: A Critical Edition, with English Translation, Introduction, and Notes,* of De multiplicatione specierum *and* De speculis comburentibus, Oxford: Oxford University Press, 1983, hereafter cited as 'Lindberg (Bacon)'). See also D. Lindberg, 'Roger Bacon on Light, Vision, and the Universal Emanation of Force', in J. Hackett (ed.), *Roger Bacon and the Sciences: commemorative essays*, Leiden: E.J. Brill, 1997, 243–274.

117 See Lindberg (Bacon), pp. xxxv–liii.

118 Lindberg (Bacon), I.1.75–76, pp. 6–7.

119 Lindberg (Bacon), I.2.220 ff., pp. 32 ff.

120 Lindberg (Bacon), pp. lviii–lix.

121 Lindberg (Bacon), I.3.52, pp. 46–47.

122 Quoted in Lindberg (Bacon), p. lxiii.

123 Lindberg (Bacon), p. lxiii. See also Bacon, *Opus Majus*, part 4, 2nd distinction, ch. 1 (pp. 129–130).

124 Bacon, *Opus Majus*, part 4, 2nd distinction, ch. 2 (p. 131): 'Every multiplication is either with respect to lines, or angles, or figures'.

125 See also Crombie, *Origins*, p. 144.

126 O. Boulnois, *Être et Représentation: une généalogie de la métaphysique moderne à l'époque de Duns Scot*, Paris: Presses Universitaires de France, 1999, p. 72: 'Avicennisme ou augustinisme? En tout cas, la lignée franciscaine rejette l'espèce intelligible: le corps ne peut rien produire par lui-même directement sur l'esprit, pas même une espèce intelligible

simple et spirituelle, qui convienne à la simplicité de son sujet, puisque sa causalité est celle d'un être corporel et composé'.

127 See Boulnois, *Être et Représentation*, pp. 67 ff. and 88 ff.

128 Boulnois, *Être et Représentation*, p. 89; cf. R. Bacon, *Compendium of the Study of Theology*, ed. and trans. T. S. Maloney, Leiden : E.J. Brill, 1988, p. 72.

129 Boulnois, *Être et Représentation*, p. 90: 'L'être de la chose pensée ne peut plus se rabattre sur l'être de la chose en elle-même. Le plan de l'objectité s'est décollé de la surface du monde'.

130 C. Pickstock, *After Writing: On the Liturgical Consummation of Philosophy*, Oxford: Blackwell, 1998, pp. 130–131. See also É. Alliez, *Capital Times: Tales from the Conquest of Time*, trans. G. van den Abbeele, Minneapolis: University of Minnesota Press, 1996, pp. 238 ff. For knowledge as representation in Avicenna, see Chapter 5.

131 See E. Gilson, 'Avicenne et le point de départ de Duns Scot', *Archives d'histoire doctrinale et littéraire du moyen âge* 2, 1927.

132 Bacon, *Opus Majus*, part 6, ch. 1 (p. 585).

133 Bacon, *Opus Majus*, part 6, chs 2–14 (pp. 587–634).

134 Bacon, *Opus Majus*, part 6, ch. 2 (p. 587). See also part 6, ch. 12, 3rd prerogative (p. 632): 'This science, moreover, knows how to separate the illusions of magic and to detect all their errors in incantations, invocations, conjurations, sacrifices, and cults'.

135 Bacon, *Opus Majus*, part 6, ch. 2 (p. 588 ff.). This is Bacon's clearest description of an experimental procedure.

136 Bacon, *Opus Majus*, part 6, ch. 12, 3rd prerogative (p. 629).

137 Ibid. (p. 632).

138 R. Bacon, *Epistola fratris Rogerii Baconis de secretis operibus artis et naturae, et de nullitate magiae*, 6, in *idem*, *Opera quaedam hactenus inedita*, ed. J. S. Brewer, London: Longman, Green, Longman and Roberts, 1859, cited in L. Daston and K. Park, *Wonders and the Order of Nature, 1150–1750*, New York: Zone Books, 1998, p. 94. See also A. G. Molland, 'Roger Bacon as Magician', in *idem*, *Mathematics and the Medieval Ancestry of Physics*, Aldershot: Variorum, 1995, ch. 11.

139 L. Thorndike, *History of Magic and Experimental Science vol. II*, New York, 1923, p. 618; *idem*, 'Roger Bacon and Experimental Method in the Middle Ages', *Philosophical Review* 23, 1914, 271–298 (see especially pp. 281ff.).

140 See Daston and Park, *Wonders and the Order of Nature*.

141 Daston and Park, *Wonders and the Order of Nature*, pp. 215 ff.

142 Daston and Park, *Wonders and the Order of Nature*, p. 215.

143 N. Fisher and S. Unguru, 'Experimental Science and Mathematics in Roger Bacon's Thought', p. 364.

144 R. Boyle, 'A Disquisition on the Final Causes of Natural Things', in T. Birch (ed.), *The Works of the Honourable Robert Boyle*, London: J & F Rivington, 1772, p. 401, quoted in Shapin and Schaffer, *Leviathan and the Air-pump*, p. 319. Shapin and Schaffer write that, 'The form of persuasion used in experimentation was a powerful weapon against the atheists: the products of such experimentation would reinforce proper theology' (*Leviathan and the Air-pump*, p. 209).

145 'When I wrote my treatise about our Systeme [the *Principia*] I had an eye upon such Principles as might work wth considering men for the beleife of a Deity & nothing can rejoyce me more than to find it usefull for that purpose', H. W. Turnbull, J. F. Scott, A. R. Hall and L. Tilling (eds), *The Correspondence of Issac Newton*, vol. 3, Cambridge: Cambridge University Press, 1961, p. 233.

4 St Thomas Aquinas: the God of motion

1 The nature of *sacra doctrina* and its relation to other sciences is complex and a lengthy, detailed discussion is beyond the immediate purview of the present chapter. For further analysis, see Per Erik Persson, *Sacra Doctrina: reason and revelation in Aquinas*, trans. E.

Mackenzie, Oxford: Blackwell, 1970; V. White, *Holy Teaching: the Idea of Theology according to St. Thomas Aquinas* (the Aquinas Society of London, Aquinas Paper 33), London, 1958; M. D. Jordan, 'Theology and Philosophy', in N. Kretzmann and E. Stump, *The Cambridge Companion to Aquinas*, Cambridge: Cambridge University Press, 1993, pp. 232–251; J. Milbank and C. Pickstock, *Truth in Aquinas*, London: Routledge, 2001, ch. 2; J. Montag, 'Revelation: The false legacy of Suárez', in J. Milbank, C. Pickstock and G. Ward (eds), *Radical Orthodoxy: a new theology*, London: Routledge, 1999, pp. 38–63.

2　Aquinas, *Summa Theologiae*, 1a.1.8.responsio. Hereafter this work is cited as '*ST*'.

3　St Augustine, *The Trinity* ('De Trinitate'), trans. E. Hill, New York: New City Press, 1991, XIV.3, p. 371, quoted in *ST*, 1a.1.2.responsio. See also 1a.1.1.responsio.

4　Aquinas, *ST*, 1a.1.1.ad 2. Unless otherwise indicated, all translations are from the Blackfriars edition, ed. T. Gilby, London: Eyre and Spottiswoode, 1963–1975.

5　Aquinas, *ST*, 1a.1.3.ad 1.

6　Aquinas, *ST*, 1a.1.5.ad 2.

7　Aquinas, *ST*, 1a.1.8.ad 2.

8　Quotations are from Aquinas, *Faith, Reason and Theology: Questions I-IV of his Commentary on the De Trinitate of Boethius*, trans. A. Maurer, Toronto: Pontifical Institute of Mediaeval Studies, 1987. Hereafter this work is cited as '*In Boeth. De Trin*'.. On this matter, see also Aquinas, *ST*, 1a.12.11.ad 3: 'We see everything in God and judge everything by him in the sense that it is by participating in his light that we are able to see and judge, for the natural light of reason is a sort of participation (*participatio*) in the divine light'.

9　Aquinas, *In Boeth. De Trin.*, 2.3.responsio.

10　Aquinas, *ST*, 1a.1.2.responsio and 1a.1.4.ad 2.

11　Aquinas, *ST*, 1a.1.1.responsio.

12　Aquinas, *In Boeth. De Trin.*, 2.4.ad 5. See also M. D. Jordan, *Ordering Wisdom: The Hierarchy of the Philosophical Discourses in Aquinas*, Notre Dame, IN: University of Notre Dame Press, 1986, p. 194: 'In Thomas, the lower studies are preparations for the higher; the human mind is not satisfied with any study below the highest. Moreover, no philosophical study is complete with the study of metaphysics – and metaphysics is not complete, except with the vision of God. In both these ways, I take Thomas to be reestablishing something like the Platonic hierarchy of study'.

13　Aristotle, *Physics*, II.1.193b. Unless otherwise indicated, all translations are by P. H. Wicksteed and F. M. Cornford in the Loeb Classical Library edition (2 vols), Cambridge, MA: Harvard University Press, 1929 and 1934.

14　Ibid.

15　Aristotle, *Physics*, III.1. 201a. On this definition of motion, see L. A. Kosman, 'Aristotle's Definition of Motion', *Phronesis* 14, 1969, 40–62. See Chapter 2.

16　Aristotle, *Physics*, III.1.201a. See also *Physics*, V.1.

17　See pp. 41 ff. of this work.

18　Aristotle, *De Anima*, II.5.417a21 ff.; Aquinas, *Commentary on Aristotle's De Anima* ('In Aristotelis Librum De Anima Commentarium'), trans. R. Pasnau, New Haven, CT: Yale University Press, 1999, II.11.§358 ff. Hereafter, this work is cited as '*In De Anima*'.

19　See Kosman, 'Aristotle's Definition of Motion'.

20　Aquinas, *Commentary on Aristotle's Physics* ('In Octo Libros Physicorum Aristotelis Expositio'), trans. R. J. Blackwell *et al.*, London: Routledge & Kegan Paul, 1963, V.1.§641. Hereafter this work is cited as '*In Physics*'.

21　Lectures 12, 13 and 14 of Book 2 contain Aquinas's comments on Aristotle's understanding of natural teleology. For Aristotle's claim that teleology is fundamental to nature, see *Physics*, II.8.

22　See p. 35 of this work; Aristotle, *Physics*, IV.8.515a; Aquinas, *In Physics*, IV.11.§524.

23　See pp. 36 ff. of this work. Aquinas, *In Physics*, VII.1.1. Anneliese Maier has written extensively on the meaning of the principle *omne quod movetur ab alio movetur* and its overthrow as part of the rise of modern physics. See A. Maier, *On the Threshold of Exact Science: Selected Writings of Anneliese Maier on Late Medieval Natural Philosophy*, trans. and ed. S. D.

Sargent, Philadelphia: University of Pennsylvania Press, 1982, especially pp. 52 ff. By contrast, James Weisheipl, OP has argued that this principle is consonant with later developments in the physics of projectiles because it does not entail a commitment to 'constant conjunction' between a mover and that which it moves. See J. Weisheipl, *Nature and Motion in the Middle Ages*, ed. W. E. Carroll, Washington, DC: Catholic University Press of America, 1985, particularly chs 4 and 5. For an interpretation of the Aristotelian mover-causality principle which is opposed to that of Weisheipl, see also W. Wallace, OP, 'Newtonian Antinomies Against the *Prima Via*', in *idem, From a Realist Point of View*, Washington, DC: University Press of America, 1979, pp. 359–364. For the view that the Aristotelian principle and the principle of inertia are compatible, see T. J. McLaughlin, 'Aristotelian Mover-Causality and the Principle of Inertia', *International Philosophical Quarterly* 38, 1998, 137–151. My own view on this matter will become clear below.

24 See Aquinas, *In Physics*, VIII.8; Aristotle, *Physics*, VIII.4.245b ff.

25 Aristotle, *Physics*, VIII.4.255b17–31; Aquinas, *In Physics*, VIII.8.§1035.

26 See Aristotle, *Physics*, VIII.4.255b8–13; Aquinas, *In Physics*, VIII.8.§1033.

27 Aquinas, *In Physics*, VII.1.§885.

28 See Aquinas, *De Veritate*, 24.1.responsio.

29 On the vexed question of self-motion in Aristotle, see M. L. Gill and J. G. Lennox (eds), *Self-Motion: From Aristotle to Newton*, Princeton, NJ: Princeton University Press, 1994, part I.

30 See, for example, Aristotle, *De Motu Animalum*, 698a7–10 and 700a7–12; Aquinas, *In De Anima*, II.5.83–103; Aquinas, *De Veritate*, 24.

31 One might think that the soul is moved by, for example, an object which it desires in such a way that the object becomes the final cause of motion. Aristotle does consider the possibility that the soul can always be said to be moved such that no *per se* self-motion can occur in nature. However, he is clear that all external motions of the soul are insufficient to account for the motion of living things. Aristotle claims that, 'That which has self-movement as part of its essence cannot be moved by anything else except incidentally: just as that which is good in itself is not good for the sake of anything else', *De Anima*, I.3.406b. Translation is by W. S. Hett *On the Soul/Parva Naturalia/On Breath* (Loeb Classical Library), Cambridge, MA: Harvard University Press, 1957.

32 Aquinas, *In Physics*, VIII.12.§1069.

33 Aquinas does not mean that the soul and God are in the same way causes of any motion. The first cause is 'more cause' than any created thing, for without the divine there would be nothing. For Aquinas, the more general a cause the higher the cause and the more immanent and potent that cause. See Aquinas, *De Potentia*, III.7.16 and *Commentary on the Book of Causes* ('In Librum De Causis Expositio'), proposition 1.

34 Aristotle, *Physics*, V.1.225a.

35 See, for example, Aquinas, *Summa Contra Gentiles*, II.17. Hereafter, this work is cited as '*SCG*'.

36 My emphasis. Unless otherwise indicated, all translations are from St Thomas Aquinas, *Summa Contra Gentiles*, trans. A. C. Pegis, J. F. Anderson, V. J. Bourke, C. J. O'Neil, Notre Dame, IN: University of Notre Dame Press, 1975 edition.

37 Weisheipl, *Nature and Motion in the Middle Ages*, pp. 88–89.

38 Weisheipl, *Nature and Motion in the Middle Ages*, p. 21.

39 See Weisheipl, *Nature and Motion in the Middle Ages*, pp. 75 ff. and especially p. 80.

40 Weisheipl, *Nature and Motion in the Middle Ages*, pp. 99–120.

41 I use the word 'consonance' rather than 'agreement'. Weisheipl is quite clear that the Aristotelian understanding of motion is in agreement with the theory of impetus, yet the principle of inertia constitutes a more radical shift in the understanding of projectile motion. In particular, inertia rejects the distinction between natural and violent motion and a teleological understanding of nature, both of which are preserved by impetus theory and are, of course, central to Aristotle's view. Weisheipl is anxious, however, to argue that Aristotelian natural philosophy can account for projectile motion in a fashion which may rival inertia while not being wholly divorced from the spirit of early modern science.

42 See Weisheipl, *Nature and Motion in the Middle Ages*, pp. 72–73, 99–120 *et passim*.
43 D. Twetten, 'Back to Nature in Aquinas', *Medieval Theology and Philosophy* 5, 1996, 205–244.
44 Aristotle, *Physics*, VIII.10.266b27–267a21; Aquinas, *In Physics*, VIII.22. Concerning the relationship between this theory, the later theory of impetus and the principle of inertia, see Weisheipl, *Nature and Motion in the Middle Ages*, ch. 2.
45 Aristotle, *Physics*, VIII.10.267a2 ff.
46 For a lengthier explanation of this definition in relation to a journey, see Chapter 2, pp. 42 f.
47 Aquinas, *Super Librum De Causis Expositio*, proposition 1. Unless otherwise indicated, all translations are from St Thomas Aquinas, *Commentary on the* Book of Causes ('Super Librum De Causis Expositio'), trans. V. A. Guagliardo, C. R. Hess and R. C. Taylor, Washington, DC: Catholic University of America Press, 1996.
48 Aquinas, *De Potentia*, III.7.
49 Aquinas, *SCG*, III.82.6. Cf. Aristotle, *Physics*, VIII.7.260b5 ff. Crucially, this is *not* to claim that all motion is reducible to local motion. See Chapter 2, p. 6.
50 Aquinas, *Sententia de caelo et mundo*, I.6.§58 ff.; Aristotle, *De Caelo*, I.3.269b ff.
51 Aquinas, *Commentary on the* Book of Causes, VII; *Sententia de caelo et mundo*, I.5.§55 ff.
52 Aquinas, *SCG*, III.82.6.
53 Aquinas, *Commentary on Aristotle's* Metaphysics ('In Duodecim Libros Metaphysicorum Aristotelis'), trans. J. P. Rowan, Chicago: Henry Regnery, 1964, V.8.§873. Hereafter, this work is cited as '*In Metaphysics*'.
54 In a further important comment, Aquinas states that the unity which is the measure of things is not univocal with the unity which is interchangeable with being. The latter involves indivisibility, whereas the former is merely a measure. Thus unity is predicated in a secondary and analogical way of those things which are the measure of their genus, and in a literal and primary way of transcendental unity. See Aquinas, *In Metaphysics*, V.8.§875.
55 Aquinas, *SCG*, III.82.6. See also Aristotle, *Physics*, VIII.8.261b27 ff.
56 Aquinas, *Sententia de caelo et mundo*, II.20.§485.
57 Aquinas, *In De Anima*, II.7.§314.
58 Aquinas, *ST*, 1a.115.3.responsio. See also *De Veritate*, 5.9.
59 Aquinas, *Sententia de caelo et mundo*, I.4.§36 and §49; *In Metaphysics*, XII.6.§2511.
60 The notion that the celestial bodies were composed of a unique kind of matter, or even of pure form (Averroës), has a long and complex history. Galileo lined up with Aquinas against such figures as Aegidius and Ockham; the latter denied that celestial matter was of a different kind. See E. Grant, *Planets, Stars and Orbs: the Medieval Cosmos, 1200–1687*, Cambridge: Cambridge University Press, 1994, especially pp. 250 ff.
61 On these 'thought experiments' (or 'idealised experiments') as logically limiting cases to phenomena such as motion, see A. Funkenstein, *Theology and the Scientific Imagination: from the Middle Aages to the seventeenth century*, Princeton, NJ: Princeton University Press, 1986, pp. 152 ff.
62 On the separation of motion and time, see E. Alliez, *Capital Times: Tales from the Conquest of Time*, trans. G. Van Den Abbeele, Minneapolis: University of Minnesota Press, 1996, pp. 222 ff.
63 Aquinas, *Sententia de caelo et mundo*, II.4.§332. See also *Commentary on the* Book of Causes where Aquinas suggests a *via negativa* for arriving at knowledge of the heavenly bodies.
64 Aquinas, *De Potentia*, V.8.sed contra.
65 Aquinas, *Sententia de caelo et mundo*, II.1.291 ff.
66 See Aquinas, *SCG*, III.23; *De Potentia*, V.5; *ST*, I.70.3. The meaning of 'contact of power' will be described below. For a description of Aquinas's view in relation to others in the Middle Ages, see E. Grant, *Planets, Stars and Orbs*, especially pp. 475 ff.
67 Aquinas, *ST*, 1a.70.3.responsio. The celestial bodies are a 'fifth' element.
68 Aquinas, *SCG*, III.21; *ST*, 1a.70.3.responsio.
69 Aquinas, *De Veritate*, 8.15.

70 Ibid., 8.15.responsio: 'From the moment of their creation, however, the intellects of angels are perfected by innate forms giving them all the natural knowledge to which their intellectual powers extend, just as the matter of celestial bodies is completely terminated by its form'. Unless otherwise indicated, all translations are from Aquinas, *Truth*, vol. 1, trans. R. W. Mulligan, SJ, Indianapolis, IN: Hackett, 1994 edition. See also *ST*, 1a.55.2.resposio.

71 Ibid., quoting Pseudo-Dionysius, *De divinis nominibus*, VII.3.

72 Aquinas considers whether the separate substances are joined to the celestial bodies as mover to moved or after the fashion of a soul to a body. Although he concludes that there is little to decide between these opinions, it is more accurate to describe the relationship between the two as akin to that between a mover and that which is moved, namely 'a contact of power'. This is because the intellective nature of the angels does not require a body in the way in which a soul requires a body (i.e. for sensation, nutrition and so on). Rather, the angelic spirits only require a relation to the celestial bodies in order to impart their knowledge through motion. See *De spiritualibus creaturis*, VI.responsio.

73 Aquinas, *ST*, 1a.70.3.responsio.

74 Aquinas, *SCG*, III.24.2: 'the species of things caused and intended by the intellectual agent exist beforehand in his intellect, as the forms of artefacts pre-exist in the intellect of the artist and are projected from there into their products'.

75 Aquinas, *SCG*, III.24.3 ff.

76 Aquinas, *ST*, 1a.8.1; *SCG*, III.65–67.

77 Aquinas, *ST*, 1a.8.2.responsio.

78 Aquinas, *SCG*, III.65.5.

79 Aquinas, *ST*, 1a.105.5.responsio (my emphasis). Aquinas continues: 'It also follows that God acts interiorly in all things, because the form of anything is within it and the more so the more basic and universal the form is; for all things God is properly the universal cause of *esse*, which is innermost in things'.

80 Aquinas, *ST*, 1a.105.2.ad 1.

81 Aquinas, *De Potentia*, III.7.

82 See Proclus, *The Elements of Theology*, trans. E. R. Dodds, Oxford: Clarendon Press, 1963, propositions 12, 56 ff. and 145.

83 Aristotle, *De Anima*, III.8.435a17 ff.; Aquinas, *In De Anima*, III.18.§865ff.

84 Ibid.

85 Aristotle, *De Anima*, II.11.423b24 f.: 'From this it is clear that that which is perceptive of what is touched is within. Thus would occur what is true in the other cases; for when objects are placed on the other sense organs no sensation occurs, but when they are placed on the flesh it does; hence the medium of the tangible is flesh'.

86 Aristotle, *De Anima*, II.11.423b26 ff.: 'This is why we have no sensation of what is as hot, cold, hard or soft as we are, but only of what is more so, which implies that the sense is a sort of mean between the relevant sensible extremes'.

87 See Aquinas, *In De Anima*, II.23.§548.

88 Milbank and Pickstock, *Truth in Aquinas*, pp. 72 ff.

89 Aquinas, *De Potentia*, I.3.7.responsio.

90 One must stress here that, although God lacks all corporality, his knowledge of matter is still more intimate than ours because his knowledge is the cause of matter. So Aquinas can state that 'our mind has immaterial knowledge of material things [because matter cannot literally enter the mind], whereas the divine and angelic minds [the latter, because they have a greater proximity to the divine ideas] have a knowledge of the same material things in a way at once more immaterial and yet more perfect', *De Veritate*, 10.4.responsio.

91 D. Burrell, *Aquinas: God and Action*, London: Routledge & Kegan Paul, 1979, p. 17.

92 Because Aquinas is merely describing some principles for referring to God appropriately, Burrell is quite correct in his insistence that Aquinas is not providing a 'doctrine of God' (see *Aquinas: God and Action*, p. 13).

93 See also D. Turner, *How to be an Atheist: Inaugural Lecture Delivered at the University of Cambridge, 12 October 2001*, Cambridge: Cambridge University Press, 2002.

94 Aquinas, *ST*, 1a.2.1.sed contra and responsio; Psalms 14.1.

95 See D. Knowles, *The Evolution of Medieval Thought*, second edition, London: Longman, 1988, especially chs 18 and 24; E. Gilson, *History of Christian Philosophy in the Middle Ages*, London: Sheed and Ward, 1980, part 9; A. de Libera, *La philosophie médiévale*, 3ième édition, Paris: Presses Universitaires de France, 1995, pp. 413–417. In Lent 1267, just months after Aquinas began his *Summa* in Rome, Bonaventure, since 1257 the Minister-General of the Franciscans, launched the first of a series of attacks against pagan learning which culminated in the 'Parisian Condemnations' of 1277 and the attempt to return to a more conservative theological consensus.

96 A much more detailed analysis of the 'Five Ways' and their context is given in V. White, 'Prelude to the Five Ways', in *idem, God the Unknown and other essays*, London: Harvill Press, 1956. White is careful to point out that Aquinas does not refer to 'proving God's existence' in any modern, pseudo-scientific sense as if existence were a property possessed by God. Aquinas only refers to the making evident of the proposition *Deus esse* or *Deus est*.

97 R. te Velde, 'Natural Reason in the *Summa Contra Gentiles*', *Medieval Philosophy and Theology* 4, 1994, 42–70. Te Velde also comments that 'If reason were justified in its claim to autonomy, the only way Christianity could affirm its faith would be by rejecting reason, by excluding rational reflection based on philosophy' (p. 58). So granting reason's autonomy would render reason impotent in relation to Christian faith.

98 Aquinas, *ST*, 1a.2.introduction.

99 Aquinas, *ST*, 1a.2.3.ad 2: 'Since nature works for a determinate *end* under the direction of a higher agent, whatever is done by nature must needs be traced back to God, as to its first cause' (my translation and emphasis). The 'fifth way' refers more particularly to teleology.

100 Aquinas, *ST*, 1a.2.3.responsio (my translation).

101 Ibid. (my translation).

102 Aquinas, *ST*, 1a.2.3.responsio; *SCG*, I.13.11–15. See also E. Gilson, *The Christian Philosophy of St. Thomas Aquinas*, trans. L. K. Shook, Notre Dame, IN: University of Notre Dame Press, 1956, pp. 59–66, 1994 edition.

103 It will be clear that I concur with David Twetten's 'metaphysical' reading of the *prima via* (see D. Twetten, 'Clearing a "Way" for Aquinas: How the Proof from Motion Concludes to God', *Proceedings of the American Catholic Philosophical Quarterly* 70, 1996, 259–278). He rejects the so-called 'physical' and 'existential' readings of Aquinas's proof on the grounds that these misconstrue what is understood by motion. If one keeps in mind that motion is understood not only physically but metaphysically, and most generally as reduction from potency to act (a distinction which divides being, the subject of metaphysics), it becomes clear that the *prima via* concludes to a being which has no potency, namely a being which is uncaused.

104 See Gilson, *The Christian Philosophy of St. Thomas Aquinas*, p. 65.

105 Aquinas, *ST*, 1a.3.introduction.

106 Aquinas, *ST*, 1a.9.1.

107 Aquinas, *ST*, 1a.9.2.ad 1.

108 Aquinas, *ST*, 1a.10.4.responsio.

109 Ibid.

110 Aquinas, *ST*, 1a.12.5.responsio. See also 12.11.ad 3: 'for the natural light of reason is a sort of participation (*participatio*) in the divine light'.

111 Aquinas, *ST*, 1a.12.11.responsio. Note that, as with Grosseteste, Aquinas does seem to concede that a vision of the divine essence is at least possible in this life under exceptional circumstances, for example in 'dreams and ecstasies'.

112 Cf. Aquinas, *ST*, 1a.4.3.

113 Ibid.

114 Aquinas, *ST*, 1a.4.3.ad 3: 'Likeness of creatures to God is not affirmed on account of agreement in form according to the formality of the same genus or species, but solely according to analogy, inasmuch as God is essential being, whereas other things are beings by participation' (my translation). Elsewhere, Aquinas quotes Dionysius's use of the example of the sun as that which possesses in itself without diversity the many qualities and substances of the things we sense and thus has a universal, and thereby more potent, causal power. God, who is the source of being, the most general require-ment for existence itself ('Being is said of everything that is', *SCG*, II.15), therefore possesses in his simplicity the diverse forms of creation and, in being the most universal cause, is therefore the most potent cause. See *ST*, 1a.4.3.ad 1; cf. Pseudo-Dionysius, *On the Divine Names*, V.8.

115 'Thus whatever is said of God and creatures is said according to the relation of a crea-ture to God as its principle and cause, wherein all perfections of things pre-exist excellently. Now this mode of community of idea is a mean between pure equivocation and simple univocation', Aquinas, *ST*, 1a.13.5.responsio.

116 Aquinas, *ST*, 1a.13.4 and 5.

117 Aquinas, *ST*, 1a.13.5.ad 1 (my translation).

118 J. Milbank, *The Word Made Strange: Theology, Language, Culture*, Oxford: Blackwell, 1997, p. 16.

119 Aquinas, *ST*, 1a.13.5.ad 3: 'God is not a measure that is proportionate to what is measured; so it does not follow that he and his creatures belong to the same order'. See also 1a.13.1.ad 4 and, importantly, *De Veritate*, 2.11.responsio.

120 On *convenientia* in relation to Aquinas's aesthetics, see G. Narcisse, *Les Raisons de Dieu: Argument de convenance et esthétique théologique selon saint Thomas d'Aquin et Hans Urs von Balthasar*, Éditions Universitaires Fribourg Suisse, 1997, pp. 184 ff. Narcisse points to *SCG*, I.42, where Aquinas explains that *omnia in esse conveniunt*. For an exposition of *conve-nientia* in relation to touch and truth (with particular reference to the incarnation), see Milbank and Pickstock, *Truth in Aquinas*, particularly chs 1 and 3.

121 See Narcisse, *Les Raisons de Dieu*, pp. 202 ff. and pp. 223 ff. Narcisse traces the link between *convenientia* and analogy to Aquinas's teacher Albert the Great and the Arabs, particularly Avicenna and al-Ghazali.

122 Aquinas, *ST*, 1a.45.2.ad.2.

123 For example, see Aquinas, *SCG*, I.13.10. See also *ST*, 1a.19.1.ad 3, on the entirely subsistent movement of the divine will.

124 Aquinas, *ST*, 1a.45: 'De modo emanationis rerum a primo principio'.

125 For a discussion of divine emanation and motion in relation to Aquinas's understanding of the perfections of being, life and knowing, see R. A. te Velde, *Participation and Substantiality in Thomas Aquinas*, Leiden: E.J. Brill, 1995, pp. 272–279.

126 Aquinas, *SCG*, IV.11.

127 Aquinas, *SCG*, IV.11.1.

128 Aquinas, *SCG*, IV.11.3.

129 Aquinas, *SCG*, IV.11.4.

130 See Aquinas, *SCG*, II. 60.

131 See Aquinas, *De Veritate*, 1.9.

132 See Aquinas, *SCG*, I.45.

133 Aquinas, *ST*, 1a.15.1.

134 Aquinas, *ST*, 1a.15.2.responsio.

135 Aquinas, *SCG*, IV.15 ff.

136 Aquinas, *SCG*, IV.19.1 ff.

137 Aquinas, *SCG*, IV.19.2.

138 Aquinas, *SCG*, IV.19.3.

139 Aquinas, *SCG*, IV.19.4.

140 Aquinas, *SCG*, IV.19.7.

141 Ibid.; I John 4.16.

142 Aquinas, *SCG*, IV.19.8.
143 Aquinas, *SCG*, IV.19.12.
144 Aquinas, *ST*, 1a.14.8.responsio.
145 Ibid.
146 Aquinas, *SCG*, IV.19.12.
147 Aquinas, *SCG*, IV.20.3.
148 W. Hankey, *God in Himself: Aquinas's Doctrine of God as Expounded in the* Summa Theologiae, Oxford: Oxford University Press, 1987, p. 56.
149 Ibid., p. 103 (my emphasis).
150 Aquinas, *ST*, 1a.18.3.responsio. See also article 1 of the same question.
151 Aquinas, *ST*, 1a.18.4.responsio. At article 3 of question 18, Aquinas notes that, in attributing perfect intellect and therefore perfect life to God, he is following Aristotle (*Metaphysics*, XII.7.1072b).
152 Aquinas, *ST*, 1a.18.3.ad.1.
153 On Aristotle's distinction between *energeia* and *kinesis*, see *Metaphysics*, IX.6 and ch.2 above, pp. 41 ff.
154 Aquinas, *SCG*, IV.20.6.
155 Ibid. Aquinas mentions John 6.64 and Ezekiel 37.5.
156 Aquinas, *ST*, 1a.59.4.reponsio; 1a.83.1.responsio.
157 Aquinas, *ST*, 1a.62.4; 1a.64.2.responsio. See also 1a.62.5.responsio: 'Now, as we have already seen, it is characteristic of and proper to the angelic nature to reach its natural completeness in a single act and not by a gradual process'.
158 Aquinas, *ST*, 1a2ae.1.2.responsio.
159 Aquinas, *ST*, 1a2ae.10.2.responsio. See also 1a2ae.9 on that which moves the will.
160 Ibid.: 'If, on the other hand, the will is offered an object that is not good from every point of view, it will not tend to it of necessity. And since lack of any good whatever is non-good, consequently, that good alone which is perfect and lacking in nothing is such a good that the will cannot not will it' (my translation). See also *De Veritate*, 22.10.ad 6.
161 Aquinas, *ST*, 1a2ae.9.3.
162 Aquinas, *ST*, 1a2ae.12.3.repsonsio: 'Now a man intends at the same time, both the proximate and the last end; as the mixing of a medicine and the giving of health'. See also *De Veritate*, 22.14.responsio.
163 Aquinas, *ST*, 1a2ae.14.1.responsio.
164 Aquinas, *ST*, 1a2ae.14.1.ad 2; 1a2ae.17.2.responsio and ad 1. See also *De Veritate*, 22.15.
165 Aquinas, *ST*, 1a2ae.13.1; 1a.83.3.responsio; *De Veritate*, 22.15.
166 Aquinas, *ST*, 1a2ae.49.4.reponsio (my translation). See also 1a2ae.49.2.responsio; 1a2ae.50.6.ad 1; 1a2ae.54.3. See Aristotle, *Nicomachean Ethics*, II.4.
167 Aquinas, *ST*, 1a2ae.50.4 and 5.
168 Aquinas, *ST*, 1a2ae.51.3.sed contra and responsio.
169 Aquinas, *ST*, 1a2ae.52.1 and 2. These passages invite brief comment on the nature of Aquinas's Platonism. He appears to be clear that a form cannot exist apart from matter or without belonging to a subject. This is apparently different from pure Platonism. However, throughout this passage, Aquinas refers to subjects participating, but participating *in what?* The notion of participation does seem to imply a sense of otherness, namely the otherness of the form in which a subject participates. Put in more Aristotelian terms, Aquinas requires an exemplar form which reduces any particular form/matter compound from potency to act. Of course, any form/matter compound which is being moved participates in the form held more perfectly and actually by its mover. Eventually, this series, in not tending to infinity, will lead to the unmoved mover and the forms which are held therein, namely in the mind of God (*ST*, 1a.15). Therefore, it remains the case that there are not 'free-floating' forms – they remain in the divine mind as the one simple divine nature. Yet Aquinas's position is clearly Platonic with Aristotelian qualifications in the sense that forms are not merely found in

particulars, although his view frequently requires elucidation and interpretation in the light of other comments elsewhere in his work. A very cursory reading might suggest a purer Aristotelianism and a rejection of Plato.

170 Aquinas, *ST*, 1a2ae.52.1.responsio: '"Things which are given the name of triangle or a circle, are accordingly triangles and circles": therefore, because indivisibility is essential to the notion of such, whatever participates their nature must participate in its indivisibility' (my translation), quoting Aristotle, *Categories*, 6.11a7–9. One might still regard this as a characteristically Platonic notion. In the *Timaeus*, complexity arises from the combination of the four elements which are described in terms of geometrical solids composed of different kinds of triangle. These shapes cannot be resolved into anything more simple. See Plato, *Timaeus*, 53c ff.

171 Ibid. See also 1a2ae.52.2.responsio: 'in forms which admit of variation in intensive magnitude growth and decay sometimes occur not in the form on its own account but through the diverse participation thereof by the subject' (my translation).

172 Aquinas, *Disputed Questions on Virtue* ('Quaestio Disputata de Virtutibus In Communi' and 'Quaestio Disputata de Virtutibus Cardinalibus'), trans. R. McInerny, South Bend, IN: St Augustine's Press, 1999, article 1.responsio: 'From all of which it can be seen that we need the habit of virtues ... in order that perfect activity might be pleasantly accomplished. This results from habit which, since it acts in the manner of a kind of nature, makes the activity proper to it, as it were, natural and, consequently, delightful'.

173 Aquinas, *ST*, 1a2ae.54.4.responsio. On *habitus* as the unity of a science, see A. Maurer, *Being and Knowing: Studies in Thomas Aquinas and Later Medieval Philosophers*, Toronto: Pontifical Institute of Mediaeval Studies, 1990, ch.1.

174 Aquinas, *ST*, 1a2ae.55.1 and 4.responsio.

175 Aquinas, *ST*, 1a2ae.59.1.responsio: 'Moral virtue, however, is not a movement, but rather a principle of movement for an appetite, being a habit'.

176 Aquinas, *ST*, 1a2ae.57.

177 Aquinas, *ST*, 1a2ae.57.1.responsio.

178 Ibid.

179 Aquinas, *ST*, 1a2ae.58.1.responsio.

180 Aquinas, *ST*, 1a2ae.23.4; *De Veritate*, 26.1.

181 Aquinas, *ST*, 1a2ae.64.1.responsio. See Aristotle, *Nichomachean Ethics*, II.2, 6–9.

182 Aquinas, *ST*, 1a2ae.90.1.responsio.

183 Aquinas, *ST*, 1a2ae.90.2.ad 2 and 90.3.

184 Aquinas, *ST*, 1a2ae.93.1.

185 Aquinas, *ST*, 1a2ae.93.4.responsio.

186 Aquinas, *ST*, 1a2ae.93.5.responsio. Aquinas goes on to comment that, 'In this fashion even non-rational creatures are subject to it through being moved by divine Providence, though they are not like rational creatures, who are subject through some understanding of the divine command'.

187 Aquinas, *ST*, 1a.2ae.91.2.responsio.

188 Aquinas, *ST*, 1a2ae.94.2.responsio.

189 Aquinas, *ST*, 1a2ae.95.1.responsio.

190 See also Aquinas, *ST*, 1a2ae.97 on changes which may be made to human law, particularly in accord with its 'spirit'. Aquinas's view of law is not one of unswerving and specific regulations. In fact, laws are not all-encompassing: it is only those virtues which pertain to an individual's relationship to the community that come under that sway of human law.

191 Aquinas, *ST*, 1a2ae.99.2.ad 2.

192 Aquinas, *ST*, 1a2ae.101–105.

193 Aquinas, *ST*, 1a2ae.99.2.responsio.

194 Aquinas, *ST*, 1a2ae.98.5.responsio.

195 Aquinas, *ST*, 1a2ae.109.1.responsio; *De Veritate*, 27.2.responsio.

196 Cf. Aquinas, *ST*, 1a.12.5.responsio; Aristotle, *Metaphysics*, I.1.

197 Aquinas, *ST*, 1a2ae.109.1.responsio.
198 Aquinas, *ST*, 1a2ae.109.2.responsio (my translation).
199 Aquinas, *De Veritate*, 27.1.responsio. See also *SCG*, III.150.3.
200 Ibid.
201 Aquinas, *ST*, 1a2ae.110.2.reponsio.
202 Ibid. See also ad 1 where Aquinas describes grace not as an efficient cause but a formal cause as 'justice makes someone just'.
203 Aquinas, *ST*, 1a2ae.110.1.responsio.
204 Aquinas, *ST*, 1a2ae.111.2.responsio.
205 Ibid.
206 Aquinas, *ST*, 1a2ae.111.3.responsio.
207 Aquinas, *ST*, 1a2ae.3.ad 2.
208 Aquinas, *ST*, 1a2ae.111.4 and 5.
209 Aquinas, *ST*, 1a2ae.61.1.responsio.
210 Aquinas, *ST*, 1a2ae.62.3.responsio.
211 Aquinas, *ST*, 1a2ae.62.4.responsio. See also 2a2ae.23.8.
212 Aquinas, *ST*, 2a2ae.1.1.responsio; 2a2ae.4.3.responsio; 2a2ae.1.2.sed contra.
213 Aquinas, *De Veritate*, 14.1.responsio.
214 Aquinas, *ST*, 2a2ae.1.6.responsio: 'Therefore faith, as regards the assent which is the chief act of faith, is from God moving man inwardly by grace'.
215 Aquinas, *De Veritiate*, 14.1.responsio.
216 Aquinas, *ST*, 2a2ae.4.1.responsio (my translation).
217 Aquinas, *ST*, 2a2ae.17.1.responsio (my translation); see also 2a2ae.17.5.
218 Aquinas, *ST*, 2a2ae.17.8.responsio and ad 2.
219 Aquinas, *ST*, 1a2ae.108.1.responsio.
220 Milbank and Pickstock, *Truth in Aquinas*, pp. 61 ff.
221 Aquinas, *ST*, 3a.1.2.responsio.
222 Ibid.. See Aquinas, *ST*, 3a.1.1.responsio.
223 See above, n. 123.
224 See particularly Aquinas, *SCG*, IV.54 and 55.
225 Aquinas, *SCG*, IV.55.2.
226 Aquinas, *SCG*, IV.55.5.
227 Aquinas, *ST*, 3a.7.9.responsio.
228 Aquinas, *ST*, 3a.7.9.ad 3.
229 Aquinas, *SCG*, IV.54.2 and 55.14.
230 Aquinas, *SCG*, IV.54.7.
231 Aquinas, *ST*, 1a2ae.113.5.responsio.
232 Aquinas, *ST*, 1a2ae.113.1.responsio.
233 Aquinas, *ST*, 1a2ae.113.6.responsio.
234 Milbank and Pickstock, *Truth in Aquinas*, p. 60.
235 Aquinas, *ST*, 3a.46.2.ad 3.
236 Aquinas, *ST*, 3a.1.2.ad 2.
237 Aquinas, *ST*, 3a.46.2.ad 3.
238 Aquinas, *De Veritate*, 27.7.ad 10: 'After the passion of Christ the condition of mankind has been much changed, because, with the debt of human nature paid, men can fly unrestrained to their heavenly home'.
239 Aquinas, *ST*, 3a.8.1.
240 Aquinas, *De Veritate*, 29.7.ad 10.
241 Aquinas, *ST*, 3a.8.1.responsio.
242 Aquinas, *ST*, 1a2ae.108.1.responsio.
243 Aquinas, *ST*, 3a.60.2.responsio.
244 Aquinas, *ST*, 3a.60.4.reponsio; Aquinas, *SCG*, IV.56.3.
245 Aquinas, *SCG*, IV.56.4.
246 Aquinas, *SCG*, IV.56.5.

247 Aquinas, *SCG*, IV.61.3; Aquinas, *ST*, 3a.66.
248 Aquinas, *ST*, 3a.73.3.
249 Aquinas, *ST*, 3a.75.3.ad 1.
250 Aquinas, *ST*, 3a.75.4.responsio.
251 Aquinas, *ST*, 3a.77.1.responsio.
252 Aquinas, *ST*, 3a.73.3.ad 2.
253 Milbank and Pickstock, *Truth in Aquinas*, p. 61.
254 See Aquinas, *ST*, 3a.74.

5 The isolation of physics

 1 In this section, 'proof' indicates a variety of related projects. It is clear that both
    Avicenna and Averroës distinguish 'proof' from 'demonstration', the latter being
    primarily syllogistic. My views concerning Aquinas's 'proofs' of God's existence were
    made clear in the previous chapter.
 2 Aristotle, *Metaphysics*, IV.1.1003a.
 3 See Avicenna, *Al-Shifa: Al-Ilahiyyat (La métaphysique du Shifa)* (2 vols), trans. G. Anawati,
    Paris: Librairie Philosophique J. Vrin, 1978–1985. For a further text which treats
    similar issues, see Avicenna, *The Metaphysica of Avicenna (ibn Sina): a critical translation-
    commentary and analysis of the fundamental arguments in Avicenna's* Metaphysica *in the* Book
    of Scientific Knowledge, trans. P. Morewedge, London: Routledge & Kegan Paul, 1973.
 4 See A. de Libera, *La philosophie médiévale*, Paris: Presses Universitaires de France, 1993,
    pp. 116 ff. For de Libera, the position represented by Avicenna 'expresses the irre-
    ducibility of the point of view of the theologian and that of the physicist'.
 5 See H. A. Davidson, *Proofs for Eternity, Creation and the Existence of God in Medieval
    Islamic and Jewish Philosophy*, Oxford: Oxford University Press, 1987, p. 284. Hereafter,
    this work is cited as '*Proofs*'. I am indebted to Davidson's clear description of Avicenna's
    position concerning the nature of metaphysics and the proof of God's existence.
 6 H. A. Davidson, *Alfarabi, Avicenna and Averroës on Intellect: their cosmologies, theories of the
    active intellect and theories of the human intellect*, Oxford: Oxford University Press, 1992, pp.
    77 ff. Hereafter, this work is cited as '*Intellect*'. Although Davidson refers to Avicenna
    'adding' proofs from existence to Aristotelian proofs from motion, it seems more accurate
    to suggest that, for Avicenna, metaphysical proofs supplant those of physics.
 7 See Avicenna, *al-Najat*, Cairo: Muhyi al-Din al Kurdi Press, 1938, p. 278, cited in
    Davidson, *Intellect*, p. 77, n. 13.
 8 Davidson, *Proofs*, p. 285.
 9 See N. Kretzmann, A. Kenny and J. Pinborg (eds), *The Cambridge History of Later
    Medieval Philosophy*, Cambridge: Cambridge University Press, 1988, pp. 386 ff.
10 In particular see Aquinas, *In librum Boethii De Trinitate expositio*, 5.4.
11 For a detailed analysis of the points of agreement between Aquinas and Avicenna on the
    division of the sciences, see J. Wippel, 'Commentary of Boethius' *De Trinitate*: Thomas
    Aquinas and Avicenna on the Relationship between First Philosophy and the Other
    Theoretical Science', *The Thomist* 37, 1973, 133–154.
12 See Averroës, *Tahafut al-tahafut* ('The incoherence of the incoherence'), trans. S. van den
    Bergh, London: Luzac, 1954.
13 As Davidson points out (*Proofs*, p. 313), there is an apparent problem in the characteri-
    sation of the subject of metaphysics. Averroës, following Aristotle, describes the subject
    matter of metaphysics both as incorporeal being and as being *qua* being. These two
    understandings of metaphysics are reconciled in Averroës' stipulation that the charac-
    teristics of being *qua* being are at once those of beings devoid of matter and motion.
14 Averroës, *Commentary on the Physics (Aristotelis opera cum Averrois commentaries, IV)*, Venice,
    1562, I, com.83, ff.47rb–48va (cited and quoted in Kretzmann *et al.* (eds), *The
    Cambridge History of Later Medieval Philosophy*, p. 386).

15 Averroës, ed. M. Bouyges, *Long Commentary on Metaphysics* (*Tasfir ma ba'd al-Tabi'a*), Beyrouth: Imprimerie Catholique, 1938–1948, XII.5 (cited in Davidson, *Proofs*, p. 315).
16 E. Gilson, *History of Christian Philosophy in the Middle Ages*, London: Sheed and Ward, 1980, p. 220. Hereafter, this work is cited as '*History*'.
17 Averroës, *Long Commentary on Metaphysics*, VI.1. See also E. Gilson, 'Avicenne et le point de départ de Duns Scot', *Archives d'histoire doctrinale et littéraire du moyen âge* 2, 1927, pp. 89–149, especially pp. 93 ff.
18 For a more detailed analysis of Aquinas's position, in particular an explanation of why his position is not ontotheological, see J. Milbank and C. Pickstock, *Truth in Aquinas*, London: Routledge, 2001, pp. 35 ff.
19 See Avicenna, trans. Morewedge, *Metaphysica*, §38, pp. 76 ff.
20 See D. Burrell, *Knowing the Unknowable God: Ibn-Sina, Maimonides, Aquinas*, Notre Dame, IN: University of Notre Dame Press, 1986, p. 26: 'And the reason for denying quiddity to what necessarily exists stems from his [Avicenna's] manner of conceiving existence as "coming to" the nature. So he must ward off a conception of necessary existence as something *attributed* to the necessary existent'. This lack of 'essence' in God betokens a certain nihilism according to Conor Cunningham, *Genealogy of Nihilism*, London: Routledge, 2002, pp. 12–13. For more detailed discussion, see E. M. Macierowski, 'Does God Have a Quiddity According to Avicenna?', *The Thomist* 54, 1988, pp. 79–87.
21 Gilson, *History*, p. 214.
22 See L. Gardet, *La pensée religieuse d'Avicenne*, Paris: Librairie Philosophique J. Vrin, 1951, p. 45.
23 Davidson, *Intellect*, p. 75; *idem, Proofs*, pp. 290 ff.
24 Avicenna, ed. G. Anawati *et al.*, *Shifa: Ilahiyyat*, Cairo, 1960, p. 405, quoted in Davidson, *Intellect*, p. 75.
25 See S. H. Nasr and O. Leaman (eds), *History of Islamic Philosophy, part I*, London: Routledge, 1996, p. 242, and E. Gilson, *History*, pp. 213 ff.
26 See, for example, Plotinus, *Enneads*, V.13.15.
27 Avicenna, *al-Najat*, pp. 256–257, cited in S. H. Nasr, *An Introduction to Islamic Cosmological Doctrines: conceptions of nature and methods for its study by Ikhwān al-Safā, al-Bīrūnī, and Ibn Sīnā*, Cambridge, MA: The Belknap Press of Harvard University Press, 1964, p. 203.
28 The medieval debate concerning the nature of the agent intellect, which will be examined briefly below with reference to Aquinas, finds its origins in Aristotle, *De Anima*, III.4–5.
29 Gilson, *History*, p. 214.
30 See Avicenna, *Metaphysica*, IX.3–5, cited in Gilson, *History*, p. 215. See also A.-M. Goichon, *La distinction de l'essence et de l'existence d'après Ibn-Sina*, Paris: Desclée de Brouwer, 1937, pp. 395 ff.
31 Cunningham, *Genealogy of Nihilism*, p.9.
32 See S. van Riet (ed.), *Avicenna Latinus (vol. 2): Liber de anima: seu, Sextus de naturalibus*, Louvain: Peeters, 1968–1972, V, pp. 125 ff. For a detailed analysis of Avicenna and Aquinas on the agent intellect, see P. Lee, 'St. Thomas and Avicenna on the Agent Intellect', *The Thomist* 45, 1981, 41–61. I am indebted to Lee in the analysis which follows.
33 See pp. 79 ff. of this work; C. Pickstock, *After Writing: on the Liturgical Consummation of Philosophy*, Oxford: Blackwell, 1998, pp. 129 ff.; É. Alliez, *Capital Times: Tales from the Conquest of Time*, trans. G. van den Abbeele, Minneapolis: University of Minnesota Press, 1996, pp. 207, 238.
34 See Lee, 'St. Thomas and Avicenna on the Agent Intellect', p. 46, and E. Gilson, 'Pourquoi Saint Thomas a critiqué Saint Augustin', *Archives d'histoire doctrinale et littéraire du moyen age* 1, 1926.
35 See O. Boulnois, *Être et Représentation: une généalogie de la métaphysique moderne à l'époque de Duns Scot*, Paris: Presses Universitaires de France, 1999, pp. 69 ff.
36 See A.-M. Goichon, 'The Philosopher of Being', in *Iran Society, Avicenna Commemoration Volume*, Calcutta: Iran Society, *c*.1956, pp. 109–110, cited in Cunningham, *Genealogy of Nihilism*, p. 35, n. 53.

37 See Aquinas, *Summa Contra Gentiles*, II.74.5–7; Lee, 'St. Thomas and Avicenna on the Agent Intellect', pp. 51–52.

38 Aquinas, *Questions on the Soul* ('Quaestiones de Anima'), trans. J. H. Robb, Milwaukee, WI: Marquette University Press, 1984.

39 See Aquinas, *De Veritate*, 2.6; *Summa Contra Gentiles*, II.76 ff.; *Summa Theologiae*, 1a.79.4. ad 4 and 1a.84.6.responsio.

40 I am here indebted to Pickstock's description and interpretation of Aquinas's view. See J. Milbank and C. Pickstock, *Truth in Aquinas*, pp. 14 ff. See Aquinas, *De Veritate*, 2.6.responsio.

41 Aquinas, *De Veritate*, 4.1.responsio, cited in Milbank and Pickstock, *Truth in Aquinas*, p. 116, n. 54.

42 See Lee, 'St. Thomas and Avicenna on the Agent Intellect', p. 58, citing Aquinas, *Summa Theologiae*, 1a.85.1.ad 2.

43 J. Weisheipl, 'The Concept of Nature: Avicenna and Aquinas', in V. Brezik, *Thomistic Papers, 1*, Houston, TX: Center for Thomistic Studies, 1984, pp. 65–82. See Carra de Vaux, *Avicenne*, Paris: F. Alcan, 1900, pp. 184–185.

44 Weisheipl, 'The Concept of Nature: Avicenna and Aquinas', p. 75.

45 Ibid.

46 Cunningham, *Genealogy of Nihilism*, p. 10.

47 John Philoponus (Ioannis Philoponi), *In Aristotelis Phisicorum libros quinque posteriores commentaria*, ed. G. Vitelli, Berolini: Typ. et impensis G. Reimeri, 1888, IV.8, pp. 636 ff., cited in J. Weisheipl, *Nature and Motion in the Middle Ages*, ed. W. E. Carroll, Washington, DC: Catholic University Press of America, 1985, p. 29, n. 18. The theory of impetus has been understood both as a natural development of Aristotelian physics which remains quite different to early modern understandings of motion, and as a precursor to Newtonian physics. On the one hand, Pierre Duhem has argued that impetus theory marks a genuine break from Aristotle's physics for, in projectile motion, an impetus is given not to air around the body, but to the body itself. By contrast, Anneliese Maier has argued that Duhem overplayed the importance of impetus and this theory remains firmly within the Aristotelian tradition which was prevalent until the time of Galileo. A. Maier, *On the Threshold of Exact Science: Selected Writings of Anneliese Maier on Late Medieval Natural Philosophy*, ed. and trans. S. D. Sargent, Philadelphia: University of Pennsylvania Press, 1982, pp. 76 ff.

48 See the MSS edited by A. Maier, *Zwei Grundprobleme der scholastischen Naturphilosophie: das Problem der intensiven Grösse die Impetustheorie*, Roma: Edizioni di storia e letteratura, 1951, pp. 166–170.

49 See J. Buridan, *Quaestiones super octo Phisicorum libros Aristotelis*, Parisiis: Venum exponutur in edibus Dionisii Roce, 1509, VIII.12. See also J. Buridan, *Iohannis Buridani Quaestiones super libris quattuor de caelo et mundo*, ed. E. A. Moody, Cambridge, MA: The Mediaeval Academy of America, 1942, II.12–13 and III.2. For reproductions of texts central to Buridan's understanding of motion, see E. Grant (ed.), *Sourcebook in Medieval Science*, Cambridge, MA: Harvard University Press, 1974, pp. 275–284. Hereafter, this work is cited as 'Grant, *Sourcebook*'. Concerning the transmission of impetus theory, Duhem has argued that it reached the medieval scholastics via the Spanish–Arabian natural philosopher Alpetragius, whose *Theorica planetarum* was translated into Latin by Michael the Scot in 1217. Further theories have linked the transmission of impetus theory to Avicenna's commentary on the *Physics*. However, Anneliese Maier has argued that the scholastics developed the theory of impetus independently from the thought of Philoponus. See Weisheipl, *Nature and Motion in the Middle Ages*, p. 30.

50 J. Buridan, *Iohannis Buridani Quaestiones super libris quattuor de caelo et mundo*, ed. E. A. Moody, reproduced and translated in M. Clagett, *The Science of Mechanics in the Middle Ages*, Madison, WI: University of Wisconsin Press, 1959, pp. 557–562. This text is also reproduced in Grant, *Sourcebook*, p. 282.

51 See Grant, *Sourcebook*, p. 277.

52 Ibid.: 'Hence by the amount more there is of matter, by that amount can the body receive more of that impetus and more intensely (*intensius*). Now in a dense and heavy body, other things being equal, there is more of prime matter than in a rare and light one. Hence a dense and heavy body receives more of that impetus and more intensely, just as iron can receive more calidity than wood or water of the same quantity'.

53 I am indebted to Edward Grant in the following analysis. See his *The Foundations of Modern Science in the Middle Ages: Their religious, institutional and intellectual contexts*, Cambridge: Cambridge University Press, 1996, pp. 96 ff. and *idem*, *God and Reason in the Middle Ages*, Cambridge: Cambridge University Press, 2001, pp. 164 ff.

54 Grant, *The Foundations of Modern Science*, pp. 97–98.

55 See Buridan, *Quaestiones super libris quattuor de caelo et mundo*, II.12; Grant, *Sourcebook*, pp. 280–283.

56 Grant, *Sourcebook*, p. 282.

57 Grant, *Sourcebook*, p. 282.

58 See Duns Scotus, *Lectura*, I.17.2.3 ff.; M. Clagett (ed. and trans.), *Nicole Oresme and the Medieval Geometry of Qualities and Motions: A Treatise on the Uniformity and Difformity of Intensities Known as* 'Tractatus de configurantionibus qualitatum et motuum', Madison, WI: University of Wisconsin Press, 1968. See also E. D. Sylla, 'Medieval Concepts of the Latitude of Forms: The Oxford Calculators', *Archives d'histoire doctrinale et littéraire du moyen-âge* 40, 1973, pp. 223–283; Grant, *The Foundations of Modern Science*, pp. 99 ff.; J. E. Murdoch and E. D. Sylla, 'The Science of Motion', in D. Lindberg (ed.), *Science in the Middle Ages*, Chicago: University of Chicago Press, 1978, pp. 206–264, especially pp. 232 ff.

59 Grant, *The Foundations of Modern Science*, p. 100.

60 See M. Clagett, 'The *Liber de motu* of Gerald of Brussels and the Origins of Kinematics in the West', *Osiris*, 1st series, 12, 1956, pp. 73–175.

61 See Murdoch and Sylla, 'The Science of Motion'; J. E. Murdoch, '*Mathesis in philosophiam scholasticam introducta*: The Rise and Development of the Application of Mathematics in Fourteenth Century Philosophy and Theology', in *Arts libéraux et philosophie médiévale*, *Université de Montréal*, Montreal: Institut d'études médiévales, 1969, pp. 215–254; M. Clagett, *The Science of Mechanics in the Middle Ages*, Madison, WI: University of Wisconsin Press, 1959.

62 See Alliez, *Capital Times*, p. 226; Galileo, *Dialogue Concerning Two Chief Systems of the World*, trans. S. Drake, Berkeley, CA: University of California Press, 1967, pp. 163 ff., *et passim*.

6 Newton: God without motion

1 Some of Newton's papers, many of them including detailed theological reflections, have only recently been made available to students of his work. The Keynes manuscripts, the eighth of which features twelve creedal articles by Newton, was made available in the late 1940s, the Yahuda manuscripts in the early 1970s, the Bodmer manuscripts in 1991 and Sotheby's Lot 255 in 2000.

2 For an account of the role of Newton's alchemical studies within his wider intellectual endeavours, see B. J. T. Dobbs, *The Janus Face of Genius: The Role of Alchemy in Newton's Thought*, Cambridge: Cambridge University Press, 1991. Dobbs has expressed more forcefully than most the view that Newton's varied intellectual pursuits form an 'ultimate unity'. However, the nature of any methodological link between these areas is not obvious. It is possible that the unity consists in a single goal, namely the demonstration of divine providence.

3 The lectures were published as R. Bentley, *A Confutation of Atheism from the Origin and Frame of the World*, London, 1693, reproduced in I. B. Cohen (ed.), *Isaac Newton's Papers and Letters on Natural Philosophy and Related Documents*, Cambridge, MA: Harvard University Press, 1978, pp. 313–394.

4 H. W. Turnbull, J. F. Scott, A. R. Hall and L. Tilling (eds), *The Correspondence of Isaac Newton*, vol. 3, Cambridge: Cambridge University Press, 1961, p. 233.

5 Turnbull *et al.*, *The Correspondence of Isaac Newton*, vol. 3, p. 234. Many of Newton's contemporaries saw the value of the *Principia* for natural theology. See, for example, W. Whiston, *New Theory of the Earth, from Its Original, to the Consummation of All Things*, London, 1696, and *idem, Astronomical Principles of Religion, Natural and Reveal'd*, London, 1717. On Whiston's use of Newton, see J. Force, 'Newton's "Sleeping Argument" and the Newtonian Synthesis of Science and Religion', in N. J. W. Thrower, *Standing on the Shoulders of Giants: A Longer View of Newton and Halley*, Berkeley, CA: University of California Press, 1990, pp. 109–127.

6 I. Newton, trans. I. B. Cohen and A. Whitman, *The* Principia: *Mathematical Principles of Natural Philosophy* ('Philosophiae Naturalis Principia Mathematica', third edition, 1726), Berkeley, CA: University of California Press, 1999, p. 943. Hereafter, this work is cited as 'Newton (Cohen and Whitman)'. All page references are to this edition and translation of the *Principia*.

7 See Descartes, *The Philosophical Writings of Descartes vol. 1*, trans. J. Cottingham, R. Stoothoff and D. Murdoch, Cambridge: Cambridge University Press, 1984–1991, 'Principles of Philosophy', II.36, pp. 240 ff.

8 R. Westfall, *Never at Rest: A Biography of Isaac Newton*, Cambridge: Cambridge University Press, 1980, p. 315.

9 Newton, Yahuda MS, var. 1, 14, f. 25 (The Jewish National and University Library, Jerusalem), reproduced in Westfall, *Never at Rest*, pp. 315–316. Note also the tenth of Newton's twelve statements.

10 Most commentators on Newton's life and work point out that one possible factor motivating this more intense interest in matters theological was the requirement that any Fellow of Trinity College, Cambridge be ordained into the Church of England within seven years of being admitted to the degree of MA. Newton was admitted to the MA in October 1667. After considerable anxiety that he may have to resign his Fellowship in order to avoid ordination into what he believed was an idolatrous Church, Newton received a royal dispensation from the requirement of ordination in April 1675. See Westfall, *Never at Rest*, pp. 179, 309 ff. and 332 ff.

11 See Newton, *Two Letters of Sir Isaac Newton to Mr Le Clerc*, London, 1754.

12 See R. Westfall, '*Isaac Newton's* "Theologiae Gentilis Origines Philosophicae"', in W. W. Wager (ed.), *The Secular Mind: Transformations of Faith in Modern Europe*, London: Holmes and Meier, 1982. Westfall notes that this treatise, which was begun in the early 1680s, now only exists in scattered notes and drafts but that much of the material appears in a sanitised form in Newton's *Chronology of Ancient Kingdoms Amended*, London, 1728. This latter work is the only example of Newton's theological writings which he prepared for publication himself. He feared the controversy that would arise if his heretical views became public, hence most of his theological writing is unpublished with only a small amount appearing posthumously in print.

13 See F. Manuel, *Isaac Newton, Historian*, Cambridge, MA: The Belknap Press of Harvard University Press, 1963, pp. 112 ff., Dobbs, *The Janus Face of Genius*, pp. 150 ff. and R. Markley, 'Newton, Corruption, and the Tradition of Universal History', in J. E. Force and R. H. Popkin (eds), *Newton and Religion: Context, Nature and Influence*, Dordrecht: Kluwer Academic, 1999, pp. 121–144. As one might expect, Newton counted the doctrine of transubstantiation a particularly despicable idolatry: 'If there be a transubstantiation, never was Pagan Idolatry so bad as the Roman', Yahuda MS, var. 1, 14, f. 20 verso. On Newton's view of the Catholic Church, see R. Iliffe, 'Those "Whose Business It Is To Cavill": Newton's Anti-Catholicism', in Force and Popkin (eds), *Newton and Religion*, pp. 97–120.

14 See Newton, Yahuda MS, var. 1, 15, f. 102 verso.

15 See J. E. McGuire and P. M. Rattansi, 'Newton and the "Pipes of Pan"', *Notes and Records of the Royal Society* 21, 1966, 108–143, especially pp. 126–127.

16 See R. Westfall, 'Newton's Theological Manuscripts', in Z. Bechler (ed.), *Contemporary Newtonian Research*, Dordrecht: Reidel, 1987, pp. 130 ff. Westfall quite correctly maintains that it is impossible to identify whether Newton's Arianism preceded his theological voluntarism or vice versa. They are, for Newton, two mutually enhancing aspects of one theological vision.

17 Library of the Royal Society, London, quoted in J. E. McGuire, 'Force, Active Principles and Newton's Invisible Realm', *Ambix* 15, 1968, 154–208, quotation appearing on p. 190 (my emphasis).

18 See H. Alexander (ed.), *Leibniz-Clark Correspondence, Together with Extracts from Newton's* Principia *and* Opticks, Manchester: Manchester University Press, 1956, pp. 11–14, 17 ff., 88 ff., *et passim*. See also D. Kubrin, 'Newton and the Cyclical Cosmos: Providence and the Mechanical Philosophy', *Journal of the History of Ideas* 28, 1967, 325–346 and S. Schaffer, 'Comets & Idols: Newton's Cosmology and Political Theology', in P. Theerman and A. F. Seelf (eds), *Action and Reaction: Proceedings of a Symposium to Commemorate the Tercentenary of Newton's* Principia, London: Associated University Presses, 1993, pp. 206–231. Kubrin argues convincingly that Newton saw divine action as a series of necessary repairs to a universe in a state of dissolution (see Query 31 of the *Opticks*). Comets were regarded as an important tool in God's replenishing of motion in the universe.

19 S. D. Snobelen, '"God of gods, Lord of lords": The Theology of Isaac Newton's *General Scholium* to the *Principia*', *Osiris*, 16, 2001, 169–208, especially pp. 191 ff. I am very grateful to Dr Snobelen for granting me access to his essay prior to publication and for a long and edifying conversation concerning Newton's theology.

20 Snobelen, '"God of gods, Lord of lords": The Theology of Isaac Newton's *General Scholium* to the *Principia*', p. 194.

21 J. Crell, *De Deo ejus attributes* (n.p. 1631), col. 100, cited and quoted in Snobelen, '"God of gods, Lord of lords": The Theology of Isaac Newton's *General Scholium* to the *Principia*', p. 192, n. 91.

22 Snobelen, '"God of gods, Lord of lords": The Theology of Isaac Newton's *General Scholium* to the *Principia*', pp. 178 ff.

23 Newton (Cohen and Whitman), pp. 940 ff.

24 See Snobelen, '"God of gods, Lord of lords": The Theology of Isaac Newton's *General Scholium* to the *Principia*', and *idem*, 'Isaac Newton, Heretic: The Strategies of a Nicodemite', *British Journal of the History of Science* 32, 1999, 381–419. See also J. E. Force, 'Newton's God of Dominion: The Unity of Newton's Theological, Scientific, and Political Thought', in J. E. Force. and R. H. Popkin, *Essays on the Context, Nature and Influence of Isaac Newton's Theology*, Dordrecht: Kluwer Academic, 1990, pp. 75–102.

25 See Newton (Cohen and Whitman), pp. 794–795.

26 See R. Iliffe, '"Making a Shew": Apocalyptic Hermeneutics and the Sociology of Christian Idolatry in the Work of Isaac Newton and Henry Moore', in J. E. Force and R. H. Popkin, *The Books of Nature and Scripture*, Dordrecht: Kluwer Academic, 1994, pp. 55–88.

27 Snobelen, 'The Theology of Isaac Newton's *General Scholium* to the *Principia*', p. 204; *idem*, 'Isaac Newton, heretic: the strategies of a Nicodemite', pp. 389 ff.; Newton, Keynes MSS, 3.

28 Turnbull *et al.* (eds), *The Correspondence of Isaac Newton*, vol. 2, pp. 485 and 492, cited in Snobelen, 'On Reading Isaac Newton's *Principia* in the Eighteenth Century', *Endeavour* 22, 1998, 159–163.

29 Cf. Snobelen, 'On Reading Isaac Newton's *Principia* in the Eighteenth Century', p. 159. A number of 'popularised' versions of the *Principia* soon appeared, for example William Whiston's *New Theory of the Earth* (1696).

30 F. Manuel, *The Religion of Isaac Newton*, Oxford: Clarendon Press, 1974, p. 19: 'That Newton was conscious of his special bond to God and that he conceived of himself as the man destined to unveil the ultimate truth about God's creation does not appear in so

many words in anything he wrote. But peculiar traces of his inner conviction crop up in unexpected ways. More than once Newton uses *Jeova sanctus unus* as an anagram for *Isaacus Neuutonus*. ... The downgrading of Christ in Newton's theology ... makes room for himself as a substitute'.

31 On early modern scientists as 'priests' of nature who revealed God by demonstrating nature's marvels, see P. Harrison, *The Bible, Protestantism and the Rise of Natural Science*, Cambridge: Cambridge University Press, 1998, p. 198: 'The book of nature and those natural philosophers who interpreted it ... assumed part of the role previously played by the sacraments and the ordained priesthood'.

32 Newton (Cohen and Whitman), p. 415.

33 Newton (Cohen and Whitman), p. 408. On the separation of time from motion in the work of Duns Scotus, see Éric Alliez, *Capital Times: Tales from the Conquest of Time*, trans. G. Van Den Abbeele, Minneapolis: University of Minnesota Press, 1996, pp. 196ff.

34 Newton (Cohen and Whitman), p. 408.

35 Newton (Cohen and Whitman), p.409.

36 Newton (Cohen and Whitman), p. 409.

37 Newton (Cohen and Whitman), p. 411.

38 Newton (Cohen and Whitman), p. 412.

39 See R. Rynasiewicz, 'By Their Properties, Causes and Effects: Newton's *Scholium* on Time, Space, Place and Motion', *Studies in History and Philosophy of Science* 26, 1995, 133–153 and 295–321, and S. Toulmin, 'Criticism in the History of Science: Newton on Absolute Space, Time, and Motion', *Philosophical Review* 68, 1959, 1–29 and 203–227. Both Rynasiewicz and Toulmin take issue with the traditional interpretation and criticism of Newton's *Scholium* which sees this part of the *Principia* as a search for, and proof of, the existence of absolute motion, time and space. On this standard interpretation of the *Scholium* see E. Mach, *The Science of Mechanics: A Critical and Historical Account of Its Development*, sixth edition, trans. T. J. McCormack, La Salle, IL: Open Court, 1960, pp. 226 ff., and G. Freudenthal, *Atom and the Individual in the Age of Newton: On the Genesis of the Mechanistic World View*, Dordrecht: Reidel, 1986, pp. 13 ff.

40 Newton (Cohen and Whitman), p. 414.

41 Newton (Cohen and Whitman), p. 414.

42 E. Mach, *The Science of Mechanics*, pp. 279 ff.: '*All* masses and *all* velocities, and consequently all forces, are relative. There is no decision about relative and absolute which we can possible meet, to which we are forced, or from which we can obtain any intellectual or other advantage. When quite modern authors let themselves be led astray by the Newtonian arguments which are derived from the bucket of water, to distinguish between relative and absolute motion, they do not reflect that the system of the world is only given *once* to us, and the Ptolemaic or Copernican view is *our* interpretation, but both are equally actual. Try to fix Newton's bucket and rotate the heaven of fixed stars and then prove the absence of centrifugal forces'.

43 S. Toulmin, 'Criticism in the History of Science: Newton on Absolute Space, Time, and Motion', and R. Rynasiewicz, 'By Their Properties, Causes and Effects: Newton's *Scholium* on Time, Space, Place and Motion'.

44 Newton (Cohen and Whitman), p. 405.

45 M. J. Buckley, *At the Origins of Modern Atheism*, New Haven, CT: Yale University Press, 1987, p. 114, quoting R. S. Westfall, *Force in Newton's Physics: The Science of Dynamics in the Seventeenth Century*, New York: American Elsevier, 1971, p. 323.

46 Newton (Cohen and Whitman), p. 416 (my emphasis). Newton was not the first to describe the principle of inertia, although its formulation in his first law is regarded as a classic statement of the principle. Galileo, for example, enunciates a similar view of inertia in his *Dialogue Concerning Two Chief Systems of the World – Ptolemaic and Copernican*, trans. S. Drake, Berkeley and Los Angeles: University of California Press, 1953, which first appeared in 1632. See *The Second Day*, p. 147 of the Drake edition.

47 H. Butterfield, *Origins of Modern Science, 1300–1800*, London: G. Bell and Sons, 1957, p. 7.

48 This is revealed particularly in the employment of the principle of inertia in Newton's calculation that the earth attracts the moon in the inverse proportion of its distance. As Weisheipl has shown, in making his calculation the principle of inertia allows Newton to assume that the moon would continue to move in a right line at a constant rate were it not for the attractive force of the earth. In other words, the actual movement of the moon or the observer are ignored as irrelevant to the other quantities in Newton's equation. This inertial system is therefore one in which motion can be discounted in an examination of nature. It is force, which instigates changes in motion or rest, which is the only significant factor. Within Newton's calculation, it is as though the bodies are at rest. See J. Weisheipl, *Nature and Motion in the Middle Ages*, ed. W. E. Carroll, Washington, DC: Catholic University Press of America, 1985, pp. 45 ff.

49 J. E. McGuire, 'Natural Motion and Its Causes: Newton and the *"Vis insita"* of Bodies', in M. L. Gill and J. G. Lennox (eds), *Self-Motion: From Aristotle to Newton*, Princeton, NJ: Princeton University Press, 1994, pp. 305–329.

50 Newton (Cohen and Whitman), p. 404.

51 McGuire argues that these forces within a body are ontologically prior in Newton's scheme to any impressive force that may attempt to change the state of that body. The fact that a body is in a continuing state is a necessary requirement for that state to be changed. Yet it is still the case that this state is merely a numerical quantity rather than also a quality of motion. Within Newton's system it is only the change of state, rather than the state itself, which reveals anything about the nature of a body.

52 Aristotle, *Physics*, Book II.1.192b18.

53 McGuire, 'Natural Motion and Its Causes: Newton and the *"Vis insita"* of Bodies', p. 312.

54 McGuire, 'Natural Motion and Its Causes: Newton and the *"Vis insita"* of Bodies', p. 313.

55 J. Zabarella, 'neque Aristotelis absolute ibi negat gravia et levia moveri a seipsis sed negat eo modo quo a se moventur animalia', *De Rebus Naturalibus, libri XXX* (Book 4, *Liber De Natura*), Cloniae, 1602, ch. 9, p. 250, cited in McGuire, 'Natural Motion and Its Causes: Newton and the *"Vis insita"* of Bodies', p. 319. My page references differ from McGuire's as I quote from an earlier edition of *De Rebus Naturalibus*.

56 See pp. 36 ff., 39 ff., and 91 ff. of this work.

57 See Aristotle, *Physics*, VIII.4.

58 The question of whether the *vis insita* or *vis inertiae* of a body can be regarded as a cause of its state of motion has been much debated. It has been argued that these are regarded merely as *properties* by which motion is *conserved* (see, for example, T. J. McLaughlin, 'Aristotelian Mover-Causality and the Principle of Inertia', *International Philosophical Quarterly* 38, 1998, 137–151, especially p. 150). However, other commentators regard the *vis insita* both as a power to resist a change of state (in which case it is merely potential until an opposing force is exerted on the body concerned) and also a force which *causes* the conservation of a state of motion (see, for example, Z. Bechler, *Newton's Physics and the Conceptual Structure of the Scientific Revolution*, Dordrecht: Kluwer Academic, 1991, p. 275). I refer to Newtonian inertial motion as a kind of 'self-motion' for the reason given above: that is, it is a state which does not necessarily require for its explanation any reference to a being other than the moving body. It is only changes in the state of motion which require reference to another being which impresses a force upon the body in question. Therefore, unlike Aristotle and Aquinas, Newton does not distinguish between the living and the non-living with reference to motion.

59 My emphasis.

60 See D. Kubrin, 'Newton and the Cyclical Cosmos: Providence and the Mechanical Philosophy'. Kubrin argues that Newton did not view the universe as in a completely constant state of operation (a view which would lead one to some kind of deism). Creation required periodic divine reparation.

61 I. Newton, *Unpublished Scientific Manuscripts of Isaac Newton: A selection from the Portsmouth collection in the University Library*, Cambridge, ed. and trans. A. R. Hall and M. B. Hall,

Cambridge: Cambridge University Press, 1962, pp. 108 and 142. This work is cited below as 'Newton (Hall and Hall)'.

62 See J. E. McGuire, 'Neoplatonism and active principles: Newton and the *Corpus Hermeticum*', in R. S. Westman and J. E. McGuire, *Hermeticism and the Scientific Revolution: Papers Read at a Clark Library Seminar*, Los Angeles: Clark Memorial Library, University of California, 1977, pp. 93–142, and J. E. McGuire, 'The Fate of the Date: The Theology of Newton's *Principia* Revisited', in M. J. Osler (ed.), *Rethinking the Scientific Revolution*, Cambridge: Cambridge University Press, 2000, pp. 271–295.

63 Newton (Hall and Hall), pp. 89–156. See also another unpublished manuscript on this subject, *Tempus et Locus*, reproduced in J. E. McGuire, 'Newton on Place, Time, and God: an unpublished source', *British Journal for the History of Science* 11, 1978, 114–129.

64 Newton (Hall and Hall), p. 132.

65 Ibid.

66 Ibid., p. 136.

67 Ibid., p. 137.

68 Ibid., p. 140.

69 See H. Alexander (ed.), *Leibniz–Clark Correspondence, Together with Extracts from Newton's* Principia *and* Opticks, pp. 11 ff., 16 ff., 21 ff., *et passim*; and A. Koyré and I. B. Cohen, 'The Case of the Missing Tanquam: Leibniz, Newton and Clarke', *Isis* 52, 1961, 555–566.

70 I. Newton, *Opticks, or A Treatise on the Reflections, Refraction, Inflections & Colours of Light* (based on the fourth edition, 1730), New York: Dover, 1952.

71 T. F. Torrance, *Space, Time and Incarnation*, Edinburgh: T & T Clark, 1997, p. 39.

72 J. E. McGuire, 'The Fate of the Date: The Theology of Newton's *Principia* Revisited', in M. J. Osler (ed.), *Rethinking the Scientific Revolution*, Cambridge: Cambridge University Press, 2000.

73 A. Gabbey, 'Newton, active powers, and the mechanical philosophy', in I. B. Cohen and G. E. Smith (eds), *The Cambridge Companion to Newton*, Cambridge: Cambridge University Press, 2002, pp. 329–357.

74 Gabbey, 'Newton, active powers, and the mechanical philosophy', p. 340. 'Alchemy' is a very broad term referring not only to the attempt to make gold from base metals, but also to more general occult transmutations of power. See B. J. T. Dobbs, *The Janus Face of Genius*.

75 Ibid.

76 Ibid.

77 Newton (Cohen and Whitman), p. 943.

78 Newton, 'Conclusio', quoted in Westfall, *Never at Rest*, pp. 388–389.

79 Gabbey, 'Newton, active powers, and the mechanical philosophy', p. 341.

80 On the influence of the Cambridge Platonists on Newton, particularly regarding the ancient provenance of the new science, see J. E. McGuire and P. M. Rattansi, 'Newton and the "Pipes of Pan"', *Notes and Records of the Royal Society* 21, 1966, 108–143.

81 McGuire and Rattansi, 'Newton and the "Pipes of Pan"', p. 132.

82 R. Cudworth, *True Intellectual System of the Universe*, London, 1743. See also *idem, A Treatise Concerning Eternal and Immutable Morality*, ed. S. Hutton, Cambridge: Cambridge University Press, 1996.

83 Cudworth, *A Treatise Concerning Eternal and Immutable Morality*, pp. 22–27.

84 Cudworth, *A Treatise Concerning Eternal and Immutable Morality*, p. 149.

85 J. C. Maxwell, *A Dynamical Theory of the Electromagnetic Field*, ed. T. F. Torrance, Edinburgh: Scottish Academic Press, 1982. Hereafter, this work is cited as Maxwell (Torrance).

86 A. Einstein, 'Maxwell's Influence on the Development of the Conception of Physics Reality', in Maxwell (Torrance), p. 29.

87 See Newton's letter to Bentley, 25 February, 1694 (in Turnbull *et al.* (eds), *The Correspondence of Isaac Newton*, vol. 3, pp. 253–254).

88 D. Park, *The Fire within the Eye: a historical essay on the nature and meaning of light*, Princeton, NJ: Princeton University Press, 1997, pp. 272 ff. In the following discussion,

I am much indebted to Park's clear analysis. See also J. T. Cushing, *Philosophical Concepts in Physics: the historical relation between philosophy and scientific theories*, Cambridge: Cambridge University Press, 1998, pp. 185 ff.

89 A longitudinal wave (what might be called a 'pressure wave') is one in which the particle displacement takes place parallel to the direction of the propagation of the wave. Sound is transmitted by longitudinal waves. A transverse wave, such as that which occurs when two people hold a rope and one person moves their arm up and down in a vertical direction, is a wave in which the particle displacement is at right angles to the direction of the propagation of the wave.

90 M. Faraday, *Experimental Researches in Electricity*, in R. M. Hutchins (ed.), *Great Books of the Western World*, vol. 45, Chicago: Encyclopaedia Britannica, 1952, p. 816.

91 M. Faraday, *Experimental Researches in Electricity*, p. 530.

92 Park, *The Fire within the Eye*, p. 276.

93 See J. C. Maxwell, *A Treatise on Electricity and Magnetism*, Oxford: Clarendon Press, 1873, vol. 1, pp. viii ff.

94 Park, *The Fire within the Eye*, p. 281.

95 J. C. Maxwell, 'On Physical Lines of Force', in W. D. Niven (ed.), *The Scientific Papers of James Clerk Maxwell*, Cambridge: Cambridge University Press, 1890, vol. 1, p. 156, quoted in M. J. Nye, *Before Big Science: The Pursuit of Modern Chemistry and Physics, 1800–1940*, Cambridge, MA: Harvard University Press, 1996, p. 71.

96 Park, *The Fire within the Eye*, p. 283.

97 Maxwell (Torrance), p. 86.

98 J. C. Maxwell, 'Are there Real Analogies in Nature?', in L. Campbell and W. Garnett (eds), *The Life of James Clerk Maxwell*, London: Macmillan, 1882, pp. 235–244.

99 Maxwell, 'Are there Real Analogies in Nature?', p. 241.

100 Ibid.

101 Nye, *Before Big Science*, p. 100.

102 Ibid.

103 See, for example, J. C. Taylor, *Hidden Unity in Nature's Laws*, Cambridge: Cambridge University Press, 2001, ch. 8; J. Gribben, *Science: A History, 1543–2001*, London: Allen Lane, 2002, pp. 508 ff.; Nye, *Before Big Science*, pp. 116 ff.

104 In the description of quantum theory which follows, I am indebted to Wolfgang Smith, 'From Schrödinger's Cat to Thomistic Ontology', *The Thomist* 63, 1999, 49–63. I will comment further below on Smith's view that Thomist ontology is more fitting to quantum theory than Cartesian dualism.

105 Smith, 'From Schrödinger's Cat to Thomistic Ontology', p. 51.

106 R. Penrose, *The Emperor's New Mind*, Oxford: Oxford University Press, 1989, p. 256, quoted in Smith, 'From Schrödinger's Cat to Thomistic Ontology', p. 54.

107 A. N. Whitehead, *The Concept of Nature*, Cambridge: Cambridge University Press, 1964, ch. 2.

108 This distinction is, in turn, similar to that found in John Locke between primary qualities, such as hardness, extension, motion and rest, which belong to a body itself, and secondary qualities, such as colour, sound and taste, which belong only within the perceiving mind as caused in us by primary qualities. Thus, for example, colour does not belong in a thing itself, but only in the mind of the person perceiving. See J. Locke, *An Essay Concerning Human Understanding*, ed. J. W. Yolton, London: J. M. Dent & Sons, 1977, II.8.§9, pp. 58 ff.

109 Smith, 'From Schrödinger's Cat to Thomistic Ontology', p. 57.

110 Smith, 'From Schrödinger's Cat to Thomistic Ontology', pp. 58 ff.

111 Smith, 'From Schrödinger's Cat to Thomistic Ontology', pp. 61 ff.

112 E. Gilson, *The Christian Philosophy of St. Thomas Aquinas*, Notre Dame, IN: University of Notre Dame Press, 1994, p. 184.

Bibliography

Aertsen, J., *Nature and Creature: Thomas Aquinas' Way of Thought*, Leiden: E.J. Brill, 1988.

Allen, R. E. (ed.), *Studies in Plato's Metaphysics*, London: Routledge & Kegan Paul, 1965.

Alliez, E., *Capital Times: Tales from the Conquest of Times*, trans. G. Van Den Abbeele, Minneapolis: University of Minnesota Press, 1996.

Annas, J., *Aristotle's* Metaphysics *Books M and N*, Oxford: Oxford University Press, 1976.

Anton, J. P. (ed.), *Science and the Sciences in Plato*, New York: Caravan Books, 1980.

Aquinas, St Thomas, *On the Power of God* ('De Potentia Dei'), trans. English Dominican Fathers, Westminster, MD: Newman, 1952.

—— *Commentary on Aristotle's* Physics ('In Octo Libros Physicorum Aristotelis Expositio'), trans. R. J. Blackwell *et al.*, London: Routledge & Kegan Paul, 1963.

—— *Commentary on Aristotle's* Metaphysics ('In Duodecim Libros Metaphysicorum Aristotelis'), trans. J. P. Rowan, Chicago: Henry Regnery, 1964.

—— *Opera Omnia*, Turin: Marietti, 1950–1965.

—— *On Being and Essence* ('De Ente et Essentia'), trans. A. Maurer, Toronto: Pontifical Institute of Mediaeval Studies, 1968.

—— *Summa Theologiae* (Blackfriars edition, general ed. T. Gilby), London: Eyre and Spottiswoode, 1963–1975.

—— *Summa Contra Gentiles*, trans. A. C. Pegis *et al.*, London: University of Notre Dame Press, 1975.

—— *Questions on the Soul* ('Quaestiones de Anima'), trans. J. H. Robb, Milwaukee, WI: Marquette University Press, 1984.

—— *The Division and Method of the Sciences: Questions V–VI of his Commentary on the* De Trinitate *of Boethius*, trans. A. Maurer, Toronto: Pontifical Institute of Mediaeval Studies, 1986.

—— *Faith, Reason and Theology: Questions I–IV of his Commentary on the* De Trinitate *of Boethius*, trans. A. Maurer, Toronto: Pontifical Institute of Mediaeval Studies, 1987.

—— *Truth* ('De Veritate'), trans. R. W. Mulligan SJ, *et al.*, Indianapolis, IN: Hackett, 1994.

—— *Commentary on* The Book of Causes ('In Librum De Causis Expositio'), trans. V. A. Guagliardo *et al.*, Washington, DC: Catholic University Press of America, 1996.

—— *On Matter and Form and The Elements* ('De Principiis Naturae' and 'De Mixtione Elementorum'), trans. J. Bobik, Notre Dame, IN: University of Notre Dame Press, 1998.

—— *Commentary on Aristotle's* De Anima ('In Aristotelis Librum De Anima Commentarium'), trans. R. Pasnau, New Haven, CT: Yale University Press, 1999.

Aristotle, *Nichomachean Ethics* (Loeb Classical Library), trans. H. Rackham, Cambridge, MA: Harvard University Press, 1934.

—— *Physics* (2 vols) (Loeb Classical Library), trans. P. H. Wicksteed and F. M. Cornford, Cambridge, MA: Harvard University Press, 1929 and 1934.

—— *Eudemian Ethics* (Loeb Classical Library), trans. H. Rackham, Cambridge, MA: Harvard University Press, 1935.

—— *Metaphysics* (2 vols) (Loeb Classical Library), trans. H. Tredennick, Cambridge, MA: Harvard University Press, 1933 and 1935.

—— *Parts of Animals* (Loeb Classical Library), trans. A. L. Peck, Cambridge, MA: Harvard University Press, 1937.

—— *Prior Analytics* (Loeb Classical Library), trans. H. P. Cooke and H. Tredennick, Cambridge, MA: Harvard University Press, 1938.

—— *On the Heavens* (Loeb Classical Library), trans. W. K. C. Guthrie, Cambridge, MA: Harvard University Press, 1939.

—— *On Coming-to-be and Passing-away/On the Cosmos* (Loeb Classical Library), trans. E. S. Forster and D. J. Furley, Cambridge, MA: Harvard University Press, 1955.

—— *On the Soul/Parva Naturalia/On Breath* (Loeb Classical Library), trans. W. S. Hett, Cambridge, MA: Harvard University Press, 1957.

—— *Posterior Analytics/Tropica* (Loeb Classical Library), trans. H. Tredennick and E. S. Foster, Cambridge, MA: Harvard University Press, 1960.

—— *Generation of Animals* (Loeb Classical Library), trans. A. L. Peck, Cambridge, MA: Harvard University Press, 1963.

Armstrong, A. H. (ed.), *The Cambridge History of Later Greek and Early Medieval Philosophy*, Cambridge: Cambridge University Press, 1970.

Ashbaugh, A. F., *Plato's Theory of Explanation: A Study of the Cosmological Account in the Timaeus*, Albany, NY: State University of New York Press, 1988.

Aubenque, P., *Le Problème de l'Être chez Aristote*, Quadrige: Presses Universitaires de France, 1962.

Averroës, *Tahafut al-tahafut ('The incoherence of the incoherence')*, trans. S. van den Bergh, London: Luzac, 1954.

Avicenna, *al-Najat*, Cairo: Muhyi al-Din al Kurdi Press, 1938.

—— *The Metaphysica of Avicenna (ibn Sina): a critical translation-commentary and analysis of the fundamental arguments in Avicenna's* Metaphysica *in the* Book of Scientific Knowledge, trans. P. Morewedge, London: Routledge & Kegan Paul, 1973.

—— *Al-Shifa: Al-Ilahiyyat (La métaphysique du Shifa)* (2 vols), trans. G. Anawati, Paris: Librairie Philosophique J. Vrin, 1978–1985.

—— *Avicenna Latinus* (5 vols), ed. S. van Riet, Louvain: Peeters, 1968–1992.

Ayers, M. and Garber, D. (eds), *The Cambridge History of Seventeenth Century Philosophy* (2 vols), Cambridge: Cambridge University Press, 1998.

Bacon, R., *Opus Majus*, trans. R. B. Burke, Philadelphia: University of Pennsylvania Press, 1928.

—— *Compendium of the Study of Theology* ('Compendium studii theologiae'), trans. T. Maloney, Leiden: E.J. Brill, 1988.

—— *Three Treatments of Universals*, trans. T. Maloney, Binghamton, NY: Center for Medieval and Early Renaissance Studies, State University of New York at Binghamton, 1989.

Ballew, L., *Straight and Circular: A Study of Imagery in Greek Philosophy*, Assen: Van Gorum, 1979.

Bamborough, R. (ed.), *New Essays on Plato and Aristotle*, London: Routledge & Kegan Paul, 1965.

Barnes, J. (ed.), *The Cambridge Companion to Aristotle*, Cambridge: Cambridge University Press, 1995.

Barnes, J., Schofield, M. and Sorabji, R. (eds), *Articles on Aristotle*, London: Duckworth, 1975–1979.

Barry, J., *Measures of Science: Theological and Technological Impulses in Early Modern Thought*, Evanston, IL: Northwestern University Press, 1996.

Baur, L. (ed.), *Die philosophischen Werke des Robert Grosseteste, Bischofs von Lincoln* (BGPM Bd. IX), Münster i. W., 1912, available at: http://www.grosseteste.com/baurframe.htm> (accessed 28 July 2004).

Bechler, Z. (ed.), *Contemporary Newtonian Research*, Dordrecht: Reidel, 1987.

—— *Newton's Physics and the Conceptual Structure of the Scientific Revolution*, Dordrecht: Kluwer Academic, 1991.

Benson, R. G. and Ridyard, S. J. (eds), *Man and Nature in the Middle Ages* (Sewanee Mediaeval Studies 6), Sewanee: University of the South Press, 1995.

Bernstein, H. R., 'Leibniz and the *Sensorium Dei*', *Journal of the History of Philosophy* 15, 1977, 171–182.

Blackwell, R. J., *Galileo, Bellarmine and the Bible*, London: University of Notre Dame Press, 1991.

Blair, G., 'Unfortunately It is a Bit More Complex: Reflections on Ενεργεια', *Ancient Philosophy* 15, 1995, 565–580.

Blanchette, O., *The Perfection of the Universe According to Aquinas: A Teleological Cosmology*, Philadelphia: Pennsylvania State University Press, 1992.

Bos, E. P. (ed.), *John Duns Scotus: Renewal of Philosophy*, Amsterdam: Rodopi, 1998.

Boulnois, O., *Être et Représentation: une généalogie de la métaphysique moderne à l'époque de Duns Scot*, Paris: Presses Universitaires de France, 1999.

Brock, S. L., *Action and Conduct: Thomas Aquinas and the Theory of Action*, Edinburgh: T&T Clark, 1998.

Brown, J. R. and Okrahlik, K., *The Natural Philosophy of Leibniz*, Dordrecht: Reidel, 1985.

Brisson, L., *Le même et l'autre dans la structure ontologique du Timée de Platon*, Paris, 1974.

Brisson, L. and Meyerstein, F. W., *Inventing the Universe: Plato's* Timaeus, *the Big Bang and the Problem of Scientific Knowledge*, Albany, NY: State University of New York Press, 1995.

Buckley, M. J., *Motion and Motion's God*, Princeton, NJ: Princeton University Press, 1971.

—— *At the Origins of Modern Atheism*, New Haven, CT: Yale University Press, 1987.

Buridan, J., *Quaestiones super octo Phisicorum libros Aristotelis*, Parisiis: Venum exponutur in edibus Dionisii Roce, 1509.

—— *Iohannis Buridani Quaestiones super libris quattuor de caelo et mundo*, ed. E. A. Moody, Cambridge, MA: The Mediaeval Academy of America, 1942.

Burrell, D. B., *Aquinas: God and Action*, London: Routledge & Kegan Paul, 1979.

—— *Knowing the Unknowable God: Ibn-Sina, Maimonides, Aquinas*, Notre Dame, IN: University of Notre Dame Press, 1986.

Butterfield, H., *The Origins of Modern Science 1300–1800*, London: G. Bell & Sons, 1957.

Butts, R. E. and Davis, J. W. (eds), *The Methodological Heritage of Newton*, Oxford: Basil Blackwell, 1970.

Calvo, T. and Brisson, L. (eds), *Interpreting the* Timaeus–Critias: *Proceedings of the IV Symposium Platonicum; selected papers* (International Plato Studies vol. 7), Sankt Augustin: Academia Verlag, 1997.

Camille Dumont, S. I., 'Enseignement de la théologie et méthode scientifique', *Gregorianum* 71, 1990, 441–463.

Campbell, L. and Garnett, W. (eds), *The Life of James Clerk Maxwell*, London: Macmillan, 1882.

Cantor, G. and Lindberg, D., *The Discourse of Light: from the Middle Ages to the Enlightenment*, Los Angeles: William Clark Andrews Memorial Library, University of California, 1985.

Carra de Vaux, B., *Avicenne*, Paris: F. Alcan, 1900.

Charles, E., *Roger Bacon: Sa vie, ses ouvrages, ses doctrines*, Paris: L. Hachette, 1861.

Chenu, M.-D., *Introduction à l'Étude de Saint Thomas d'Aquin*, Montréal: Institut d'Études Médiévales, 1993.

—— *Man, Nature and Society in the Twelfth Century: Essays on New Theological Perspectives in the Latin West*, ed. and trans. J. Taylor and L. K. Little, Toronto: University of Toronto Press, 1997.

Clagett, M., 'The *Liber de motu* of Gerald of Brussels and the Origins of Kinematics in the West', *Osiris*, 1st series, 12, 1956, 73–175.

—— *The Science of Mechanics in the Middle Ages*, Madison, WI: University of Wisconsin Press, 1959.

—— (ed. and trans.), *Nicole Oresme and the Medieval Geometry of Qualities and Motions: A Treatise on the Uniformity and Difformity of Intensities Known as* 'Tractatus de configurantionibus qualitatum et motuum', Madison, WI: University of Wisconsin Press, 1968.

Claghorn, G. S., *Aristotle's Criticism of Plato's* Timaeus, The Hague: Martinus Hijhoff, 1954.

Cohen, I. B., *Isaac Newton's Papers and Letters On Natural Philosophy and related documents*, Cambridge, MA: Harvard University Press, 1978.

Cohen, I. B. and Koyré, A., 'The Case of the Missing *Tanquam*: Leibniz, Newton and Clarke', *Isis* 52, 1961, 555–566.

Cohen, I. B. and Smith, G. E. (eds), *The Cambridge Companion to Newton*, Cambridge: Cambridge University Press, 2002.

Cohen, I. B. and Westfall, R. S. (eds), *Newton: texts, backgrounds, commentaries*, London: W. W. Norton, 1995.

Cohen, S. M., *Aristotle on Nature and Incomplete Substance*, Cambridge: Cambridge University Press, 1996.

Corbin, M., *Le Chemin de la Théologie chez Thomas d'Aquin*, Paris: Bibliothèque des Archives de Philosophie, 1974.

Cornford, F. M., *The Laws of Motion in Ancient Thought*, Cambridge: Cambridge University Press, 1931.

—— *Plato's Cosmology: The* Timaeus *of Plato translated with a running commentary*, London: Routledge & Kegan Paul, 1937.

Crombie, A., *Robert Grosseteste and the Origins of Experimental Science 1100–1700*, Oxford: Clarendon Press, 1953.

—— *Science, Art and Nature in Medieval and Modern Thought*, London: The Hambledon Press, 1996.

Cudworth, R., *True Intellectual System of the Universe*, London, 1743.

—— *A Treatise Concerning Eternal and Immutable Morality*, ed. S. Hutton, Cambridge: Cambridge University Press, 1996.

Cunningham, C., *Genealogy of Nihilism*, London: Routledge, 2002.

Cushing, J. T., *Philosophical Concepts in Physics: The Historical Relation between Philosophy and Scientific Theories*, Cambridge: Cambridge University Press, 1998.

Dales, R., 'The Text of Robert Grosseteste's *Questio de fluxu et refluxu maris* with an English Translation', *Isis* 57, 1966, 455–474.

Damerow, P., Freudenthal, G., McLaughlin, P. and Renn, J., *Exploring the Limits of Preclassical Mechanics*, New York: Springer Verlag, 1992.

Dante Alighieri, *The Divine Comedy (La Commedia Divina)*, trans. P. Dale, London: Anvil Press Poetry, 1996.

Dauphinais, M. A., 'Loving the Lord your God: the *Imago Dei* in Saint Thomas Aquinas', *The Thomist* 63, 1999, 241–267.

Davidson, H. A., *Proofs for Eternity, Creation and the Existence of God in Medieval Islamic and Jewish Philosophy*, Oxford: Oxford University Press, 1987.

——— *Alfarabi, Avicenna, and Averroes, on Intellect: their cosmologies, theories of the active intellect and theories of the human intellect*, Oxford: Oxford University Press, 1992.

Davies, B., *The Thought of Thomas Aquinas*, Oxford: Oxford University Press, 1992.

Debrock, G. and Scheurer, P. B. (eds), *Newton's Scientific and Philosophical Legacy*, Dordrecht: Kluwer Academic, 1988.

Demos, R., 'Plato's Doctrine of the Psyche as a Self-Moving Motion', *Journal of the History of Philosophy* 6, 1968, 133–145.

Devereux, D. and Pellegrin, P. (eds), *Biologie, Logique at Métaphysique chez Aristote* (Actes du Séminaire C.N.R.S.-N.S.F. Oléron 28 juin–3 juillet 1987), Paris: Editions du Centre National de la Recherche Scientifique, 1990.

Di Liscia, D. A., Kessler, E. and Methuen, C. (eds), *Method and Order in Renaissance Philosophy of Nature: The Aristotelian Commentary Tradition*, Aldershot: Ashgate, 1997.

Dobbs, B. J. T., *The Janus Faces of Genius: The role of alchemy in Newton's thought*, Cambridge: Cambridge University Press, 1991.

Donati, S., 'The Anonymous Questions on *Physics* II–IV of MS Philadelphia, Free Library, Lewis Europ. 53 (ff. 71ra–85rb) and Roger Bacon', *Viarium* 35, 1997, 177–221.

Donohoo, L. J., 'The Nature and Grace of Sacra Doctrina in St. Thomas's *Super Boetium De Trinitate*', *The Thomist* 63, 1999, 341–401.

Duhem, P., *Le Système du Monde: histoires des doctrines cosmologiques de Platon à Copernic* (10 vols), Paris: Hermann, 1913–1959.

——— *Medieval Cosmology: Theories of Infinity, Place, Time, Void, and the Plurality of Worlds*, ed. and trans. R. Ariew, Chicago: University of Chicago Press, 1985.

——— *The Aim and Structure of Physical Theory*, trans. P. P. Wiener, Princeton, NJ: Princeton University Press, 1991.

——— *Essays in the History and Philosophy of Science*, trans. R. Ariew and P. Barker, Indianapolis: Hackett, 1996.

Duns Scotus, John, *Opera Omnia*, ed. C Balić *et al.*, Vatican City: Vatican Polyglot Press, 1950.

——— *Philisophical writings*, trans. A. Wolter, Indianapolis, IN: Hackett Publishing Co., 1987.

Düring, I. (ed.), *Naturphilosophie bei Aristoteles und Theophrast: Verhandlungen des 4. Symposium Aristotelicum veranstaltet in Göteborg, August 1966*, Heidelberg: Verlag, 1969.

Düring, I. and Owen, G. E. L. (eds), *Aristotle and Plato in the mid-Fourth Century: papers of the Symposium Aristotelicum held at Oxford in August, 1957*, Göteborg: Elanders Boktryckeri Aktiebolag, 1960.

Dutton, B. D., 'Nicholas of Autrecourt and William of Ockham on Atomism, Nominalism, and the Ontology of Motion', *Medieval Philosophy and Theology* 5, 1996, 63–85.

Easterling, H. J., 'Causation in the Timaeus and Laws X', *Eranos* 65, 1967, 25–38.

Eastwood, B. S., 'Grosseteste's "Quantitative" Law of Refraction: a chapter in the history of non-experimental science', *Journal of the History of Ideas* 28, 1967, 403–414.

Eco, U., *The Aesthetics of Thomas Aquinas*, London: Radius, 1988.

Elders, L. J., *Aristotle's Theory of the One: A Commentary on* Book X *of the* Metaphysics, Assen: Van Gorcum, 1960.

——— *Aristotle's Cosmology: A Commentary on the* De Caelo, Assen: Van Gorcum, 1965.

——— *Aristotle's Theology: A Commentary on* Book Λ *of the* Metaphysics, Assen: Van Gorcum, 1972.

——— *La Philosophie de la Nature de Saint Thomas d'Aquin*, Rome: Libreria Editrice Vaticana, 1982.

——— *The Metaphysics of Being of St. Thomas Aquinas in a Historical Perspective*, Leiden: E.J. Brill, 1993.

Elders, L. J. and Hedwig, K. (eds), *Lex et Libertas: Freedom and Law According to St. Thomas Aquinas*, Rome: Libreria Editrice Vaticana, 1987.

Ellis, G. F. R. and Murphy, N., *On the Moral Nature of the Universe*, Minneapolis: Fortress Press, 1996.

Ferrari, G. R. F., *Listening to the Cicadas: A Study of Plato's* Phaedrus, Cambridge: Cambridge University Press, 1987.

Feyerabend, P., *Farewell to Reason*, London: Verso, 1987.

—— *Against Method*, London: Verso, 1993.

Fisher, N. W. and Unguru, S., 'Experimental Science and Mathematics in Roger Bacon's Thought', *Traditio* 27, 1971, 353–378.

de Finance, J., *Être et Agir Dans La Philosophie de Saint Thomas*, Rome: Presses de L'Université Grégorienne, 1965.

FitzPatrick, P. J., *In Breaking of Bread: The Eucharist and Ritual*, Cambridge: Cambridge University Press, 1993.

Folkerts, M. and Hogendijk, J. P. (eds), *Vestigia Mathemaitca: studies in medieval and early modern mathematics in honour of H. L. L. Busard*, Amsterdam: Rodopi, 1993.

Force, J. E. and Katz, D. S., *Everything Connects: In conference with Richard H. Popkin*, Leiden: E.J. Brill, 1999.

Force, J. E. and Popkin, R. H. (eds), *Essays on the Context, Nature and Influence of Isaac Newton's Theology*, Dordrecht: Kluwer Academic, 1990.

—— *The Books of Nature and Scripture: recent essays on natural philosophy, theology, and biblical criticism in the Netherlands of Spinoza's time and the British Isles of Newton's time*, Dordrecht: Kluwer Academic, 1994.

—— *Newton and Religion: Context, Nature and Influence*, Dordrecht: Kluwer Academic, 1999.

—— *The Millenarian Turn: millenarian contexts of science, politics, and everyday Anglo-American life in the seventeenth and eighteenth centuries (Millenarianism and messianism in early modern European culture vol. 3)*, Dordrecht: Kluwer Academic, 2001.

Frank, E., 'The Fundamental Opposition of Plato and Aristotle', *American Journal of Philology* 61, 1940, 34–53 and 166–185.

Frank, W. A., 'Duns Scotus on Autonomous Freedom and Divine Co-Causality', *Medieval Philosophy and Theology* 2, 1992, 142–164.

French, R. and Cunningham, A., *Before Science: The Invention of the Friars' Natural Philosophy*, London: Scolar Press, 1996.

Freudenthal, G., *Atom and the Individual in the Age of Newton: On the Genesis of the Mechanistic World View*, Dordrecht: Reidel, 1986.

Funkenstein, A., *Theology and the Scientific Imagination*, Princeton, NJ: Princeton University Press, 1986.

Gadamer, H. G., *Dialogue and Dialectic: Eight Hermeneutical Studies on Plato*, trans. P. C. Smith, New Haven, CT: Yale University Press, 1980.

—— *The Idea of the Good in Platonic–Aristotelian Philosophy*, trans. P. C. Smith, New Haven, CT: Yale University Press, 1986.

de Gandt, F. and Souffrin, P. (eds), *La Physique d'Aristote et les Conditions d'Une Science de la Nature*, Paris: Librairie Philosophique J. Vrin, 1991.

Gardet, L., *La pensée religieuse d'Avicenne (Ibn Sīnā)*, Paris: Librairie Philosophique J. Vrin, 1951.

Gerson, L. P., *God and Greek Philosophy: Studies in the Early History of Natural Philosophy*, London: Routledge, 1990.

—— (ed.), *The Cambridge Companion to Plotinus*, Cambridge: Cambridge University Press, 1996.

Gill, M. L. and Lennox, J. G. (eds), *Self-motion: from Aristotle to Newton*, Princeton, NJ: Princeton University Press, 1994.

Gilson, E., 'Pourquoi Saint Thomas a critiqué Saint Augustin', *Archives d'histoire doctrinale et littéraire du moyen age* 1, 1926.

—— 'Avicenne et le point de départ de Duns Scot', *Archives d'histoire doctrinale et littéraire du moyen âge* 2, 1927, 89–149.

—— *History of Christian Philosophy in the Middle Ages*, London: Sheed and Ward, 1980.

—— *From Aristotle to Darwin and Back Again: A Journey in Final Causality, Species, and Evolution*, trans. J. Lyon, London: Sheed and Ward, 1984.

—— *The Christian Philosophy of St. Thomas Aquinas*, Notre Dame, IN: University of Notre Dame Press, 1994 edition.

Goddu, A., *The Physics of William of Ockham*, Leiden: E.J. Brill, 1984.

Goichon, A.-M., *La distinction de l'essence et de l'existence d'après Ibn-Sina (Avicenne)*, Paris: Desclée de Brouwer, 1937.

—— *The Philosophy of Avicenna*, trans. M. S. Kahn, Delhi: Delhi Motil al Banarsidass, 1969.

Goldish, M., *Judaism in the Theology of Sir Isaac Newton*, Dordrecht: Kluwer Academic, 1998.

Goodfield, J. and Toulmin, S., *The Fabric of the Heavens*, London: Hutchinson, 1961.

Gotthelf, A. (ed.), *Aristotle on Nature and Living Things: Philosophical and Historical Studies Presented to David M. Balme on His Seventieth Birthday*, Pittsburgh: Mathesis Publications, 1985.

Graham, D., *Aristotle's Physics: book VIII*, Oxford: Oxford University Press, 1999.

Grant, E. (ed.), *A Sourcebook in Medieval Science*, Cambridge, MA: Harvard University Press, 1974.

—— *Studies in Medieval and Natural Philosophy*, London: Variorum Reprints, 1981.

—— *The Foundations of Modern Science in the Middle Ages: their religious, institutional and intellectual contexts*, Cambridge: Cambridge University Press, 1996.

—— *God and Reason in the Middle Ages*, Cambridge: Cambridge University Press, 2001.

Grant, E. and Murdoch, J. E. (eds), *Mathematics and Its Application to Science and Natural Philosophy in the Middle Ages*, Cambridge: Cambridge University Press, 1987.

Gribbin, J., *Science: A History, 1543–2001*, London: Allen Lane, 2002.

de Groot, J., 'Form and Succession in Aristotle's "Physics"', *Proceedings of the Boston Area Colloquium in Ancient Philosophy* 10, 1994, 1–23.

Grosseteste, R., *Robert Grosseteste on Light* (De Luce), intro. and trans. C. C. Reidl, Milwaukee: Marquette University Press, 1942.

—— *Commentarius in Posteriorum Analyticorum Libros*, ed. P. Rossi, Firenze: Leo S. Olschiki Editore, 1981.

—— *On the Six Days of Creation* ('Hexaëmeron'), trans. C. F. J. Martin, Oxford: Oxford University Press, 1996.

Hackett, J. (ed.), *Roger Bacon and the Sciences: commemorative essays*, Leiden: E.J. Brill, 1997.

—— 'Roger Bacon and Aristotelianism: introduction', *Viarium* 35, 1997, 129–135.

—— 'Roger Bacon, Aristotle, and the Parisian Condemnations of 1270, 1277', *Viarium* 35, 1997, 283–314.

Hall, A. R. and Hall, M. B. (eds), *Unpublished Scientific Papers of Isaac Newton*, Cambridge: Cambridge University Press, 1962.

Hall, D. C., *The Trinity: An Analysis of St. Thomas Aquinas' Expositio of the De Trinitate of Boethius*, Leiden: E.J. Brill, 1992.

Hall, P. M., 'Towards a Narrative Understanding of Thomistic Natural Law', *Medieval Philosophy and Theology* 2, 1992, 53–73.

Harrison, P., 'Newtonian Science, Miracles, and the Laws of Nature', *Journal of the History of Ideas* 56, 1995, 531–553.

—— *The Bible, Protestantism, and the rise of Natural Science*, Cambridge: Cambridge University Press, 1998.

Heidegger, M., *The Question Concerning Technology and Other Essays*, trans. W. Lovitt, New York: Harper & Row, 1977.

Honnefelder, L., Wood, R. and Dreyer, M. (eds), *John Duns Scotus: Metaphysics and Ethics*, Leiden: E.J. Brill, 1996.

Hussey, E., *Aristotle's* Physics: *books III and IV*, Oxford: Oxford University Press, 1983.

Hutchins, R. M. (ed.), *Great Books of the Western World, vol. 45*, Chicago: Encyclopaedia Britannica, 1952.

Jacob, M. C., *The Newtonians and the English Revolution 1689–1720*, Hassocks, Sussex: Harvester Press, 1976.

Jacobi, K. (ed.), *Argumentationstheorie: scholastische Forschungen zu den logischen and semantischen Regeln korrekten Folgerns*, Leiden: E.J. Brill, 1993.

Jenkins, J., 'Expositions of the Text: Aquinas's Aristotelian Commentaries', *Medieval Philosophy and Theology* 5, 1996, 39–62.

Johnson, M. F., 'Another Look at the Plurality of the Literal Sense', *Medieval Philosophy and Theology* 2, 1992, 117–141.

Jordan, M. D., 'The Order of Lights: Aquinas on Immateriality as Hierarchy', *Proceedings of the American Philosophical Association* 52, 1978, 112–120.

—— *Ordering Wisdom: The Hierarchy of the Philosophical Discourses in Aquinas*, Notre Dame, IN: University of Notre Dame Press, 1986.

—— *The Alleged Aristotelianism of Thomas Aquinas*, Toronto: Pontifical Institute of Mediaeval Studies, 1992.

Judson, L., *Aristotle's Physics: A Collection of Essays*, Oxford: Oxford University Press, 1991.

Kardaun, M. and Spruyt, J. (eds), T*he Winged Chariot: Collected Essays on Plato and Platonism in Honour of L. M. de Rijk*, Leiden: E.J. Brill, 2000.

Kerr, F., *After Aquinas: versions of Thomism*, Oxford: Blackwell, 2002.

Kirwan, C., *Aristotle* Metaphysics *Books Γ, Δ, and E*, Oxford: Oxford University Press, 1993.

Klein, J., *Greek Mathematical Thought and the Origin of Algebra*, Cambridge, MA: MIT Press, 1968.

Knasas, J. F. X., 'Thomistic Existentialism and the Proofs *Ex Motu* at *Contra Gentiles* I,C. 13', *The Thomist* 59, 1995, 591–615.

Kondoleon, T. J., 'The Argument from Motion and the Argument for Angels: a reply to John F. X. Knasas', *The Thomist* 62, 1998, 269–290.

Kosman, L. A., 'Aristotle's Definition of Motion', *Phronesis* 14, 1969, 40–62.

Koyré, A., 'The Origins of Modern Science: A New Interpretation', *Diogenes* 16, 1956, 1–22.

—— *Newtonian Studies*, London: Chapman and Hall, 1965.

—— *Metaphysics and Measurement: Essays in the Scientific Revolution*, London: Chapman and Hall, 1968.

—— *Galileo Studies*, trans. J. Mepham, Hassocks, Sussex: Harvester Press, 1978.

Krämer, H. J., *Plato and the Foundations of Metaphysics: A Work on the Theory of the Principles and Unwritten Doctrines of Plato with a Collection of the Fundamental Documents*, ed. and trans. J. R. Catan, Albany, NY: State University of New York Press, 1990.

Kraut, R. (ed.), *The Cambridge Companion to Plato*, Cambridge: Cambridge University Press, 1992.

Kretzmann, N. (ed.), *Infinity and Continuity in Ancient and Medieval Thought*, Ithaca, NY: Cornell University Press, 1982.

Kretzmann, N. and Stump, E. (eds), *The Cambridge Companion to Aquinas*, Cambridge: Cambridge University Press, 1993.

Kretzmann, N., Kenny, A. and Pinborg, J. (eds), *The Cambridge History of Later Medieval Philosophy*, Cambridge: Cambridge University Press, 1988.

Kubrin, D., 'Newton and the Cyclical Cosmos: Providence and the Mechanical Philosophy', *Journal of the History of Ideas* 28, 1967, 325–346.

Kuhn, T. S., *The Essential Tension: Selected Studies in Scientific Tradition and Change*, London: University of Chicago Press, 1977.

—— *The Structure of Scientific Revolutions*, Chicago: University of Chicago Press, 1996 edition.

Kupfer, J., 'The Father of Empiricism: Roger not Francis', *Vivarium* 22, 1974, 52–62.

Lang, H. S., *Aristotle's* Physics *and Its Medieval Varieties*, Albany, NY: State University of New York Press, 1992.

—— 'Aristotle's *Physics* IV, 8: A Vexed Argument in the History of Ideas', *Journal of the History of Ideas* 56, 1995, 353–376.

—— 'Thomas Aquinas and the Problem of Nature in *Physics* II,1', *History of Philosophy Quarterly* 13, 1996, 411–432.

—— *The Order of Nature in Aristotle's* Physics: *Place and the Elements*, Cambridge: Cambridge University Press, 1998.

Laporte, J.-M., 'The Motion of Operative and Cooperative Grace: Retrievals and Explorations', *Lonergan Workshop* 13, 1997, 79–94.

Lawrence, C. and Shapin, S. (eds), *Science Incarnate: Historical Embodiments of Natural Knowledge*, Chicago: University of Chicago Press, 1998.

Lear, J., 'Aristotle's Philosophy of Mathematics', *Philosophical Review* 91, 1982, 161–192.

—— *Aristotle: the desire to understand*, Cambridge: Cambridge University Press, 1988.

Leclerc, I., 'Motion, Action, and Physical Being', *International Philosophical Quarterly* 21, 1981, 17–28.

Lee, P., 'St. Thomas and Avicenna on the Agent Intellect', *The Thomist* 45, 1981, 41–61.

Lentz, W., 'The Problem of Motion in the Sophist', *Apeiron* 30, 1997, 89–108.

Lewis, F. A. and Bolton, R. (eds), *Form, Matter and Mixture in Aristotle*, Oxford: Blackwell, 1996.

de Libera, A., *La philosophie médiévale*, Paris: Presses Universitaires de France, 1993.

Lindberg, D. C., 'Roger Bacon's Theory of the Rainbow: Progress or Regress?', *Isis* 57, 1966, 235–248.

—— (ed.), *Science in the Middle Ages*, Chicago: University of Chicago Press, 1978.

—— 'On the Applicability of Mathematics to Nature: Roger Bacon and his Predecessors', *British Journal for the History of Science* 15, 1982, 3–25.

—— *Roger Bacon's Philosophy of Nature: a critical edition, with English translation, introduction, and notes, of* De multiplicatione specierum *and* De speculis comburentibus, Oxford: Oxford University Press, 1983.

—— 'The Genesis of Kepler's Theory of Light: Light Metaphysics from Plotinus and Kepler', *Osiris* 2nd series, 1986, 5–42.

—— 'Science as Handmaiden: Roger Bacon and the Patristic Tradition', *Isis* 78, 1987, 518–536.

—— *The Beginnings of Western Science: The European Scientific Tradition in Philosophical, Religious, and Institutional Context, 600 B.C. to A.D. 1450*, Chicago: University of Chicago Press, 1992.

—— *Roger Bacon and the Origins of* Perspectiva *in the Middle Ages: a critical edition and English translation of Bacon's* Perspectiva *with introduction and notes*, Oxford: Clarendon Press, 1996.

Link-Salinger, R., *A Straight Path: Studies in Medieval Philosophy and Culture: Essays in Honor of Arthur Hymann*, Washington, DC: Catholic University Press of America, 1988.

Lintott, A., *Violence, Civil Strife and Revolution in the Classical City 750–330BC*, London: Croom Helm, 1982.

Little, A., *The Platonic Heritage of Thomism*, Dublin: Golden Eagle Books, 1949.

Little, A. G., *Roger Bacon: essays, contributed by various writers on the occasion of the commemoration of the seventh century of his birth*, Oxford: Clarendon Press, 1914.

Livesey, S. J., *Theology and Science in the Fourteenth Century: three questions on the unity of subalternation of the sciences from John of Reading's* 'Commentary on the Sentences', Leiden: E.J. Brill, 1989.

Lloyd, G. E. R., 'Plato as Natural Scientist', *Journal of Hellenistic Studies*, 1968, 78–92.

—— *Magic, Reason and Experience: studies in the origins and development of Greek science*, Cambridge: Cambridge University Press, 1979.

—— *The Revolutions of Wisdom: studies in the claims and practice of ancient Greek science*, Los Angeles: University of California Press, 1987.

—— *Methods and Problems in Greek Science*, Cambridge: Cambridge University Press, 1991.

—— *Aristotelian Explorations*, Cambridge: Cambridge University Press, 1996.

Lloyd, S. A., *Ideals as interests in Hobbes's* Leviathan: *The power of mind over matter*, Cambridge: Cambridge University Press, 1992.

Locke, J., *An Essay Concerning Human Understanding*, ed. J. W. Yolton, London: J. M. Dent & Sons, 1977.

Lonergan, B. J. F., *Grace and Freedom: Operative Grace in the Thought of St. Thomas Aquinas*, London: Darton, Longman and Todd, 1971.

Long, R. J. (ed.), *Philosophy and the God of Abraham: Essays in Memory of James A. Weisheipl*, Toronto: Pontifical Institute of Mediaeval Studies, 1991.

—— 'Roger Bacon on the Nature and Place of Angels', *Viarium* 35, 1997, 266–282.

Loose, P., 'Roger Bacon on Perception: a reconstruction and critical analysis of the theory of visual perception expounded in the Opus Majus', unpublished PhD thesis, Ohio State University, 1979.

Machamer, P. K., 'Aristotle on Natural Place and Natural Motion', *Isis* 69, 1978, 377–387.

Machamer, P. K. and Turnbull, R. G., *Motion and Time, Space and Matter: Interrelations in the History and Philosophy of Science*, Columbus, OH: Ohio State University Press, 1976.

Macierowski, E. M., 'Does God Have a Quiddity According to Avicenna?', *The Thomist* 54, 1988, 79–87.

MacIntosh, J. J., 'St. Thomas on Angelic Time and Motion', *The Thomist* 59, 1995, 547–575.

Mackinnon, E., 'Motion, Mechanics and Theology', *Thought* 36, 1961, 344–370.

Madigan, A., *Aristotle's* Metaphysics Books B and K 1–2, Oxford: Oxford University Press, 1999.

Maier, A., *Zwei Grundprobleme der scholastischen Naturphilosophie: das Problem der intensiven Grösse die Impetustheorie*, Roma: Edizioni di storia e letteratura, 1951.

—— *On the Threshold of Exact Science: selected writings of Anneliese Maier on late medieval natural philosophy*, ed. and trans. S. D. Sargent, Philadelphia: University of Pennsylvania Press, 1982.

Maloney, T., *Three Treatments of Universals by Roger Bacon: a translation with introduction and notes*, Binghamton, NY: State University of New York at Binghampton, 1989.

Manuel, F. E., *The Religion of Isaac Newton*, Oxford: Oxford University Press, 1974.

Marenbon, J. (ed.), *Aristotle in Britain during the Middle Ages*, Turnhout: Brepolis, 1996.

Marmo, C., 'Bacon, Aristotle (and the others) on Natural Inferential Signs', *Viarium* 35, 1997, 136–154.

Marrone, S. P., *William of Auvergne and Robert Grosseteste: New Ideas of Truth in the Early Twelfth Century*, Princeton, NJ: Princeton University Press, 1983.

—— *The Light of Thy Countenance: Science and Knowledge of God in the Thirteenth Century* (2 vols), Leiden: E.J. Brill, 2001.

Martens, R., *Kepler's Philosophy and the New Astronomy*, Princeton, NJ: Princeton University Press, 2000.

Marvin, F. S. (ed.), *Science and Civilization*, London, 1923.

Maurer, A., *Being and Knowing: Studies in Thomas Aquinas and Later Medieval Philosophers*, Toronto: Pontifical Institute of Mediaeval Studies, 1990.

Maxwell, J. C., *A Treatise on Electricity and Magnetism* (2 vols), Oxford: Clarendon Press, 1873.

—— *A Dynamical Theory of the Electromagnetic Field*, ed. T. F. Torrance, Edinburgh: Scottish Academic Press, 1982.

McEvoy, J., *The Philosophy of Robert Grosseteste*, Oxford: Clarendon Press, 1982.

—— (ed.), *Robert Grosseteste: new perspectives on his thought and scholarship*, Turnhout: Brepolis, 1995.

—— *Robert Grosseteste*, Oxford: Oxford University Press, 2000.

McGuire, J. E., *Tradition and Innovation: Newton's Metaphysics of Nature*, Dordrecht: Kluwer Academic, 1995.

McGuire, J. E. and Rattansi, P. M., 'Newton and the "Pipes of Pan"', *Notes and Records of the Royal Society of London* 21, 1966, 108–143.

McGuire, J. E. and Tamney, M., *Certain Philosophical Questions: Newton's Trinity Notebook*, Cambridge: Cambridge University Press, 1983.

McInerny, R., *Aquinas and Analogy*, Washington, DC: Catholic University Press of America, 1996.

McKeon, R. (ed. and trans.), *Selections from Medieval Philosophers* (2 vols), New York: Charles Scribner's Sons, 1957.

—— *On Knowing – The Natural Sciences*, ed. D. B. Owen and Z. K. McKeon, Chicago: University of Chicago Press, 1994.

McLaughlin, T. J., 'Aristotelian Mover-Causality and the Principle of Inertia', *International Philosophical Quarterly* 38, 1998, 137–151.

McMullin, E., *The Inference that Makes Science* (The Aquinas Lecture, 1992), Milwaukee: Marquette University Press, 1992.

Menn, S., 'The Origin of Aristotle's Concept of Ενεργεια: Ενεργεια and Δυναμισ', *Ancient Philosophy* 14, 1994, 73–114.

Metzger, D., *The Lost Cause of Rhetoric: The Relation of Rhetoric and Geometry in Aristotle and Lacan*, Carbondale, IL: Southern Illinois University Press, 1995.

Milbank, J., *Theology and Social Theory: Beyond Secular Reason*, Oxford: Blackwell, 1990.

—— *The Word Made Strange: Theory, Language, Culture*, Oxford: Blackwell, 1997.

Milbank, J. and Pickstock, C., *Truth in Aquinas*, London: Routledge, 2001.

Milbank, J., Pickstock, C. and Ward, G. (eds), *Radical Orthodoxy: a new theology*, London: Routledge, 1999.

Mohr, R., 'The Mechanism of Flux in Plato's "Timaeus"', *Apeiron* 14, 1980, 96–114.

—— 'The Sources of Evil Problem and the αρχη κινησεως Doctrine in Plato', *Apeiron* 14, 1980, 41–56.

—— *The Platonic Cosmology*, Leiden, 1985.

Molland, G., 'Roger Bacon and the Hermetic Tradition in Medieval Science', *Vivarium* 31, 1993, 140–160.

—— *Mathematics and the Medieval Ancestry of Physics*, Aldershot: Variorum, 1995.

Moravcsik, J., *Plato and Platonism: Plato's Conception of Appearance and Reality in Ontology, Epistemology, and Ethics, and its Modern Echoes*, Oxford: Blackwell, 1992.

Moreno, A., 'The Law of Inertia and the Principle *Quidquid Movetur Ab Alio Movetur*', *The Thomist* 38, 1974, 306–331.

—— 'Time and Relativity: Some Philosophical Considerations', *The Thomist* 45, 1981, 62–79.

de Muralt, A., *L'Enjeu de la Philosophie Médiévale: études thomistes, scotistes, occamiennes et grégoriennes*, Leiden: E.J. Brill, 1991.

Murdoch, J. E., '*Mathesis in philosophiam scholasticam introducta*: The Rise and Development of the Application of Mathematics in Fourteenth Century Philosophy and Theology', in *Arts libéraux et philosophie médiévale, Université de Montréal*, Montréal: Institut d'études médiévales, 1969, pp. 215–254.

Murdoch, J. E. and Sylla, E. Dudley (eds), *The Cultural Context of Medieval Learning* (Proceedings of the First International Colloquium on Philosophy, Science, and Theology in the Middle Ages – September 1973), Dordrecht: Reidel, 1975.

Narcisse, G., *Les Raisons de Dieu: Argument de convenance et Esthétique théologique selon saint Thomas d'Aquin et Hans Urs von Balthasar*, Fribourg: Editions Universitaires Fribourg Suisse, 1997.

Nasr, S. H., *An Introduction to Islamic Cosmological Doctrines: conceptions of nature and methods for its study by Ikhwān al-Safā, al-Bīrūnī, and Ibn Sīnā*, Cambridge, MA: The Belknap Press of Harvard University Press, 1964.

Nasr, S. H. and Leaman, O. (eds), *History of Islamic Philosophy part I*, London: Routledge, 1996.

Newton, I., *The Chronology of Ancient Kingdoms Amended*, London, 1728.

—— *Observations upon the Prophecies of Daniel, and the Apocalypse of St. John*, London, 1733.

—— *Two Letters of Isaac Newton to Mr. Le Clerc*, London, 1754.

—— *Opticks*, New York: Dover, 1952.

—— *The Principia: Mathematical Principles of Natural Philosophy*, trans. I. B. Cohen and A. Whitman, London: University of California Press, 1999.

Niven, W. D. (ed.), *The Scientific Papers of James Clerk Maxwell*, vol. 1, Cambridge: Cambridge University Press, 1890.

Nodé-Langlois, M., 'L'intuitivité de l'intellect selon Thomas d'Aquin', *Revue Thomiste* 100, 2000, 179–203.

Noone, T., 'Roger Bacon and Richard Rufus on Aristotle's Metaphysics: a search for the grounds of disagreement', *Viarium* 35, 1997, 251–265.

Nussbaum, M., *Aristotle's De Motu Animalum: text and translation, commentary, and interpretative essays*, Princeton, NJ: Princeton University Press, 1985.

Nye, M. J., *Before Big Science: The Pursuit of Modern Chemistry and Physics, 1800–1940*, Cambridge, MA: Harvard University Press, 1996.

Oakley, F., 'Christian Theology and the Newtonian Science: The Rise of the Concept of the Laws of Nature', *Church History* 30, 1961, 433–457.

Oakley, F. and O'Conner, D. (eds), *Creation: the impact of an idea*, New York: Charles Scribner's Sons, 1969.

O'Mera, D. J. (ed.), *Platonic Investigations*, Washington, DC: Catholic University of America Press, 1985.

Onians, R. B., *The Origins of European Thought about the Body, the Mind, the Soul, the World, Time and Fate*, Cambridge: Cambridge University Press, 1951.

O'Rourke, F., *Pseudo-Dionysius and the Metaphysics of Aquinas*, Leiden: E.J. Brill, 1992.

Osler, M. J., *Rethinking the Scientific Revolution*, Cambridge: Cambridge University Press, 2000.

Osler, M. J. and Farber, L. P. (eds), *Religion, Science and Worldview: essays in honour of Richard Westfall*, Cambridge: Cambridge University Press, 1985.

Owen, G. E. L., 'The Platonism of Aristotle', *Proceedings of the British Academy* 51, 1965, 125–150.

Park, D., *The Fire within the Eye: a historical essay on the nature and meaning of light*, Princeton, NJ: Princeton University Press, 1997.

Pavelich, A., 'Descartes' Eternal Truths and Laws of Motion', *Southern Journal of Philosophy* 35, 1997, 517–537.

Pegis, A. C., 'St. Thomas and the Coherence of the Aristotelian Theology', *Mediaeval Studies* 35, 1973, 67–117.

Penrose, R., *The Emperor's New Mind*, Oxford: Oxford University Press, 1989.

Philoponus, J., *In Aristotelis Physicorum libros quinque posteriores commentaria*, ed. G. Vitelli, Berolini: Typ. et impensis G. Reimeri, 1888.

Pickstock, C. J., *After Writing: On the Liturgical Consummation of Philosophy*, Oxford: Blackwell, 1998.

Plato, *Euthyphro/Apology/Crito/Phaedo/Phaedrus* (Loeb Classical Library), trans. H. N. Fowler, Cambridge, MA: Harvard University Press, 1914.

—— *Sophist* (Loeb Classical Library), trans. H. N. Fowler, Cambridge, MA: Harvard University Press, 1914.

—— *Timaeus/Critias/Cleitophon/Menexenus/Epistles* (Loeb Classical Library), trans. R. G. Bury, Cambridge, MA: Harvard University Press, 1929.

—— *Republic* (2 vols) (Loeb Classical Library), trans. P. Shorey, Cambridge, MA: Harvard University Press, 1963.

—— *Laws* (Loeb Classical Library), trans. R. G. Bury, Cambridge, MA: Harvard University Press, 1967.

Plotinus, *Enneads* (vols VI and VII, Loeb Classical Library), trans. A. H. Armstrong, Cambridge, MA: Harvard University Press, 1988.

Proclus, *The Elements of Theology*, trans. E. R. Dodds, Oxford: Clarendon Press, 1963.

—— *Commentary on Plato's* Timaeus (2 vols), trans. T. Taylor, London: The Prometheus Trust, 1998.

Protevi, J., *Time and Exteriority: Aristotle, Heidegger, Derrida*, London and Toronto: Associated University Presses, 1994.

Pseudo-Dionysius, *The Complete Works*, trans. C. Luibheid, New York: Paulist Press, 1987.

Randall, J. H., *The Career of Philosophy: From the Middle Ages to the Enlightenment*, London: Columbia University Press, 1962.

Ray, C., *Time, Space and Philosophy*, London: Routledge, 1991.

Ross, W. D., *Aristotle's* Physics: *a revised text with introduction and commentary*, Oxford: Oxford University Press, 1936.

Rossetti, L. (ed.), *Understanding the* Phaedrus: *Proceedings of the II Symposium Platonicum*, Sankt Augustin: Academia Verlag, 1992.

Rossi, P., *The Birth of Modern Science*, trans. C. De Nardi Ipsen, Oxford: Blackwell, 2000.

Rynasiewicz, R., 'By Their Properties, Causes and Effects: Newton's *Scholium* on Time, Space, Place and Motion – I. The Text', *Studies in History and Philosophy of Science* 26, 1995, 133–153.

—— 'By Their Properties, Causes and Effects: Newton's *Scholium* on Time, Space, Place and Motion – II. The Context', *Studies in History and Philosophy of Science* 26, 1995, 295–321.

Sarasohn, L. T., 'Motion and Morality: Pierre Gassendi, Thomas Hobbes and the Mechanical World-View', *Journal of the History of Ideas* 46, 1985, 363–379.

Scaltsas, T., Charles, D. and Gill, M. L. (eds), *Unity, Identity, and Explanation in Aristotle's Metaphysics*, Oxford: Oxford University Press, 1994.

Schaffer, S., 'The Show that Never Ends: perpetual motion in the early eighteenth century', *British Journal of the History of Science* 28, 1995, 157–189.

Scheurer, P. B. and Debrock, G. (eds), *Newton's Scientific and Philosophical Legacy*, Dordrecht: Kluwer Academic, 1988.

Serene, E. F., 'Robert Grosseteste on Induction and Demonstrative Science', *Synthese* 40, 1979, 97–115.

Shapin, S., *A Social History of Truth: Civility and Science in Seventeenth Century England*, Chicago: University of Chicago Press, 1994.

Shapin, S. and Schaffer, S., *Leviathan and the Air-pump: Hobbes, Boyle and the Experimental Life*, Princeton, NJ: Princeton University Press, 1985.

Shields, C., *Order in Multiplicity: Homonymy in the Philosophy of Aristotle*, Oxford: Oxford University Press, 1999.

Sider, D. (ed.), *The Fragments of Anaxagoras*, Meisenheim am Glan: Hain, 1981.

Sigurdarson, Eríkur Smári, 'Plato's Ideal of Science', in Erik Nis Ostenfeld (ed.), *Essays on Plato's Republic*, Aarhus: Aarhus University Press, 1998.

Sirridge, M., 'As It Is, It Is an Ax: Some Medieval Reflections on De Anima II.1', *Medieval Philosophy and Theology* 6, 1997, 1–24.

Skemp, J. B., *The Theory of Motion in Plato's Later Dialogues*, Amsterdam: Adolf M. Hakkert, 1967.

Slowik, E., 'Huygens' Center-of-Mass Space-Time Reference Frame: Constructing a Cartesian Dynamics in the Wake of Newton's "De Gravitatione" Argument', *Synthese* 112, 1997, 247–269.

Smith, W., *The Quantum Enigma*, Peru, IL: Sherwood Sugden, 1995.

—— 'From Schrödinger's Cat to Thomistic Ontology', *The Thomist* 63, 1999, 49–63.

Snobelen, S. D., 'On Reading Isaac Newton's *Principia* in the Eighteenth Century', *Endeavour* 22, 1998, 159–163.

—— 'Isaac Newton, heretic: the strategies of a Nicodemite', *British Journal for the History of Science* 32, 1999, 381–419.

—— ' "God of gods, and Lord of lords". The Theology of Isaac Newton's *General Scholium* to the *Principia*', *Osiris* 16, 2001, 169–208.

Sorabji, R., *Matter, Space and Motion: Theories in Antiquity and Their Sequel*, London: Duckworth, 1988.

Speer, A., 'Reception – Mediation – Innovation: Philosophy and Theology in the Twelfth Century', in J. Hamesse (ed.), *Bilan et Perspectives des Etudes Médiévales en Europe*, Louvain-la-Neuve: Fédération Internationale des Instituts d'Etudes, 1995, pp. 129–149.

Strange, S. K., 'The Double Explanation in the *Timaeus*', *Ancient Philosophy* 5, 1985, 25–39.

Sweeney, E. C., 'From Determined Motion to Undetermined Will and Nature to Supernature in Aquinas', *Philosophical Topics* 20, 1992, 189–214.

Sylla, E. Dudley, 'Medieval Concepts of the Latitude of Forms: The Oxford Calculators', *Archives d'histoire doctrinale et littéraire du moyen-âge* 40, 1973, 223–283.

—— 'Aristotelian Commentaries and Scientific Change: The Parisian Nominalists on the Cause of the Natural Motion of Inanimate Bodies', *Vivarium* 21, 1993, 37–83.

Tachau, K. H., *Vision and Certitude in the Age of Ockham: Optics, Epistemology and the Foundations of Semantics 1250–1345*, Leiden: E.J. Brill, 1988.

Taylor, A. E., *A Commentary on Plato's* Timaeus, Oxford: Clarendon Press, 1928.

Taylor, C. C. W. (ed.), *The Atomists Leucippus and Democritus: fragments: a text and translation*, Toronto: University of Toronto Press, 1999.

Taylor, J. C., *Hidden Unity in Nature's Laws*, Cambridge: Cambridge University Press, 2001.

te Velde, R. A., 'Natural Reason in the *Summa contra Gentiles*', *Medieval Philosophy and Theology* 4, 1994, 42–70.

—— *Participation and Substantiality in Thomas Aquinas*, Leiden: E.J. Brill, 1995.

Thorndike, L., 'Roger Bacon and Experimental Method in the Middle Ages', *Philosophical Review* 23, 1914, 271–298.

Torrance, T. F., *Christian Theology and Scientific Culture*, Belfast: Christian Journals, 1980.

—— *Space, Time and Incarnation*, Edinburgh: T&T Clark, 1997.

Toulmin, S., 'Criticism in the History of Science: Newton on Absolute Space, Time, and Motion, I', *Philosophical Review* 68, 1959, 1–29.

—— 'Criticism in the History of Science: Newton on Absolute Space, Time, and Motion, II', *Philosophical Review* 68, 1959, 203–227.

—— *Cosmopolis: The Hidden Agenda of Modernity*, Chicago: University of Chicago Press, 1990.

Trifolgi, C., 'Roger Bacon and Aristotle's Doctrine of Place', *Viarium* 35, 1997, 155–176.

—— *Oxford Physics in the Thirteenth Century (ca. 1250–1270): motion, infinity, place and time*, Leiden: E.J. Brill, 2000.

Turnbull, H. W., Scott, J. F., Hall, A. R. and Tilling, L. (eds), *The Correspondence of Isaac Newton* (7 vols), Cambridge: Cambridge University Press, 1959–1977.

Turner, D., *How to be an Atheist: Inaugural Lecture Delivered at the University of Cambridge, 12 October 2001*, Cambridge: Cambridge University Press, 2002.

Twetten, D., 'Back to Nature in Aquinas', *Medieval Theology and Philosophy* 5, 1996, 205–244.

—— 'Clearing a "Way" for Aquinas: How the Proof from Motion Concludes to God', *Proceedings of the American Catholic Philosophical Quarterly* 70, 1996, 259–278.

Työrinoja, R., Lehtinen, A. I. and Føllesdal, D. (eds), *Knowledge and the Sciences in Medieval Philosophy* (Proceedings of the Eighth International Congress of Medieval Philosophy, vol. 3), Helsinki: Yliopistopaino, 1990.

Unguru, S. (ed.), *Physics, Cosmology and Astronomy, 1300–1700: Tension and Accommodation*, Dordrecht: Kluwer Academic, 1991.

van Ophuijsen, J. M. (ed.), *Plato and Platonism*, Washington, DC: Catholic University Press of America, 1999.

Vlastos, G., *Plato's Universe*, Oxford: Clarendon Press, 1975.

Wadell, P. J., *Friends of God: virtues and gifts in Aquinas*, New York: Peter Lang, 1991.

Wager, W. W. (ed.), *The Secular Mind: Transformations of Faith in Modern Europe*, London: Holmes and Meier, 1982.

Wallace, W., 'St. Thomas Aquinas, Galileo, and Einstein', *The Thomist* 24, 1961, 1–22.

—— 'Buridan, Ockham, Aquinas: Science in the Middle Ages', *The Thomist* 40, 1976, 475–483.

—— *From a Realist Point of View: Essays on the Philosophy of Science*, Washington, DC: University of America Press, 1979.

—— *Prelude to Galileo: Essays on Medieval and Sixteenth-Century Sources of Galileo's Thought*, Dordrecht: Reidel, 1981.

—— *Galileo, the Jesuits and the Medieval Aristotle*, Aldershot: Variorum, 1991.

—— *The Modelling of Nature: Philosophy of Science and Philosophy of Nature in Synthesis*, Washington, DC: Catholic University Press of America, 1996.

—— 'Thomism and the Quantum Enigma', *The Thomist* 61, 1997, 455–467.

Wardy, R., *The Chain of Change: A Study of Aristotle's* Physics VII, Cambridge: Cambridge University Press, 1990.

Waterlow, S., *Nature, Change and Agency in Aristotle's* Physics, Oxford: Oxford University Press, 1982.

—— *Passage and Possibility*, Oxford: Clarendon Press, 1982.

Weisheipl, J. A., *Friar Thomas d'Aquino: his life, thought and works*, Oxford: Basil Blackwell, 1974.

—— 'The Concept of Nature: Avicenna and Aquinas', in V. Brezik (ed.), *Thomistic Papers*, 1, Houston, TX: Center for Thomistic Studies, 1984, pp. 65–82.

—— *Nature and Motion in the Middle Ages*, ed. W. E. Carroll, Washington, DC: Catholic University Press of America, 1985.

Werner, C., *Aristote et L'Idéalism Platonicien*, Paris: F. Alcan, 1910.

Westfall, R. S., *Never at Rest: A Biography of Isaac Newton*, Cambridge: Cambridge University Press, 1980.

Westman, R. S. and McGuire, J. E., *Hermeticism and the Scientific Revolution*, Los Angeles, 1977.

Whewell, W., *History of the Inductive Science* (2 vols), New York, 1858.

Whitehead, A. N., *The Concept of Nature*, Cambridge: Cambridge University Press, 1964.

Wians, W. (ed.), *Aristotle's Philosophical Development: Problems and Prospects*, London: Rowman and Littlefield, 1996.

Williams, S. J., 'Roger Bacon and His Edition of the Pseudo-Aristotelian *Secretum secretorum*', *Speculum* 69, 1994, 57–73.

Wilson, C., 'Motion, Sensation, and the Infinite: the lasting impression of Hobbes on Leibniz', *British Journal of the History of Philosophy* 5, 1997, 339–351.

Wippel, J., 'Commentary of Boethius' *De Trinitate*: Thomas Aquinas and Avicenna on the Relationship between First Philosophy and the Other Theoretical Science', *The Thomist* 37, 1973, 133–154.

Wood, R., 'Roger Bacon: Richard Rufus' Successor as a Parisian Physics Professor', *Viarium* 35, 1997, 222–250.

Wright, M. R. (ed.), *Empedocles: the extant fragments*, New Haven, CT: Yale University Press, 1981.

Zabarella, J., *De Rebus Naturalibus*, Cloniae, 1602.

Index

Milton Keynes UK
Ingram Content Group UK Ltd.
UKHW040107071024
449327UK00019B/874